bility and Mathematical Statistics (Contir

LEHMANN • Theory of Point Estimatic
MATTHES, KERSTAN, and MECKE •
MUIRHEAD • Aspects of Multivariate {
OCHI • Applied Probability and Stocha
Physical Sciences
PRESS • Bayesian Statistics: Principles, Models, and Applications
PURI and SEN • Nonparametric Methods in General Linear Models
PURI and SEN • Nonparametric Methods in Multivariate Analysis
PURI, VILAPLANA, and WERTZ • New Perspectives in Theoretical and
Applied Statistics
RANDLES and WOLFE • Introduction to the Theory of Nonparametric
Statistics
RAO • Linear Statistical Inference and Its Applications, *Second Edition*
RAO • Real and Stochastic Analysis
RAO and SEDRANSK • W.G. Cochran's Impact on Statistics

d Statistical

esses, and

Iathematical

l Statistical

tics
lications to

A|

ting

cs

atistics

ns to the

tion
rd Edition
Manpower

plications
l Science
dentifying

n
on

urfaces
lethod for

Survey Errors and
Survey Costs

Survey Errors and Survey Costs

ROBERT M. GROVES
The University of Michigan

WILEY

JOHN WILEY & SONS

New York • Chichester • Brisbane • Toronto • Singapore

HB
849.49
.G76
1989
X

Library of Congress Cataloging-in-Publication Data:

Groves, Robert M.
  Survey errors and survey costs / Robert M. Groves.
    p.   cm.
  Bibliography: p.
  ISBN 0-471-61171-9
  1. Household surveys. 2. Social surveys. I. Title.
HB849.49.076   1989
001.4'33--dc19   89-5674
                                                               CIP

Printed in the United States of America

10 9 8 7 6 5 4 3 2 1

PREFACE

I work in a research environment that puts me in contact with statisticians, economists, psychologists, sociologists, and anthropologists. In our conversations about various survey design options for research projects, I never cease to be impressed by the variety of perspectives the different groups bring to these discussions. Even among social scientists there are many different uses of surveys, different conceptions of survey quality, and different approaches to analyzing survey data.

When I was beginning work in survey design, I attributed the resulting strained communication in these discussions to my own ignorance. It has taken me over fifteen years to be comfortable in my judgment that my problems are shared by many others. I wrote this book with these problems in mind.

The different perspectives toward surveys are mirrored in the distinct literatures created by the different disciplines. The current major texts focus either on statistical or on social science issues surrounding the methodology. For example, there are many good treatments of statistical sampling theory and its applications. However, they tend to give little attention to nonsampling errors, except through some discussion of measurement error models. There is also a growing number of books in the social sciences on questionnaire design and measurement methods. They often seek to identify the reasons for the errors and to focus on error reduction rather than error measurement.

In psychology in the late 1960's there were good summary treatments of psychological measurement theory, and they influence many survey researchers trained in psychology. Currently, there is much activity in estimation of psychometrically-based measurement error model parameters. Most of this estimation uses models based on formal assumptions about the survey measurement environment. In econometrics, issues of errors arising from failure to measure parts of the population resemble issues treated in survey sampling texts, but there is little reference to that literature in those texts. In short, the boundaries separating academic disciplines have produced different perspectives on survey research design, different languages about survey errors, and different assessments of the importance of various survey design features.

Another motivation for the book came from my perception of a current weakness in the survey methodological literature. Too many procedures to improve surveys are described without explicit treatment of their cost impact on data collection. Neither statistics nor the social sciences has treated the costs of alternative designs as seriously as it has treated errors. A careful student can learn how to reduce errors in the sampling, questionnaire design, and data collection phase, but little about how such efforts might increase the costs of the research project. This produces, in my view, a mismatch between theory and practice, one exceeding those in most fields, simply because quality improvement efforts are often discussed in isolation of the real limitations facing a researcher.

In this book I have tried to synthesize the statistical and social science perspectives to produce a more complete treatment of survey errors and costs of alternative survey designs. I start with the premise that all sample surveys are subject to various types of errors:

1. **Coverage error**, from the failure to give any chance of sample selection to some persons in the population.

2. **Nonresponse error**, from the failure to collect data on all persons in the sample.

3. **Sampling error**, from heterogeneity on the survey measures among persons in the population.

4. **Measurement error**, from inaccuracies in responses recorded on the survey instruments. These arise from:

    a. effects of **interviewers** on the respondents' answers to survey questions;

    b. error due to **respondents**, from inability to answer questions, lack of requisite effort to obtain the correct answer, or other psychological factors;

    c. error due to the weaknesses in the wording of survey **questionnaires**; and

    d. error due to effects of the **mode of data collection**, the use of face to face or telephone communication.

Most methods that attempt to reduce these various errors have cost implications for the survey. This occurs because administrative features of a survey design can affect the size of these errors (e.g., interviewer training, sample design and size, effort at persuading sample persons to cooperate). Further, these errors can be linked to one another in practice—attempting to decrease one source of error may merely increase another (e.g., reducing nonresponse by aggressively persuading sample persons to cooperate may result in larger measurement errors in the survey data).

My goals in writing this book were:

1. to consolidate the social science and statistical literatures on survey errors, drawing on the statistical sciences for insights into measuring the errors and the social sciences for explaining why they exist;

2. to explore evidence of relationships among the several types of survey errors, presenting examples of increases in one error as an unintended result of decreasing another; and

3. to present cost models that correspond to efforts to reduce various errors, making explicit the consequences of data quality maximization on research budgets.

The discussion focuses on survey methods using interviewers and largely ignores self-administered questionnaire techniques, except when they illustrate characteristics of errors or costs applicable to interviewer-dependent methods. The book also ignores errors in survey data created in steps after the data collection. These include the coding of text into numeric categories and other data processing stages of surveys. In some surveys these are large sources of error, but steps differ from others in that they can be repeated without recontact with the sample unit and with no effects on the recorded data.

Although the book uses research findings from throughout the world, it retains a distinctly U.S. flavor and focuses on many problems of surveys which are important to U.S. researchers but less so to those in other countries. Household surveys are the focus of the discussion. Use of the survey method to study businesses, organizations, and institutions is ignored. Finally, the discussion focuses on surveys involving an initial sampling and contact of households. Panel surveys and multi-wave data collections bring with them special problems of coverage, nonresponse, and measurement error, and they are not treated here.

I wrote this book for statisticians, survey methodologists, and for social scientists (and their students), especially those interested in methodological issues and those using survey research as a method of data collection. Statisticians will learn about social science approaches to the errors they treat as fixed and those they treat as stochastic or random. They will learn of the hypothesized psychological and sociological causes of survey errors. Survey methodologists will learn about statistical approaches concerning errors they have tried to eliminate in surveys. Social scientists will learn how to use formal cost and error models to aid design decisions in surveys.

This book exploits the growing realization that the different academic disciplines can contribute to each other's understanding of the survey method. In that regard, it takes advantage of the recent interest of the statistical community in social science approaches to survey error (e.g., cognitive psychological insights into measurement error).
Simultaneously, it presents to social scientists alternative statistical approaches for measuring various errors. Finally, it combines concerns about errors with concerns about limited research funds, in order to guide design decisions within cost constraints.

One of the problems I faced in writing for multiple audiences is that some use mathematics as a language more often than others. Mathematical notation is often used to describe sampling and measurement error models used by statisticians to estimate survey errors. Words are more often used by social scientists to describe errors. I have attempted to react to this difference by presenting both in words and notation the common ground of different perspectives.

I hope this book will act to reduce the barriers between researchers who now work together on different parts of survey research. I also hope that the book is used to train the next generation of researchers, as a text in graduate courses in survey design for the statistical and social sciences. I have used it in several editions of a graduate seminar for quantitative social scientists and statisticians. I have found that the introduction of language differences presented in Chapter 1 needs to be emphasized throughout the other chapters to insure comprehension. I believe that a shorter version of the course might be given using Chapter 1 (introduction to errors), Chapter 3 (coverage error), Chapter 4 (nonresponse rates), Chapter 6 (sampling error), and Chapter 7 (introduction to measurement error). Some courses devoted to sampling statistics may find it useful to include only material on nonsampling errors in Chapters 3–5 and 7. For these courses, the replacement of Chapter 6 by other readings would be desirable.

This book is the result of released creative tension that is caused by working on the margin of established academic disciplines. It is the result of searching to understand the perspectives taken by different researchers who work in the same general area. Such diversity is a designed feature of the Survey Research Center and the Institute for Social Research at The University of Michigan. My learning has been enhanced by the freedom of discourse fostered by this organization and by the tolerance of and respect for different approaches that are followed by researchers there. The sabbatical time provided to senior researchers for scholarly writing permitted this book to be written; the enlightened policy of letting this time be spread out over two different years aided its completion.

Despite its remaining deficiencies, this book was greatly improved through the merciless criticisms that only inquiring students can provide. Several cohorts of students suffered through early drafts that only I could understand, and through their patience and devotion to learning, provided insights to clearer presentation. They tried to keep me honest, and I am in their debt.

Colleagues spread throughout the world were kind enough to read chapter drafts as they matured. Frank Andrews provided encouragement throughout the early days. Barbara Bailar commented on early chapters and offered helpful criticisms. Charles Cannell provided both advice and access to his large archive of survey methodology literature. Charles Jones and others at the U.S. Census Bureau helped me locate important literature on coverage error. Leslie Kish was willing to engage in risky discussion about the philosophical underpinnings of various concepts of error. Jim Lepkowski and I discuss issues presented in this book so often that I am not sure which ideas are mine and which are his. He deserves special thanks for ideas on cost modeling, and parts of Chapters 2 and 11 are the result of earlier collaborative work with him. Lars Lyberg conscientiously read each chapter draft and mailed comments from Sweden. Stanley Presser gave advice on the telephone and even in bus stations. Howard Schuman allowed me to pick his brain on various Saturday afternoons spread over several years. There were many others who gave me advice on specific arguments and pointed me to literatures about which I was ignorant. I thank them all.

I have been blessed by working with two dedicated editors at Wiley, Beatrice Shube and Kate Roach. They provided an excellent blend of probing and freedom. I owe a large debt to Yvonne Gillies, who tirelessly managed the final manuscript preparation and typesetting.

In an effort as broad as this book, I am certain that errors of commission and omission remain. Writing it has taught me that such a book can only be ended not finished. I accept full responsibility for the remaining weaknesses. Further, there is much ongoing research which will enrich this discussion over the next few years. I encourage the reader to stay current with these developments.

ROBERT M. GROVES

Ann Arbor, Michigan
March 1989

ACKNOWLEDGMENTS

The author appreciates permissions granted to reprint copyrighted material. Corresponding Table or Figure numbers in *Survey Errors and Survey Costs* and the sources are listed below:

Tables 3.4, 4.6, 7.2, 8.3, 8.4, 8.5, 11.1, 11.4, with permission of Academic Press.

Tables 4.5, 5.1, 7.9, 8.1, 8.2, 8.9, 8.10, 8.11, 9.6, and 10.2, with permission of American Association for Public Opinion Research and the University of Chicago for the *Public Opinion Quarterly*. Copyright © (1981, 1980, 1984, 1986, 1979, 1971, 1975, 1981, 1979) by the American Association for Public Opinion Research.

Tables 5.2 and 10.9, with permission of the American Marketing Association for the *Journal of Marketing Research*. Copyright © (1974) American Marketing Association.

Table 5.12, with permission of the American Psychological Association for the *Journal of Personality and Social Psychology*. Copyright © (1978) by the American Psychological Association. Adapted by permission of publisher and author.

Table 9.8, with permission of American Psychological Association for *Psychology and Aging*. Copyright © (1989) by the American Psychological Association. Adapted by permission of publisher and author.

Tables 3.2, 4.7, 4.8, 4.9, 5.10, 6.7, 10.4, 10.5, and 10.6, with permission of American Statistical Association, for the *Journal of the American Statistical Association*.

Tables 5.8, 8.3, 8.5, 8.6, and 11.3, with permission of the American Statistical Association for the *Proceedings of the American Statistical Association*.

Table 5.5, with permission of The Gerontological Society of America for the *Journal of Gerontology: Social Sciences.* Copyright © (1988) The Gerontological Society of America.

Tables 7.10 and 10.3, with permission of Gower Publishing Group.

Table 10.7, with permission of the Institute of Statisticians for *The Statistician.* Copyright © 1980 The Institute of Statisticians.

Tables 9.7, and 10.8, with permission of Jossey-Bass.

Table 8.8, with permission of Jossey-Bass for the *Sociological Methodology.*

Tables 4.2 and 4.3, with permission of Minister of Supply and Services Canada for *Survey Methodology.*

Table 5.9, with permission of National Bureau of Economic Research for the *Annals of Economic and Social Measurement.*

Table 10.10, with permission of Royal Statistical Society for *Applied Statistics.* Reproduced by permission of the Royal Statistical Society, London.

Table 10.1, with permission of Russell Sage Foundation.

Tables 8.9 and 8.12, with permission of Sage Publications, Inc. for *Sociological Methods and Research.*

Tables 6.1 and 9.9, with permission of Seymour Sudman.

Tables 11.5, 11.6, and 11.7, and Figures 11.6, 11.7, with permission of Statistics Sweden for the *Journal of Official Statistics.*

Table 9.5, with permission of The University of Chicago for the *Journal of Labor Economics.* Copyright © (1985) by The University of Chicago. All rights reserved.

Table 3.5, with permission of John Wiley & Sons, Inc.

CONTENTS

    Error   441

9.8 Summary   445

10. **Measurement Errors Associated With the
     Questionnaire**   449

    10.1 Properties of Words in Questions   450

    10.2 Properties of Question Structure   460
         10.2.1 Length of the Question   461
         10.2.2 Open Versus Closed Questions   462
         10.2.3 Number and Order of Response Categories in
                Fixed Alternative Questions   464
         10.2.4 Explicit "Don't Know" Response Categories   468
         10.2.5 Question Structure in Measurement of
                Sensitive Topics   471
         10.2.6 Question Structure and Recall of Past Events 475

    10.3 Properties of Question Order   477
         10.3.1 Context Effects on Responses   478
         10.3.2 Position of the Question in the Questionnaire 481

    10.4 Conclusions About Measurement Error Related to the
         Questionnaire   482

    10.5 Estimates of Measurement Error from Multiple
         Indicator Models   484

    10.6 Cost Models Associated With Measurement Error
         Through the Questionnaire   490

    10.7 Summary   497

11. **Response Effects of the Mode of Data Collection**   501

    11.1 Two Very Different Questions About Mode of Data
         Collection   502

    11.2 Properties of Media of Communication   507

Survey Errors and
Survey Costs

CHAPTER 1

AN INTRODUCTION TO SURVEY ERRORS

For constantly I felt I was moving among two groups—comparable in intelligence, identical in race, not grossly different in social origin, earning about the same incomes, who had almost ceased to communicate at all, who in intellectual, moral and psychological climate had so little in common that instead of going from Burlington House or South Kensington to Chelsea, one might have crossed an ocean.

C. P. Snow, The Two Cultures, 1959, p. 2

1.1 DIVERSE PERSPECTIVES ON SURVEY RESEARCH

Survey research is not itself an academic discipline, with a common language, a common set of principles for evaluating new ideas, and a well-organized professional reference group.[1] Lacking such an organization the field has evolved through the somewhat independent and uncoordinated contributions of researchers trained as statisticians, psychologists, political scientists, and sociologists.

Such a *melange* of workers certainly breeds innovation, but it also spawns applications of the method for radically different purposes, suffers severe problems of communication, and produces disagreements about the importance of various components of survey quality. The word "survey" is not well defined. Its requisite components have not been enumerated. Surveys have not been distinguished formally from "polls," despite the latter's connotation of less rigorous methodology. Furthermore, surveys are used for myriad purposes.

Every month interviewers from the U.S. Census Bureau visit or telephone over 50,000 households and administer a short labor force survey. Years of research and analysis have been devoted to developing

---

[1] Providing a formal definition of "survey research" seems requisite in a first chapter of a book entitled *Survey Errors and Survey Costs*, but I do not provide a succinct one. By the end of this chapter the reader should have a notion of the extent of the concept as I use it.

sampling procedures that minimize the likely deviation of the unemployment rate as computed on the sample households from the corresponding rate in the full population. Thousands of Bureau staff have carefully implemented procedures to ensure that all types of residence have been given a chance of being selected for the interview. Interviewers strive to contact each sample household, enumerate all members of the household, and ask a series of questions about the employment status of household members. Complex editing procedures are used in an attempt to improve the quality of the recorded answers. Within a matter of days after the last interview is taken the monthly unemployment rate for the United States is publicly announced. The estimate is a complex statistical function of the answers obtained on the short questionnaire. Its computations reflect all the complexities of the data collection design used. The number is viewed as a *description* of an important characteristic of the society at that moment. At the time of its release no analysis has been performed to understand why the sample persons have succeeded or failed to be employed. The number is given meaning by comparing it to similar numbers available from previous editions of the survey. The next month another wave of interviewing produces that month's unemployment rate estimate.

At the same time as this survey is being conducted throughout the year hundreds of social scientists request computer tapes from data archives for various years of the National Election Studies. The existence of these large data archives permits researchers who have no direct knowledge of the data collection process to analyze the data. In this case, the data were collected in telephone and personal interviews of adults throughout an election year. Which candidate won the election is well known from published vote totals. Description of the election outcome is not the purpose of the exercise. The analysts use the data to test theories about political attitudes or electoral behavior. The theories may involve the effects of several variables on decisions to vote for one candidate or another. The theories are tested by the analyst's use of statistical models that posit the particular way various personal attributes (e.g., financial status, attention to media coverage of political events) produce the tendency to vote for one candidate or another. The analysts may be unconcerned about the fact that only about 70 percent of the sampled persons provided answers to the survey questions. Instead, efforts are made to include in the model all those variables that affect the vote decision. The analysts will argue that if the model is well formulated, these weaknesses will not impede a good test of the theory. Thousands of different models, attempting to explain several different attitudes or behaviors, are tested with the same set of data over many years of use.

These two examples illustrate how *describers* and *modelers* use survey data differently. Describers are interested in discovering some property of a fixed set of people (e.g., the unemployment rate in March 1989). The sample survey is a cost-efficient way to do so. Modelers are motivated by hypotheses about the cause of social phenomena (e.g., does having a broad set of skills reduce risks of unemployment?). Government agencies often specialize in descriptive uses of survey data, while academic social scientists specialize in analytic uses.

Because of these different purposes, different strengths and weaknesses of the survey method are of interest to the two types of user. The describers focus on whether the sample reflects the population they are seeking to describe (e.g., are people who are transient between two households disproportionately missed by the survey?). The modelers focus on whether the survey measures fully capture the concepts involved in their theory (e.g., is "years of education" a sufficiently strong indicator of skill mix to form a useful test of the theory?).

Describers and modelers form only one dichotomy among users of survey research. There is a large literature, scattered across journals of various disciplines, which identifies what components of surveys affect the quality of information obtained by them. Statistical sampling theory came into its own in the 1930s and describes how the selection of those interviewed in a survey can affect the survey quality. Studies into the properties of interviewing and question wording started about the same time but arose from a psychological perspective. Because of disciplinary boundaries there are few researchers who make major contributions to both areas.

Most *sampling statisticians* devote all their energies to the measurement and reduction of sampling error in survey estimates, the error arising because only a subset of the population was measured. They attempt to do this not only by increasing the sample size as long as resources permit but also by assuring representation of different groups in the sample (through stratification), by adjusting probabilities of selection for different subgroups to reduce sampling error, and by using clustered selections to increase cost efficiency. A separate, rarely overlapping group of researchers attempts to reduce nonresponse and measurement errors, sometimes referred to as examples of nonsampling error. They study ways of writing survey questions to seek better answers, of contacting sample persons to improve cooperation, and of training interviewers to obtain consistent behavior from respondents. Research in sampling error is unified by statistical sampling theory and is described in terms of empirical estimates of error. Research in nonsampling errors, since nonresponse and measurement problems arise from human behavior, is sometimes based on social science theories but often fails to obtain useful measures of nonsampling error. New analytic statistical models have provided social scientists with ways of acknowledging the effects of nonresponse and measurement errors in their substantive analysis. Little research exists, however, in combined estimates of sampling and nonsampling error on survey estimates.

The third important division of interest to us in this book lies within research on survey nonsampling errors. Within this literature are efforts to *reduce* the various errors afflicting surveys and efforts to *measure* the errors. On the side of reduction are scores of efforts to train interviewers in techniques of obtaining information freed from any influence they themselves might exert on the respondent, examples of question wording that improves the quality of the information collected, and efforts to increase response rates in order to improve the extent to which all eligible population members have a chance of selection in the sample. Most of this research uses an experimental design imbedded in a survey to compare two or more methods of doing the same work. Sometimes a source of information external to the survey (e.g., administrative records) is used to identify the "best" method. For example, there are studies comparing different ways of measuring reports of voting, compared to records on voting. There are studies comparing ways of asking persons about past criminal victimizations, compared to police reports. There are other studies of reports of hospital stays and doctor visits, compared to medical records on the same events. On the whole these studies are designed as one-time efforts to reduce survey error by identifying a better method than that currently in use. They offer no way of routine assessment of the size of the remaining errors. In the absence of such routine measurement the researcher has no assurance that "best" method might vary by survey topic, population studied, or other aspects of the survey design.

In sharp contrast to this research is a set of statistical models of nonsampling errors, designed to estimate their magnitudes. These include measures of interviewer effects that vary over different interviewers, variability in answers from the same respondent over repeated administrations, and the relative quality of different questions attempting to measure the same concept in the same questionnaire. All these methods yield empirical estimates of nonsampling error, but all of them require some assumptions about the data collection process, which may or may not be completely warranted.

Reducing nonsampling error requires an understanding of the causes of the errors. Measuring nonsampling errors requires a data collection design compatible with a statistical model of the errors. Unfortunately, many of the efforts to reduce error have not simultaneously attempted clear specification of statistical models of the errors. Hence, the two streams of research have rarely merged. Clearly, reduction *and* measurement of error are preferable to either singly because measurement evaluates how successfully the error was reduced.

In short, the diverse approaches to survey research can be described by a series of interlocking and overlapping perspectives, each with its set of practitioners and its own set of academic or professional organizations. They are:

1. those who use surveys to describe characteristics of a fixed population (describers) and those who seek to identify causes of phenomena constantly occurring in a society (modelers),

2. those who study sampling error (sampling statisticians) and those who study nonsampling error (field and substantive experts), and

3. those who attempt to measure errors (measurers) and those who attempt to reduce them (reducers).

These groups are not mutually exclusive. Instead, they mostly define categories of work within survey research. There are, however, many specialist researchers who concentrate their efforts on only one side of each of the dichotomies.

These groups rarely enjoy a single forum for interchange of ideas. They have, between the 1930s and the 1980s, in both the United States and Europe, chosen different problems to study, developed separate languages, and constructed distinct practices in survey methods. In general, they hold each other in mild disdain, not unlike the "intellectuals" and "scientists," or the "numerate" and "nonnumerate" encountered by Snow in Britain in the 1950s (Snow, 1959).[2]

This book has the immodest goal of comparing and contrasting the perspectives of these groups. It is addressed to the professionals in each of the groups and seeks to call their attention to noteworthy research findings by other groups. It is also addressed to serious students, now just learning their trade, in order to arm them with knowledge about other perspectives on the same issues. At the minimum, it attempts to juxtapose the various perspectives; at its most ambitious, it seeks to integrate them into one unified perspective.

---

[2] As is discussed later, the same debate about the value of quantification and the use of mathematical notation as a language of discourse is involved in the different approaches to survey research.

1.2 THE THEMES OF INTEGRATION: ERRORS AND COSTS

Despite the diversity of outlooks on the survey method, the several approaches enjoy a common concern—errors inherent in the methodology which inhibit the researchers from obtaining their goals in using surveys. In this context "error" refers to deviations of obtained survey results from those that are true reflections of the population. Some choose one error to study; others choose another. Some study ways to eliminate the error; others concentrate on measuring its impact on their work. But all of them share the preoccupation with weaknesses in the survey method.

Much of the research identified above examines survey errors relative to the costs of collecting and processing survey data. For example, there are a set of cost models tailored to different sample design features. These are used to identify the optimal characteristics of a sample design within the constraints created by a fixed research budget. Similarly, there are cost models for staff training or efforts to attain full cooperation of a sample. The value of such efforts is generally assessed with concern about the cost implications of increasing such efforts. Thus, costs of survey activities often act as limiting influences on efforts to reduce survey errors. In this book an attempt is made to link survey errors with specific design features and to discuss the costs of those features. When possible the discussion will formally balance costs and errors in order to stimulate the search for the best design features within the resources available.

The explicit inclusion of cost concerns may seem antithetical to discussions of quality of survey data. In the crudest sense it may even evoke thoughts of attempts to "cut corners." This book argues that this reaction confuses "cost efficiency" with "low cost." In contrast, the position taken here is that only by formally assessing the costs of alternative methods (jointly measuring quality) can the "best" method be identified. This logic will probably receive an easier hearing when the discussion focuses on sampling error than when it focuses on measurement error. This will not result from the inapplicability of the approach to measurement error, but rather the weakness of current knowledge about error levels resulting from different methods.

1.3 THE LANGUAGES OF ERROR

Thus far in our discussion we have avoided formal definitions of concepts of error, such as "measurement error," "nonresponse error," and "sampling error." This cannot continue without jeopardizing the goal of comparing different perspectives on survey error. Unfortunately, along with their emphasis on different properties of the survey method, the different perspectives use different terminology. We thus face the unpleasant task of language lessons, reviewing words and their meanings. The following sections are best read twice. The first reading should be devoted to learning the terminology not yet familiar. The second reading might take place as other chapters of the book are being read, to reinforce the understanding of differences and similarities of different terms.

There appear to be at least three major languages of error that are applied to survey data. They are associated with three different academic disciplines and exemplify the consequences of groups addressing similar problems in isolation of one another. The three disciplines are statistics (especially statistical sampling theory), psychology (especially psychometric test and measurement theory), and economics (especially econometrics). Although other disciplines use survey data (e.g., sociology and political science), they appear to employ similar languages to one of those three.

It is unlikely that the substantive foci of the different disciplines are the causes of the different vocabularies. Rather, as we shall see below, they may be the result of focusing on different parts of the survey research act (as exemplified by groups above). Regardless of the source of language differences, there would likely be no debate about whether researchers from different disciplines have difficulty understanding one another's discussions of survey errors.[3]

This chapter introduces the reader to these different languages of error in order to offer some definitions of the terms used throughout this book and also to contrast and compare other terms that the reader might use more regularly. It will become clear that differences across these fields are not merely a matter of different labels for similar concepts of error, but real differences in the set of factors that are seen to influence survey estimates. The different perspectives will themselves be presented in both verbal and mathematical form (the mathematical forms are presented in an appendix to this chapter). This approach is necessary because there are survey researchers who prefer one mode of presentation to another (and unfortunately often do not bother to present their thoughts in both). The reader, however, will be exposed to the complete discussion by reading either one of the treatments.

---

[3] One of the more enjoyable encounters I have had was a set of meetings with economists, sociologists, psychologists, and statisticians at the Survey Research Center, which attempted to translate from one discipline's language of errors to another's. This section profits from those discussions.

1.4 CLASSIFICATIONS OF ERROR WITHIN SURVEY STATISTICS

Four nested levels of concepts are used throughout this book to classify errors (see Figure 1.1). This perspective leans heavily on the conceptual map of Kish (1965), used also by Andersen et al. (1979). The total error of a survey statistic is labeled the **mean square error**[4]; it is the sum of all variable errors and all biases (more precisely, the sum of variance and squared bias). Errors are specific to a single statistic (e.g., the mean years of education of respondents to a single survey), not all statistics in a single survey, nor all statistics of a certain form in one survey (e.g., all means or correlation coefficients). The mean square error is rarely fully measured for any survey statistic.

Bias is the type of error that affects the statistic in all implementations of a survey design; in that sense it is a constant error (e.g., all possible surveys using the same design might overestimate the mean years of education per person in the population). A variable error, measured by the **variance** of a statistic, arises because achieved values differ over the units (e.g., sampled persons, interviewers used, questions asked) that are the sources of the error. The concept of variable errors inherently requires the possibility of repeating the survey, with changes of units in the replications (e.g., different sample persons, different interviewers). If there were no possibility of such replication, the distinction between variable errors and biases does not exist.

Variable errors and biases are therefore connected; bias is the part of error common to all implementations of a survey design, and variable error is the part that is specific to each trial. A "survey design" defines the fixed properties of the data collection over all possible implementations. Hansen et al. (1961) might refer to these as the "essential survey conditions." For example, the concept of sampling variance refers to changes in the value of some statistic over possible replications of a survey, where the sample *design* is fixed (e.g., stratification, clustering, rules for probabilities of selection), but different *individuals* are selected into different samples. For example, the computed mean years of education of respondents will vary over different subsets of the population selected using the same sample design. "Sampling variance" exists because the value of the survey statistic varies over persons in the population *and* sample surveys measure only a subset of the population. In contrast, the phrase "response variance" is used by some to denote the

---

[4]Throughout this section, when a new phrase is introduced and defined, it is set in boldface.

variation in answers to the same question if repeatedly administered to the same person over different **trials** or **replications**. The units of variation for response variance are different applications of the survey question to the same person.

Variable errors require an assumption of replicability. With regard to sampling variance, for example, it is rare that full surveys are repeated in order to estimate the change in sample values of statistics. Instead, the notion of sampling variance is based on the conceptual replicability of the survey with a different set of respondents, but with all other properties of the survey the same. In general, variable errors are based on a model of replicated implementation of a survey design, *whether or not such replication is ever actually conducted.*[5]

In some situations it is assumed that there is an observable true value of the statistic in the population and a true answer of the survey question for the respondent in the sample. For example, when measuring the proportion of the household population that visited a physician in the last 6 months, there is the assumption that there is a correct answer to the question for each population member, and there is a correct value of the proportion for the population. Bias is the difference between the average value of the statistic (obtainable with the survey design) and that true value. Variance is a summary of how the obtained values vary over conceptual replications of the survey design (e.g., over possible samples, interviewer corps, trials).

This presumed existence of an observable true value distinguishes the survey statistics viewpoint from the psychometric perspective discussed later. There are good reasons for the differences. Psychological measurements are generally taken on attributes that cannot be observed by anyone other than the respondents themselves. In the extreme, there is no evidence that the attribute really exists. An interviewer asks whether a respondent favors or opposes the way that the president is handling problems in the Middle East. The respondent answers that he favors it. How can we assess that this is a correct report of his attitude? How do we know that he really has an attitude on the issue? Contrast this to the case in which an interviewer asks the respondent to report his weight. Clearly, the appeal of a concept of bias (departures from the true value) is greater in this second case. The comparison can be clouded, however, if we inquire whether the question means the weight in the morning or the evening, weight to the nearest pound or to the nearest

---

[5] There are some exceptions to this. For example, some researchers employ a model of measurement in which persons commit different errors of response but each always commits the same error. "Response variance" is used to describe the variability over persons in these fixed errors.

Figure 1.1 The conceptual structure and language of error sources in surveys to be employed in this book.

ounce, weight measured with and without clothes, with and without shoes, and so on.

There are two types of error under both "variance" and "bias" in Figure 1.1. **Errors of nonobservation** are those arising because measurements were not taken on part of the population. **Observational errors** are deviations of the answers of respondents from their true values on the measure. In short, these are errors of omission and errors of commission. The most familiar error of nonobservation in surveys, "sampling error," stems from measuring only a subset of the population, a sample, not the full population. The estimates of population characteristics based on that subset will not be identical to the actual values. The most familiar observational error is probably that associated with respondents' inability to provide the correct answer (e.g., failure to report correctly whether she visited a physician in the last 6 months). The answer provided departs from the true value for that person. If there is a tendency to make such an error throughout the population, the overall survey proportion will depart from the true population proportion (yielding a measurement bias).

The final level of conceptual structure in Figure 1.1 concerns the alternative sources of the particular error. Errors of nonobservation are viewed as arising from three sources—coverage, nonresponse, and sampling. **Coverage error** exists because some persons are not part of the list or **frame** (or equivalent materials) used to identify members of the population. Because of this they can never be measured, whether a complete census of the frame is attempted or a sample is studied. For example, telephone surveys of the U.S. household population fail to include households without telephones. This is a potential coverage error for a statistic describing the full household population. **Nonresponse error** arises because some persons on the frame used by the survey cannot be located or refuse the request of the interviewer for the interview. For example, about 20 to 25 percent of sample persons in the General Social Survey, an ongoing face to face interview household survey in the United States, are not interviewed for one reason or another. To the extent that they have different values on the survey variables than others in the sample, the survey statistics can be affected. **Sampling error** arises because the statistic is computed on a subset of the population being studied. To the extent that different subsets exhibit different characteristics on the statistic of interest, the sample survey estimates will vary depending on which subset happens to be measured.

Observational errors are conveniently categorized into different sources—the interviewer, the respondent, the questionnaire, and the mode of data collection. **Interviewer errors** are associated with effects on respondents' answers stemming from the different ways that interviewers administer the survey. Examples of these errors include the failure to read the question correctly (leading to response errors by the respondent), delivery of the question with an intonation that influences the respondent's choice of answer, and failure to record the respondent's answer correctly. There are also effects on the quality of the respondents' answers from the wording of the question or flow of the questionnaire, labeled **instrument error**. Research has found that small, apparently innocuous, changes in wording of a question can lead to large changes in responses. Thus, **respondent error** is another source of observational error. Different respondents have been found to provide data with different amounts of error, because of different cognitive abilities or differential motivation to answer the questions well. Finally, there are observational errors associated with the **mode of data collection**. It has been found, for example, that respondents have a tendency to shorten their answers to some questions in telephone interviews relative to personal visit interviews.

The reader will note that Figure 1.1 is not a complete enumeration of all sources of error in survey data. The most notable omissions are those arising after the answers to the survey questions have been obtained by the interviewers—the coding, editing, imputation, and other data processing activities that follow the data collection phase. These items were deliberately omitted from Figure 1.1, not because they are always trivial sources of error but because they do not involve the interviewer, questionnaire, sample persons, and other survey design characteristics. Instead, they result from actions of processors and analysts of the data.

1.4.1 A More Restrictive View, Sampling Statistics

Having presented the full conceptual structure of error terms that will organize this book, we discuss alternative viewpoints on survey error. First, we note two perspectives within the field sometimes labeled "total survey error" and another taken by sampling statistics. The former incorporates the view that the given survey is one of an infinite number of possible implementations of the design on a particular sample. In sampling statistics no aspect of the design is viewed as a source of variable error other than sampling. The former view includes more sources of error in its approach. It holds, for example, that different interviewers perform their tasks differentially well. Hence, another possible implementation of the survey design would have used a different set of interviewers who would obtain different levels of error. Similarly, they view the performance of respondents in answering questions as varying over possible replications. Hence, the values of statistics will vary over potential trials of the survey (Hansen et al., 1961).

The simpler model of the sample survey is based on the assumption that the only source of variation in survey results comes from measuring different subsets of the population (Figure 1.2). Thus, **sampling variance** is the only variable error. This view is taken in most of standard statistical sampling theory. It holds that if the survey could be repeated on the same sample persons, the same results would be obtained. Biases in survey estimates, fixed errors over replications, arise from coverage error (failures of the frame to cover part of the population), nonresponse bias (failures of the interviewers to obtain measurements on all the sample people), and sampling bias (through the failure to give some persons in the population any chance of selection).[6] These errors are viewed to be constant over all possible samples, given the survey design.

It is important to note that sampling variance is conceptualized as variation among values of a survey statistic resulting from a sample design. That is, if each survey is subject to coverage or nonresponse errors of a similar sort, the variation is measured about the expected value of the statistics, each of which is subject to those errors. Sampling variance is not a function of deviations from the population value, but from the expected value of sample statistics. Thus, confidence intervals built on survey data (using sampling variance alone) do not reflect any of the coverage, nonresponse, measurement, or sampling biases that may plague the survey.

The attraction of this perspective is that it concentrates on a survey error, sampling variance, that can be measured with well-developed techniques. Partially for that reason, the perspective has dominated many design decisions in surveys. Many designs have maximized sample size (in an attempt to reduce sampling variance) while minimizing attention to questionnaire development or interviewer training (ignoring measurement errors associated with them). This viewpoint is incomplete and unsatisfactory, however, to those concerned with nonsampling error as well as sampling error. It omits from its concern the practical problems of survey administration—the fact that some interviewers obtain very high response rates and others obtain low rates; findings that suggest some respondents report more accurately than others and different questions appear better suited to obtaining correct data than others.

---

[6] The word "population" is meant to refer to the set of persons identified as eligible for the survey, the "frame population" in the language of Chapter 3. A more complete discussion of the different concepts of population can be found in that chapter.

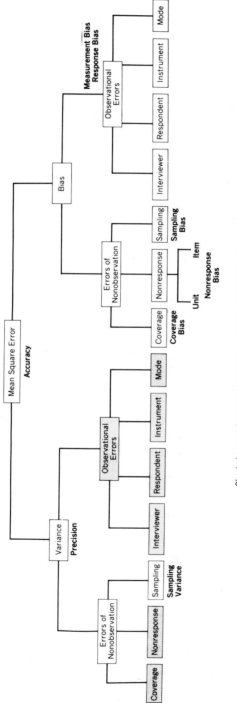

Shaded concepts are not central to viewpoint of sampling statistics.

Figure 1.2 The structure and language of errors used in sampling statistics.

14

1.4.2 Total Survey Error

A more elaborated view of survey error held by some survey statisticians comes from those interested in total survey error (e.g., Fellegi, 1964; Hansen et al., 1964; Bailar and Dalenius, 1969; Koch, 1973; Bailey et al., 1978; Lessler et al., 1981). This perspective retains the possibility of fixed errors of coverage and nonresponse. In addition to variability over samples, it acknowledges variability in errors over different trials of the survey. Underlying this is the notion that the survey at hand is only one of an infinite number of possible trials or replications of the survey design. Respondents are assumed to vary in their answers to a survey question over trials, leading to **simple response variance** (Hansen et al., 1964). They may also be viewed as varying in their decision to cooperate with the interview request over trials. The same respondent might cooperate on one trial but refuse on the next, without a well-developed rationale yielding predictable behavior. This would lead to variable nonresponse errors over trials, a nonresponse error variance term.

The interviewer is often treated as a source of error in this perspective, most often conceptualized as a source of variable error. That is, each trial of a survey is viewed to consist of both a set of sampled persons (one replication of the sample design) and a set of interviewers (one set selected to do the work, from among those eligible). Both sets are viewed to change over trials (both the sample and the interviewer corps and the assignment of interviewers to sample persons). The variable effects that interviewers have on respondent answers are sometimes labeled **correlated response variance** in this perspective (Bailey et al.,, 1978). This arises from the notion that errors in responses might be correlated among sample persons interviewed by the same person. A generalization of this perspective would permit interviewers to differ in their effects on coverage and nonresponse error also. This would lead to variation over trials in these errors because of different interviewer corps. There are very few examples, however, of this generalization.

1.4.3 Other Error Terms in Survey Statistics

There are several other phrases that are commonly encountered in treatments of error by survey statisticians. **Accuracy** is used to mean "the inverse of total error, including bias and the variance" (Kish, 1965, p. 25) or, alternatively, "the quality of a survey result that is measured by the difference between the survey figure and the value being estimated. The true value is seldom known, although it can be approximated in some instances" (U.S. Department of Commerce, 1978, p. 1). **Precision** most often refers to the sum of variable errors and is the converse of variance (high precision is equivalent to low variance); in some rare instances (e.g., Bailar, 1976) survey statisticians will use the term **reliability** to mean the same thing. (We shall see that "reliability" in psychometric usage has a very different meaning.) There appears to be no phrase that is the converse of "bias" in the terminology of survey statistics. An "accurate" estimate, in this usage, means that it is both stable over replications and close in value to the true value of a statistic. A "precise" estimate merely means that it is stable over replications.

Many treatments of survey error contrast **sampling error** versus **nonsampling errors**. This is merely a different categorization of the errors than that of Figure 1.1. The term "nonsampling errors" generally means all the sources of error in Figure 1.1 that do not arise merely because a subset of the population is measured. It is likely that the categories "sampling" and "nonsampling error" are popular because the two types of error offer different ease of measurement. The dichotomy is unfortunate, however, because it combines very different kinds of problems on the nonsampling error side.

The phrase **standard error** is used to denote the square root of the sampling variance, not the square root of the total error term, the mean square error. The square root of the mean square error is called the **root mean square error**.

Measurement error is generally equated with the term "observational errors," whether they arise from the interviewer, the instrument, or the respondent. Unfortunately, some researchers use the term **response errors** to mean the same thing (e.g., Horvitz and Koch, 1969). Those who use "measurement errors" to denote all observational errors generally reserve the term "response errors" for errors associated with the respondent, not the questionnaire or the interviewer.

All the sources of error that are normally associated with bias (nonresponse, coverage) can also be associated with variable terms. All that is necessary for this switch of category is that the magnitude of these errors vary over replications of the survey. In Figure 1.3 the measurement entities of interest that would vary over replications are (1) the persons in the sample, (2) the trial of the measurement, and (3) any administrative unit (e.g., interviewers, supervisors) that can be viewed as the result of a stochastic (i.e., involving randomization) selection process.

The other perspectives on error differ on which sources of variable error they bring into their scope of concern. By ignoring some they build theories of the data collection process that treat as uncomplicated (or fixed) some portions to which other fields devote most of their energies. This causes great difficulty when researchers discuss, for example, the "variance properties" of a survey statistic (e.g., the estimated proportion

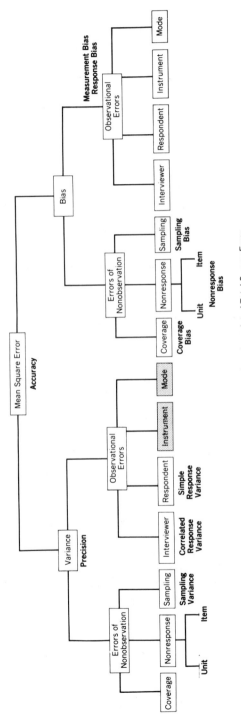

Shaded concepts are not central to viewpoint of Total Survey Error.

Figure 1.3 The structure and language of errors used in the total survey error literature.

17

of persons owning their home). Some will include more sources of error than others and possibly reach very different conclusions about the quality of the estimate.

1.5 TERMINOLOGY OF ERRORS IN PSYCHOLOGICAL MEASUREMENT

When moving from survey statistics to psychometrics, the most important change is the notion of an unobservable characteristic that the researcher is attempting to measure with a survey indicator (i.e., a question). Within survey statistics, in contrast, the measurement problem lies in the operationalization of the question (**indicator**, in psychometric terms). That is, the problem is not the impossibility of measuring the characteristic but the weakness of the measure. The psychometrician, typically dealing with attitudinal states, is more comfortable labeling the underlying characteristic (**construct**, in psychometric terms) as unobservable, something that can only be approximated with any applied measurement.[7]

There are two influential measurement models that are discussed in this section. In the first, **classical true score theory**, all observational errors are viewed as joint characteristics of a particular measure and the person to whom it is administered. Errors in responses are acknowledged. In such measurement, however, the expected value (over repeated administrations) of an indicator is the true value it is attempting to measure. That is, there is no measurement bias possible, only variable errors over repeated administrators.

Although classical true scores provide the basis for much of the language of errors in psychometrics, it is found to be overly restrictive for most survey applications. The need to acknowledge possible biases in survey measurements is strong. In psychometrics different labels are given to this kind of measurement. Most survey measures are labeled as sets of **congeneric measures** or indicators in a **multiple factor model**, where measurement errors can yield biases in indicators of underlying constructs, and indicators can be influenced by various methods of measurement. A set of congeneric measures are indicators of different underlying characteristics. The different characteristics, however, are all simple linear functions of a single construct. These models are discussed in more detail in Chapter 7.

---

[7] The existence of an observable true value is labeled an example of Platonic true scores (Bohrnstedt, 1983).

An additional change when moving to the field of psychometric measurement is the explicit use of models as part of the definition of errors. That is, error terms are defined *assuming* certain characteristics of the measurement apply. In classical true score theory (a model), the most important assumption is that if an indicator were administered to a person repeatedly (and amnesia induced between trials), the mean of the errors in the respondent's answers would be zero. That is, the indicator is an "unbiased" measure of the respondent's characteristic, in the sense used by survey statisticians. (Here the parameter of interest is the single respondent's value on the construct.) This is not as strong an assumption as it may appear to be because psychometricians often view the scale on which their measurements are made as rather arbitrary (e.g., a 0 to 100 scale in their view having all the important properties of a scale from -50 to $+50$).[8] Indeed, as Lord and Novick (1968) note, "if certain effects— for example, poor lighting, excessive noise, and the inherent properties of the particular measurement scale adopted—have a biasing effect on the observed scores, this bias is represented in the classical model by a change in the true score" (p. 43). In short, if an indicator is not an unbiased measure of one construct, it is of another.

In this perspective expectations of measures are taken over trials of administration of the measurement of a person. That is, each asking of a question is one sample from an infinite population (of trials) of such askings. The **propensity distribution** describes the variability over trials of the error for the particular person. Under the classical true score assumption the mean of that distribution is zero. As shown in Figure 1.4, the only concept of error akin to those of the survey statistician is the variance of the error term, the **error variance** (Lord and Novick, 1968). This is the dispersion of the propensity distribution. When there is interest in a population of persons, the expected value of the indicator is taken both over the many propensity distributions of the persons in the population *and* the different persons. *It is only within this context (measurement of a set of persons) that other concepts of error are defined.* This case is shown in Figure 1.5.

Recall that with most of the psychometric perspective, only variable errors exist. Two terms, "validity" and "reliability," are frequently used to label two kinds of variable error. The notion of **theoretical validity**, sometimes **construct validity**, is used to mean "the correlation between

---

[8]This fact arises because the statistics of interest to psychometricians are not generally means or totals for persons studied, but rather correlation coefficients, relative sizes of variance components, factor loadings, and standardized regression coefficients. All these statistics are functions of variance and covariance properties of measures, not of means (expected values).

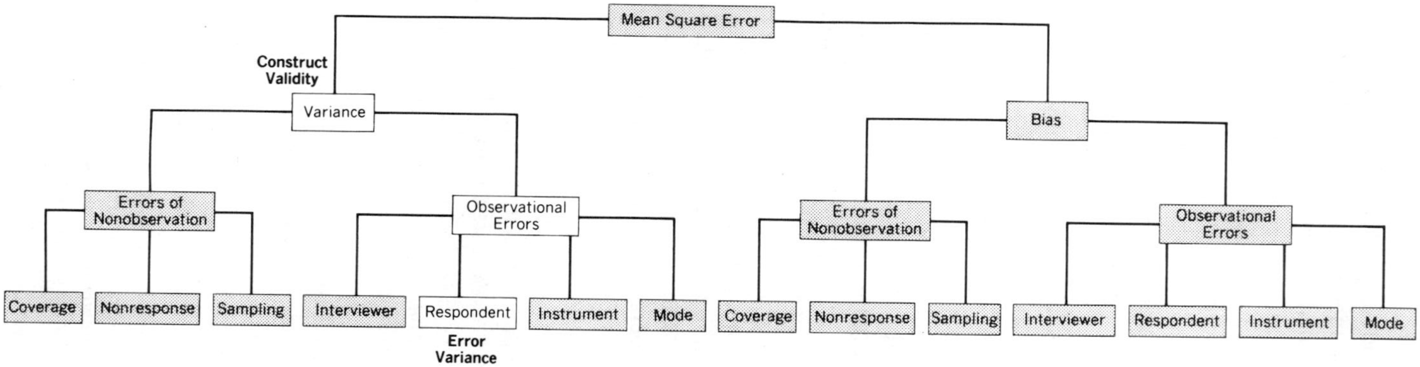

Shaded concepts are not central to viewpoint of psychometrics for individual measurement.

Figure 1.4 The structure and language of errors used in psychometric true score theory for measurement of an individual.

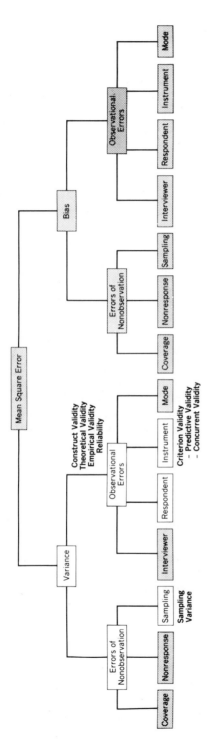

Shaded concepts are not central to viewpoint of psychometrics for population measurement.

Figure 1.5 The structure and language of errors used in psychometric theory for populations.

the true score and the respondent's answer over trials." Note well that "validity" is not to be simply equated with "unbiasedness," as used by survey statisticians. Under true score theory all the questions produce unbiased estimates of the person's true value on the trait. Instead, a completely valid question is one that has a correlation of 1.0 with the true score. Since validity is based on correlations, it is defined only on a population of persons (who vary on the true values), not on a single person. That is, there is no concept of a valid measure of a single person's attribute.

The other error concept used in psychometrics, when only one construct is under examination, is **reliability**, the ratio of the true score variance to the observed variance (Bohrnstedt, 1983). Variance refers to variability over persons in the population and over trials within a person. With this definition of reliability, it can be noted that the concept is not defined for measurements on a single person, only on a population of persons, *and* has a value specific to that population. Each population will produce its own reliability magnitude on a particular measure.

When dealing with populations of persons, true score theory adds another assumption about the errors, that their values are uncorrelated with the true values of the persons on any of the trials. With this assumption a simple mathematical relationship between validity and reliability results. The theoretical validity of a measure is merely the square root of its reliability. This relationship shows how different the concepts of reliability and validity, on one hand, are from variance and bias, on the other. Validity and reliability are both functions of variable response errors made in the measurement process. Given this definition, the traditional statement that "no measure can be valid without also being reliable, but a reliable measure is not necessarily a valid one," is not true. Given the assumptions of classical true score theory, if a measure is reliable, it is valid; if it is valid, it is reliable. Note the contrast with the survey statistics concepts of bias and variance. In that field a sample statistic may have an expected value over samples equal to the population parameter (unbiasedness) but may have very high variance from a small sample size. Conversely, a sample statistic can have very low sampling variance (from an efficient sample design) but may have an expected value very different from the population parameter (high bias).

1.5.1 Estimating Validity

Validity and reliability can be assessed only with multiple indicators. Psychometricians use a set of adjectives for "validity" in the context of multiple measures of a single construct and/or multiple constructs. Figure

1.6 illustrates this case. Bohrnstedt (1983) makes the distinction between theoretical validity, which is defined on a single indicator, and **empirical validity**, an estimation of theoretical validity that can be implemented only with multiple measures of the same construct. Empirical validity is the correlation of the measure and another observed variable that is viewed to be an indicator of the same construct. Thus, empirical validity of one indicator is always measured in relation to another measure. The concept is not defined without reference to that other measure. Sometimes **criterion validity** is used to denote that the other measure is assumed to be measured without any variable error. Lord and Novick (1968, p. 72) state the relationship between criterion validity and reliability clearly: "unless the index of reliability of a test is sufficiently high, validity cannot be high for *any* criterion. Of course, it does not follow that a high index of reliability guarantees or even suggests high validity; indeed it does not. High reliability is a necessary, but not a sufficient, condition for high validity." They might have emphasized that this means high *empirical* validity. They could have also said that high *empirical* validity is a necessary and sufficient condition for high reliability, a perspective that seems to be rarely taken.

There are at least two types of empirical validity, which differ in the characteristics of the criterion chosen. **Predictive validity** is the correlation between one measure of a construct and another (presumably with better error features) taken at a later time. Bohrnstedt gives the example of the empirical validity of a measure of job performance taken at time of hire being measured by its correlation with the evaluative rating of supervisors later. **Concurrent validity** is the correlation between a measure and some criterion measured at the same point in time. Both of these can use any number of criteria to assess validity, and implicit in the choice of the criterion is the assumption that both measures are indicators of the same construct. These error measurement procedures in psychometrics resemble the reinterview studies in surveys, used to measure response variance on the part of the sample. In these the same question is asked of the respondent at two points in time, and the response variance is measured as a function of differences in the answers.

When several constructs are being considered, another perspective on validity can be used, one that is prominent in the multitrait multimethod approach suggested by Campbell and Fiske (1959). In order to demonstrate **convergent validity**, correlations among indicators of the same construct that employ different methods of measurement should be high. This merely says that measures of the same thing should be positively correlated. Conversely, **discriminant validity** is demonstrated when the correlations between measures of different constructs are low even when they employ similar methods. Convergent

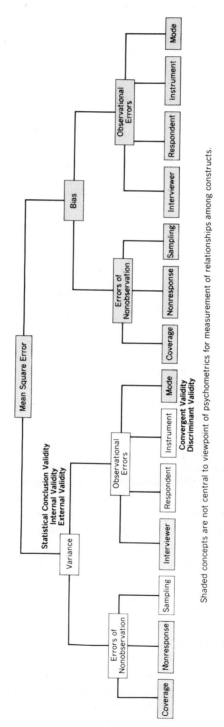

Shaded concepts are not central to viewpoint of psychometrics for measurement of relationships among constructs.

Figure 1.6 The structure and language of errors used in psychometric theory for measures of relationships between constructs.

and discriminant validity must be assessed jointly, relative to one another. One would want correlations among indicators of the same construct to be higher in general than among indicators of different constructs. These notions of validity, however, do not yield themselves to a single measure. Survey statisticians are also interested in method effects but generally do not investigate them by concurrently using two methods of measuring the same concept on the same respondent. The belief is that two measures of the same concept (e.g., two question wordings) might not yield independent measures. Specifically, if question form A were administered before question form B, the answers to B might be different if the order were changed. Such order effects are generally ignored in the multitrait multimethod approach.

Finally, **content validity** is a property of a set of measures of a construct. It applies to the situation in which the multiple measures do not all have the same expected value but are indicators of different parts of the construct in question, different "domains of meaning" (Bohrnstedt, 1983). It is based on the premise that there are systematic variations in which domain a particular indicator covers. Content validity refers to the extent to which the set of items used is a good sample of the different domains of meaning. Thus, its meaning is somewhat different from those terms that concern the extent of error variance between a measure and underlying constructs; it is better viewed as a procedure to increase construct validity (Bohrnstedt, 1983).

All the concepts above are common to classical true score theory and to most approaches to survey data used by psychologists. Relationships between reliability and validity found above are heavily dependent on the simplifying assumptions of the classical true score model. A more general model, the **congeneric measures model**, yields different concepts of error. Congeneric measures are always treated in sets of two or more. Two congeneric measures do not have to have the same expected values (as in the case in assessments of empirical validity under the classical true score model). Instead, the two expected values are themselves each linear functions of an underlying construct. This fits the survey measurement model more closely. For example, the two indicators might be affected differently by interviewer influences and thus have different expected values. Once the interviewer effect is accounted for, however, they have equal expected values.

Even more appealing is the **multiple factor model**, which posits that an indicator may simultaneously reflect characteristics on two underlying constructs. In the case of survey measures, the indicators might reflect the substantive construct of interest *and* an effect of method of measurement. The observations are thus viewed to be subject to both **systematic errors** and **random errors**. Systematic errors exist, say

Carmines and Zeller (1979), when the indicators "represent something other than the intended theoretical concept" (p. 15). These errors do not have one or more of the necessary characteristics of the errors in the classical true score model. That is, they do not have a zero expected value, they are correlated with the true score value, or they are correlated across persons. **Random measurement error**, as used by Andrews (1984, p. 412), refers to "deviations (from true or valid scores) on one measure that are statistically unrelated to deviations in any other measure being analyzed concurrently." In the language of survey statistics this would refer to lack of correlation between two variables in their response deviations. **Correlated measurement error** means "deviations from true scores on one measure that *do* relate to deviations in another measure being concurrently analyzed." Thus, "correlated measurement error" means something very different from the "correlated response variance" used by survey statisticians. The latter refers to correlations among respondents contacted by the same interviewer (or other administrative units) in deviations obtained on *one* indicator. The correlated measurement errors could arise from the fact that two indicators share the effects of a common method of measurement. Such a viewpoint is central to the multitrait multimethod approach to estimating construct validity. This alternative measurement model retains all the basic concepts of error, but necessarily alters the computational forms of error estimates (see next section).

1.5.2 Other Notions of Validity

A large set of new concepts of error is created when there is concern with a population of persons *and* relationships among constructs. Not all these types of validity yield themselves to empirical measures. Their meaning seems to lie closer to "appropriateness of research activities." **Statistical conclusion validity**, in Cook and Campbell's (1979, p. 41) terminology, concerns "inferences about whether it is reasonable to presume covariation given a specified α level and the obtained variances." The concept is closely related to the execution of statistical tests of hypotheses. Therefore, statistical conclusion validity is threatened by all the circumstances that can threaten the appropriate application of a test: for example, violation of the assumptions of the tests (generally distributional assumptions) and effect on power of sequential tests of hypotheses on the same data. The inferences here are to the process generating the relationship.

   Internal validity is "the approximate validity with which we infer that a relationship between two variables is causal or that the absence of a relationship implies the absence of cause" (Cook and Campbell, 1979, p. 37). The adjective "internal" refers to the inference to the set of respondents examined in the data collection, not to others outside the set of subjects, nor to situations outside that of the data collection itself. Threats to internal validity in a measured relationship include problems of operationalization. Cook and Campbell, in the context of quasi-experimental designs, cite problems of selection of nonequivalent persons in different treatment groups, mortality of the respondent pool between pretest and posttest, correlation of response errors with the number of measurements taken over time, and ambiguity of the causal direction, among others.

External validity is "the approximate validity with which we can infer that the presumed causal relationship can be generalized to and across alternate measures of the cause and effect and across different types of persons, settings, and times" (Cook and Campbell, 1979, p.37). This concept concerns inference to persons beyond the set of respondents actually measured. This concept is akin to some of the error concepts in survey statistics. Linked to the "external validity" are notions of target populations, failure to cover different parts of the target population, and concerns with how "representative" of the full target population the sample of respondents is. Thus, what a survey statistician might call coverage, nonresponse, or sampling bias, this approach might term threats to external validity or forces toward external invalidity.

The more one reads of errors in psychological measurement, the more types of validity that one encounters. Each author seems to feel both free to ignore phrases used in prior research and compelled to invent his/her own. For example, some use the terms "ecological" validity, "methodological" validity, and "explanatory" validity to describe the ability of the researcher to describe the limits of his findings in terms of the substantive inference, procedures of data collection, and conceptual structure. Still other terms refer to methods of assessing validity— "synthetic" validity, "job-analytic" validity, "rational" validity, and "factorial" validity. It is important for the reader of these literatures to seek definitions of these terms to separate normative statements about the desirability of analytic practice from concepts of measurement error.

1.6 THE LANGUAGE OF ERRORS IN ECONOMETRICS

One additional term is required to be conversant with most discussions of errors in survey data. This arises from the field of econometrics, whose terminology for errors arises chiefly through the language of estimation of the general linear model. To illustrate the language consider the case

(Figure 1.7) in which the estimators of interest are those in a regression equation. The basis of the language of errors in this field is typically an assumption that the model is correctly specified (i.e., all predictors included that relate to those that *are* included, no others included, functional form correct, error term correctly specified). In this field, errors in the respondents' answers for the variables in the model are allowed; termed the errors in variable case, these errors are viewed to be stochastic in nature (see Kmenta, 1971). In addition, there is concern with the error of nonobservation from sampling. The observations analyzed are viewed to be a collection of events from a random process (Malinvaud, 1966).

Recently, there has been acknowledgment of the impact of errors in estimates of regression coefficients from other errors of nonobservation. These might arise when the selection process for observations systematically excludes cases with distinctive values on the dependent variable (see Chapter 3). For example, if one is studying the effects of an educational program on its participants but has data only for those who completed the program, the measured effect of the program might be altered by that omission. This general set of errors, which could arise from exclusions due to what was called "noncoverage" or "nonresponse" earlier, is labeled **selection bias** in that literature.

The econometric literature concerning selection biases offers analysts of data a variety of methods to reduce the error of nonobservation. These require a model predicting the likelihood of a member of the population being included in the data collection. To estimate these predictive models requires some data on cases which were excluded from the data collection. The methods are being compared to traditional postsurvey weighting for nonresponse and noncoverage, and conditions under which they are preferable to those methods are being identified.

1.7 DEBATES ABOUT INFERENTIAL ERRORS

Although the review of error terminology above cites most of the key differences, it underemphasizes one striking difference between the approaches of survey statistics, on one hand, and those of psychometrics and econometrics, on the other. Survey statistics concerns itself with the estimation of characteristics of a finite population fixed at a moment in time. Psychometrics and econometrics concern themselves with the articulation of causal relationships among variables. Sometimes these two viewpoints are labeled **descriptive** versus **analytic** uses of data (Deming, 1953; Anderson and Mantel, 1983) within the survey statistics

Figure 1.7 The structure and language of errors used in econometrics for estimates of regression model parameters.

Shaded concepts are not central to viewpoint of econometrics.

field, and **applied** versus **theoretical** uses of data by those in the quasi-experimental design group (Calder et al., 1982, 1983; Lynch, 1982).

Those who build causal models using data most often concentrate on the correct specification of the form of the models and less so on various errors of nonobservation. Those who use the data for descriptive purposes are more concerned with nonresponse and noncoverage issues. The debates in psychology about the relative importance of external validity (e.g., Mook, 1983) and in statistics about **design based inference** or **model based inference** essentially revolve around how the researcher conceptualizes the impact of errors of nonobservation. These debates are discussed in more detail in Chapters 3 and 6.

1.8 IMPORTANT FEATURES OF LANGUAGE DIFFERENCES

The sections above provide the kind of information common to a glossary, but they do not highlight the several reasons for misunderstanding among the various groups mentioned in Section 1.1. It appears that the answers to three questions are sufficient to eliminate most misunderstandings:

1.   What is the statistic of interest when errors are being considered?

2.   Which features of the data collection are viewed to be variable over replications and which are fixed?

3.   What assumptions are being made about the nature of those persons not measured or about properties of the observational errors?

The answer to the first question will determine whether observational errors of constant magnitude across all persons have any effects on the statistic of interest. For example, if all respondents underestimate their own weight by 5 pounds, the mean weight of the population will be underestimated, but correlations of reported weight and other variables will not be. The impact of errors at the point of data collection varies over different statistics in the same survey. Another example concerns a measurement procedure for which each person's error is constant over trials, but the sum of the errors over all persons is zero. That yields a case of a biased indicator of the person's true value, but an unbiased estimate of a population mean. (Note that such a case would violate the assumptions of true score theory.) The variation in effects of errors across statistics creates misunderstandings between the group labeled "describers" above and those labeled "modelers." Modelers most often deal with statistics that are functions of variances and covariances of survey measures, and hence some shared or constant part of measurement errors may not affect their statistics. Describers, however, are more often affected by these.

The second question determines whether a problem is viewed as a bias or as a component of variance of the statistic. One of the most common instances of this in the experience of sampling statisticians is the following. A researcher who is the client of the sampling statistician, after having collected all the data on the probability sample survey (for simplicity, let us assume with a 100 percent response rate), observes that there are "too many" women in the sample. The researcher calls this a "biased" sample; the statistician, viewing this as one of many samples of the same design (and being assured that the *design* is unbiased), views the discrepancy as evidence of sampling variance. In the view of the sampler, the sample drawn is one of many that could have been drawn using the design, with varying amounts of error over the different samples. If properly executed the proportion of women would be correct *in expectation* over all these samples. The sampler claims the sample proportion is an unbiased estimate of the population proportion. This is a conflict of models of the research process. The sampler is committed to the view that the randomization process on the average produces samples with desirable properties; the analyst is more concerned with the ability of this single sample to describe the population.[9]

The third question determines whether some types of error in the statistic of interest are eliminated by the model assumptions. This is perhaps the most frequent source of disagreement about error in statistics. The simplest example is the use of true score theory to eliminate response biases, as we saw above. Certainly if true score theory were accepted, but statistics were calculated that were affected by expected values of measures (e.g., intercepts in regression models), then the existence of error in the statistics would be dependent on that assumption. Another example concerns the fitting of a regression model with data that a survey researcher would claim is subject to large sampling bias. The analysis regressed a measure of annoyance with airport noise on a measure of the distance from the respondent's home to the airport. The sample included (1) some persons so far from the airport that the noise was just noticeable by equipment that simulated normal hearing capabilities and (2) some persons in the neighborhood closest to the

---

[9] The solution to this problem, acceptable to both positions, is a poststratification adjustment of the sample data.

airport. The survey researcher claims that the omission of persons living in intermediate neighborhoods will bias the regression coefficient.[10] The model builder claims that the omission has nothing to do with the error in the regression coefficient. The difference in viewpoints is that the model builder, following ordinary least squares theory, asserts the unbiasedness of the coefficient, *given that the model is well specified.* She is specifying a linear model, hence, the persons in intermediate neighborhoods are not needed to obtain an unbiased estimate of the slope coefficient. In contrast, the survey researcher does not accept the assumption behind the regression estimator and claims that there may be a bias in the coefficient because people in intermediate neighborhoods may behave differently than that implied by observation of the closest and farthest away neighborhoods. In essence, the survey researcher is questioning how certain the model builder should be about the specification of the model. If the model is incorrect (if a curvilinear relationship exists), the model builder will never be able to discover the misspecification with the data in hand.

The reader will need to be armed with the three questions above when reading the literature on survey errors and errors in survey analysis because the authors of articles assume very different answers to them and often fail to communicate them to the reader.

1.9 SUMMARY: THE TYRANNY OF THE MEASURABLE

Even a careful reader (especially a careful reader?) should feel somewhat overwhelmed by the wealth of terms used to describe various kinds of error. What are the important differences in these languages? What are the strengths and weaknesses of the different approaches? Is there a synthesis possible?

The largest differences among the languages arise because of the first question that needs to be posed during debates about error terminology: "What is the statistic of interest when errors are being considered?" Psychometrics and econometrics are fields devoted to measuring relationships among measures; they are fields thus populated by "modelers" attempting to unearth how phenomena affect one another. Their languages of error concern those statistics (e.g., correlations, regression coefficients) that are used to measure how characteristics covary. Their languages are rather barren concerning statistics that are univariate in nature, those describing characteristics of a population. For

---

[10] This problem is discussed in some detail in Chapter 6.

example, in classical true score theory all measurement is considered to be unbiased (only variable errors that cancel one another over replications exist). Univariate statistics is the territory inhabited by "describers." Hence, the error terminology of survey statistics, both sampling theory and total survey error, focuses mostly on estimates of means and totals. Unfortunately, some of the measurement error terms are the same across these fields. For example, when "correlated measurement error" is mentioned, it is necessary to ask the speaker what statistic is being considered. The speaker may be referring to correlation across measures or correlations across respondents (e.g., interviewer effects).

There is another large difference among the different perspectives. The survey perspectives on error describe error sources close to the measurement process itself. They identify interviewers as error-producing units, they examine features of measurements, and they concern themselves with effects of the mode of data collection. In contrast, the psychometric approach tends to view error at a later stage, as the combined result of all these possible influences. The researcher is more removed from the creation of the errors. The errors for the most part are viewed as flowing from the subjects themselves or the instrument. Most of the literature ignores possible effects of the researcher (see Rosenthal, 1966, for the exception to this rule). Furthermore, the survey literature has discussed errors of nonobservation, coverage, and nonresponse, but the psychometric literature has tended to ignore these. Absence of concern with errors of nonobservation is typical of fields dominated by modelers. Instead of worrying whether the sample base well represents a known population, the investigator is typically concerned about whether the theoretical specification of relationships among variables is correct.

Despite the different perspectives taken by various groups who produce or use survey data, all tend to give more attention to those errors for which empirical estimates have been developed. Measurable errors may, with appropriate study or alternative remedies, form a progressively smaller portion of total error in survey statistics. Errors that elude simple empirical estimation often are ignored in survey practice. Furthermore, concentration of the reduction of the measurable errors may lead to designs that reduce them successfully while increasing others.

The easiest example of this is sampling error, the one error common to the different perspectives reviewed above. Given a statistical theory that both provides estimates of the error and a set of procedures that permit its reduction, survey researchers can balance tolerable costs and desired sampling error within small limits. In addition, so ubiquitous is the reporting of sampling errors that it is probably the case that most survey designs are determined to minimize that error alone, within the constraints of the budget.[11] In contrast, for example, nonresponse error, in the absence of empirical estimates of its effects on statistics, is often relegated to a secondary priority. Nonresponse *rates* are given some attention in technical appendixes of survey reports, but rarely do designers or analysts alter survey designs to increase sampling error in order to reduce nonresponse error. Why not? Only rarely could the designer be assured of what reduction in nonresponse error would result from the reallocation of resources from a given number of sample units to greater efforts to measure a smaller number of sample units.

Numbers have a power over design decisions that is partially legitimate. Statistical sampling theory is a useful science in measuring and reducing *sampling* error. Alternative statistical models of measurement error provide empirical estimates, given appropriate designs. However, the *art* of survey design consists of judging the importance of unmeasurable (or not cheaply measurable) sources of error relative to the measured. This requires some understanding of the causes of the various errors, their relative sizes, changes over time in their effects, and the costs of efforts to reduce them. Unfortunately, those designing surveys are often specialists in one error or another, rarely exploring the interconnected nature of the various error sources. The challenge facing the field is either (1) how to blend quantitative and nonquantitative information in minimizing survey error for fixed costs, or (2) how to develop quantitative estimates of the error sources that have previously defied measurement.

1.10 SUMMARY AND PLAN OF THIS BOOK

Diverse activities involving survey research have lead to diverse languages and concerns about survey quality. Describers use survey data to estimate characteristics of fixed populations. Modelers use the same data to test theoretical hypotheses about social processes. Biases for estimates of means and proportions, of concern to the describer, may not trouble the modeler, who is more interested in errors affecting covariance structures of data. The describer worries more about errors of nonobservation (especially coverage and nonresponse), while the modeler worries more about errors of observation (especially those arising from the respondent or questionnaire).

---

[11] An interesting survey could be done on the society's understanding of sampling error, as presented in media polls. Even such a poll among most journalists would be enlightening. Most persons, I suspect, believe that the confidence intervals cited are a measure of total error and that the survey estimate is unbiased.

Many sampling statisticians devote their energies to understanding the sampling error properties of diverse estimators from varied sample designs. Since the 1930s their work has produced the most important advances in the survey method. In unfortunate isolation of their research, however, other methodologists have attempted to reduce and sometimes construct measures of the various nonsampling errors (i.e., coverage, nonresponse, and measurement errors) which affect survey data. The total survey error approach attempts to acknowledge all sources of errors simultaneously. It attempts therefore to minimize mean square error of estimates not merely the sampling variance of them. The potential of the approach has not been realized in most applications because of the failure to provide measures of various nonsampling errors.

The conceptual error structure used in this book separates errors of nonobservation (coverage, nonresponse, sampling) from those occurring during observation (measurement errors). It discriminates variable errors from biases. Variable errors are those that take on different values over replications of the measurement design. These replications are sometimes internal to the survey (e.g., different interviewers) and sometimes only conceptual (e.g., different samples using the same design). Biases are the components of errors viewed to be constant over replications. Any bias can be related to a variable error if the design unit producing it is viewed to vary over replications. For example, coverage error differences might arise if the same set of interviewers are assigned to different sample blocks to list sample addresses.

Concepts of validity and reliability dominate the psychometric notions of error in survey data. Both of these would be called types of variable error in the total survey error perspective. Biases are irrelevant to the measures of correlation and covariance is important to the psychometric enterprise. Errors of nonobservation have been brought into the psychometric perspective through notions of external validity, but they have no mathematical formulation of practical utility. The econometric error concept of selection biases incorporates noncoverage and nonresponse errors into the framework of analysis of analytic models.

It is not likely that nonsampling errors will yield themselves to well grounded statistical theory in the near future. That does not imply that our understanding of their processes cannot improve. The acknowledgment that survey estimates are subject to many sources of error is a first step which then demands attention to coverage, nonresponse, and measurement errors.

This book is organized around different error sources. There are chapters on coverage error, nonresponse error, sampling error, and measurement errors. The first chapters discuss errors of nonobservation:

Chapter 3	—	Coverage Error
Chapters 4 and 5	—	Nonresponse Error
Chapter 6	—	Sampling Error

The latter chapters describe various properties of measurement error:

Chapter 7	—	Designs to Estimate Measurement Error
Chapter 8	—	The Interviewer as a Source of Measurement Error
Chapter 9	—	The Respondent as a Source of Measurement Error
Chapter 10	—	The Questionnaire as a Source of Measurement Error
Chapter 11	—	The Mode of Data Collection as a Source of Measurement Error

Each of the chapters first introduces the error source in conceptual terms. Then relevant social science, survey methodological, and statistical literatures are reviewed which are relevant to the error source. Sometimes these literatures overlap. For example, there may exist cognitive psychological theory which explains respondent behavior productive of a certain error in responding to a question. This theory may have motivated experiments imbedded in surveys which suggest a particular statistical model of the error-producing process.

For the most part, however, the social science, statistical, and methodological literatures are not well integrated. The discussions in the chapters review the literatures organized into sections on the measurement of a particular error source and sections describing attempts to reduce the error. Sometimes social science research is described which has not yet been applied to survey research, but which is descriptive of behaviors that appear to resemble those producing errors in surveys. In short, the social sciences are called upon to answers questions about what produces the error in question. Why do people refuse to respond to surveys? Why can't respondents provide accurate answers to questions about past events they experienced? The statistical literature is used to characterize the impact of these errors on the survey estimates.

Finally, each chapter attempts, whenever possible, to discuss the impact on costs of attempts to reduce a particular error source. This acknowledges the general point that, in many ways, cutting corners in a survey or attempting to save money often affects some error property of the estimates from the survey (e.g., attempting to contact a sample person only once typically increases nonresponse error). On the other hand, it also appears to be true that attempts to eliminate one completely greatly inflates survey costs (e.g., increasing a response rate from 90 to 100 percent typically increases costs to intolerable levels). Costs and errors are related. The nature of their relationship has been studied for some errors but not for all. Hence, our discussions of survey costs related to error sources are rich for some sources and meager for others. In order to motivate the role of costs in survey design approaches, the next chapter reviews a variety of issues in cost modeling.

Summaries of each chapter are intended both to review briefly the discussion of the chapter and to comment on the current state of understanding of a particular error source. This understanding concerns both methods to measure the error and methods to reduce it. We shall see that those two goals are not always jointly achieved.

APPENDIX FOR MATHEMATICAL PRESENTATION OF ERROR PERSPECTIVES

1.A Classifications of Error Within Survey Statistics

The discussion below presents in mathematical notation the total survey error perspective represented graphically in Figure 1.3. As with the English version, terms are defined beginning with the top of the figure. The statistic of interest is a sample mean.

Define the **mean square error** as

$$\text{Mean Square Error} = \text{Variance} + \text{Bias}^2$$

$$E_{s,t,i,a}[\bar{y}_{stia} - \overline{X}]^2 = E_{s,t,i,a}(\bar{y}_{stia} - \bar{y}_{....})^2 + (\bar{y}_{....} - \overline{X})^2 ,$$

where $E_{s,t,i,a}[\ ]$ = expectation over all samples, s, given a sample design; all trials, t; all sets of interviewers, i, chosen for the study; and all assignment patterns, a, of interviewers to sample persons;

$\bar{y}_{stia}$ = mean over respondents in the sth sample, tth trial, ith set of interviewers, ath assignment pattern of interviewers to sample persons, for y, the survey measure of the variable X in the target population; ·

$\bar{y}_{....}$ = expected value of $\bar{y}_{stia}$ over all samples of respondents, all trials, all sets of interviewers, and all assignment patterns;

$\overline{X}$ = mean of target population for true values on variable X.

The **bias** of the mean is

$$(\bar{y}_{....} - \overline{X}).$$

The **variance** of the mean is

$$E_{s,t,i,a}\left(\bar{y}_{stia} - \bar{y}_{....}\right)^2.$$

It contains as components **sampling variance**, due to the selection of different persons,

$$E_{t,i,a}\left(E_s\left(\bar{y}_{stia} - \bar{y}_{\cdot tia}\right)^2\right),$$

where $\bar{y}_{\cdot tia}$ = expected value of $\bar{y}_{stia}$ over all samples of respondents, s, given a sample design.

Nonresponse variance, due to stochastic respondent decisions, is

$$E_t\left(E_{s,i,a}\left(\bar{y}_{stia} - \bar{Y}_{stia}\right)^2\right),$$

where $\bar{Y}_{stia}$ = mean on the survey measure y for all sample persons in sample s, both respondents and nonrespondents.

There is a similar term for coverage variance.[12]

There are also variance terms for observational errors. **Simple response variance** is defined as

$$E_{s,i,a}\left(E_t\left(\bar{y}_{stia} - \bar{y}_{s\cdot ia}\right)^2\right) = \sigma_r^2.$$

Because of the correlation of errors among a group of respondents interviewed by the same interviewer (the notion of **correlated response variance**), response variance is inflated so that the total response variance is

$$\frac{\sigma_r^2}{nk}(1 + \rho(n-1)),$$

---

[12] There are also covariance terms involving coverage, nonresponse, and sampling error, but few treatments in the total survey error literature treat them empirically. Fellegi (1964) discusses the covariance term between sampling and measurement errors.

where σ_r^2 = simple response variance;

n = number of respondents interviewed by each interviewer;

k = number of interviewers employed on the survey;

ρ = intraclass correlation of response deviations for respondents interviewed by the same interviewer.

Bailar and Dalenius (1969) and Bailey et al. (1978) label $\rho\sigma_r^2$ the **correlated response variance**.

Coverage bias is

$$E_{s,t,i,a}\left((P_{nc})_{stia}\left(\overline{X}_{stia} - (\overline{X}_{nc})_{stia}\right)\right),$$

where $(P_{nc})_{stia}$ = proportion of the target population not covered by the frame during frame construction operations on the sth sample, the tth trial, with the ith set of interviewer, and the ath assignment pattern;

$\overline{X}_{stia}$ = mean of true values for those covered by the frame in the sth sample, the tth trial, ith set of interviewers, and the ath assignment pattern;

$(\overline{X}_{nc})_{stia}$ = mean of true values for those not covered by the frame in the sth sample, the tth trial, ith set of interviewers, and the ath assignment pattern.

Nonresponse bias is

$$E_{s,t,i,a}\left((P_{nr})_{stia}\left(\overline{y}_{stia} - (\overline{y}_{nr})_{stia}\right)\right),$$

where $(P_{nr})_{stia}$ = proportion of the sample that is nonrespondent for the sth sample, tth trial, ith set of interviewers, and ath assignment pattern;

$(\bar{y}_{nr})_{stia}$ = mean of true values for those nonrespondent on sth sample, the tth trial, ith set of interviewers, and the ath assignment pattern.

There is a similar definition for sampling bias and the other biases.

1.B Terminology of Errors in Psychological Measurement

Much of the mathematical notation for psychometric measurement is presented clearly in Lord and Novick (1968). In order to call attention to those aspects of psychometric measurement that differ from survey statistics, we have adapted their notation to use the same symbols as above.

The true score X_j on the **construct** X, of a person, j, on the **indicator** g is defined as the expected value of the observed score; that is,

$$X_j = E_t\,[y_{tgj}]\,,$$

where y_{tgj} = response to indicator g on the tth trial for the jth person;

$E_t[\ ]$ = expectation with respect to the **propensity distribution** over trials of the indicator's administration for the jth person.

Thus, the model of measurement is

$$\text{Response} = \text{True Score} + \text{Error}$$

$$y_{tgj} = X_j + \varepsilon_{tgj}\,,$$

where ε_{tgj} = the error for the gth indicator committed by the jth person on the tth trial.

In a population of persons it is assumed that $\text{Cov}(X,\varepsilon) = 0$; where $\text{Cov}(\ )$ is the covariance over trials and persons in the population.

Theoretical validity of the gth indicator, for the population of which the jth person is a member, is

Covariance of Indicator and True Score

(Standard Deviation of Indicator)(Standard Deviation of True Score)

$$\frac{E_{tj}\Big([y_{tgj} - \bar{y}_{\cdot g \cdot}] [X_j - \bar{X}_{\cdot}] \Big)}{\sqrt{E_{tj}[y_{tgj} - \bar{y}_{\cdot g \cdot}]^2}\sqrt{E_{tj}[X_j - \bar{X}_{\cdot}]^2}} = \rho_{yX},$$

where ρ_{yX} = correlation between true scores and observed
        values over trials and persons in the population;

$\bar{y}_{\cdot g \cdot}$ = mean over persons and trials of observed scores
        on indicator y;

$\bar{X}_{\cdot}$ = mean over persons of true values.

Reliability is the proportion of the observed score variance
associated with the true score:

Variance of True Score

Variance of Indicator

$$\frac{E_{tj}[X_j - \bar{X}_{\cdot}]^2}{E_{tj}[y_{tgj} - \bar{y}_{\cdot g \cdot}]^2} = \frac{\sigma_X^2}{\sigma_y^2} = \rho_y,$$

where σ_X^2 = variance of the true scores across the population and
        trials;

σ_y^2 = variance of the observed scores across the population;

ρ_y = index of reliability.

Reliability and theoretical validity are thus related because

Validity = Square Root of Reliability

$$\rho_{yX} = \frac{\text{Cov}(y,X)}{\sigma_y \sigma_X}$$

$$= \frac{\text{Cov}(X + \varepsilon, X)}{(\sigma_y \sqrt{\rho_y})\sigma_y}$$

$$= \frac{\sigma_X^2}{\sigma_y^2 \sqrt{\rho_y}}$$

$$= \sqrt{\rho_y}$$

where $\text{Cov}(y,X)$ = covariance of true scores and observed scores
over trials and persons in the population.

Empirical validity or **criterion validity** of y_1 in relation to y_2 is

$$\rho_{y_1 y_2}$$

where y_1 = one indicator of X;

y_2 = a second indicator of X;

$\rho_{y_1 y_2}$ = correlation of y_1 and y_2 over trials and persons
in the population.

Because

$$y_{1j} = X_j + \varepsilon_{1j}$$

and

$$y_{2j} = X_j + \varepsilon_{2j},$$

it follows that $\rho_{y_1 y_2} \le \rho_{y_1 X}$. If, in addition, $\text{Var}(\varepsilon_1) = \text{Var}(\varepsilon_2)$, y_1 and y_2
are called **parallel measures**.

If y_1 and y_2 were measured at the same time, $\rho_{y_1 y_2}$ would be termed **concurrent validity**; if measured at two points in time, **predictive validity**.

If y_1 and y_2 measure one construct X, and y_3 and y_4 measure another, X', and

$$y_{1j} = X_j + \varepsilon_{1j},$$

$$y_{2j} = X_j + \varepsilon_{2j},$$

$$y_{3j} = X'_j + \varepsilon_{3j},$$

$$y_{4j} = X'_j + \varepsilon_{4j},$$

then to the extent that $\rho_{y_1 y_2}$ and $\rho_{y_3 y_4}$ are greater than $\rho_{y_1 y_3}$, $\rho_{y_1 y_4}$, $\rho_{y_2 y_3}$, and $\rho_{y_2 y_4}$, there is evidence for **discriminant validity**. To the extent that $\rho_{y_1 y_2}$ and $\rho_{y_3 y_4}$ are high, there is evidence for **convergent validity**.

Two indicators are viewed as **congeneric measures** if

$$y_{1j} = X_j + \varepsilon_{1j}$$

and

$$y_{2j} = X_j + \varepsilon_{2j},$$

where $X_j \neq X_j$, but all other assumptions in classical true score theory above apply;

and

$$X_{1j} = \mu_1 + \beta_1 X''_j$$

$$X_{2j} = \mu_2 + \beta_2 X''_j.$$

That is, the expected value of each indicator is a linear function of the same underlying variable, X''_j.

The multifactor model views each indicator as the result of the person's value on the underlying construct and some systematic error due to another (disturbing) factor:

Response = Population Mean + Influence of True Value
+ Method Effect + Error

$$y_{gjk} = \mu_g + \beta_g X_j + \alpha_g M_{jk} + \varepsilon_{gjk} ,$$

where y_{gjk} = response on the gth indicator, for the jth person, using the kth method;

μ_g = a constant for the population;

β_g = coefficient for the indicator on the construct;

X_j = true value of the construct for the jth person;

α_g = coefficient for the indicator on the method effect variable, k;

M_{jk} = effect experienced by the jth person on the kth method.

With this measurement model, reliability is defined as

$$\frac{\text{Var}(\beta_g X_j + \alpha_g M_{jk})}{\text{Var}(y_{gjk})} ,$$

and validity is defined as

$$\frac{\text{Var}(\beta_g X_j)}{\text{Var}(y_{gjk})} .$$

Note that, with the introduction of systematic error, the simple relationship between reliability and validity is broken. Reliability will always be higher than validity, as defined immediately above. Indeed, if β_1 is zero (the indicator reflects only the method effect), then reliability can be 1.0, but validity would be zero.

There appear to be no simple mathematical descriptions of notions of **statistical conclusion validity**, **internal validity**, or **external validity**.

1.C The Language of Errors in Econometrics

Assume interest in estimating the regression equation,

Dependent Variable Response = Constant + Effect of x + Error

$$y_j = \alpha + \beta x_j + \varepsilon_j$$

where y_j = response for jth person on variable y ,

   x_j = response for jth person on variable x ,

   α, β = unknown model parameters .

Assume $\varepsilon \sim N[0, \sigma^2]$, for all persons,

   $E[\varepsilon_i, \varepsilon_j] = 0$, for all $i \neq j$

Furthermore, assume we observe

$$y_j^* = y_j + v_j \text{ and } x_j^* = x_j + w_j ,$$

where y_j^* = response on y from jth person;

   y_j = true value on y for jth person;

   v_j = error on y for jth person;

   x_j^* = response on x for jth person;

   x_j = true value on x for jth person; and

   w_j = error on x for jth person .

such that

$$E[v_i v_j] = E[w_i w_j] = E[v_j w_j] = E[v_j \varepsilon_j] = E[w_j \varepsilon_j] = 0, i \neq j .$$

That is, there are variable but uncorrelated measurement errors on both x
and y.

If a regression is run on the observed scores,

$$y_j^* = \alpha' + \beta' x_j^* + \varepsilon'_j,$$

then $\beta' \neq \beta$, the result of **errors in variables**.

By **selection bias** is meant that the observed values of y in the data collection are limited to a subset with different variance and covariance properties than those of the full population. In truncation, all observed y_j's in the sample, s, are such that $y_j < Z$, some threshold value. In that case,

$$y_{sj} = \alpha_s + \beta_s x_{sj} + \varepsilon_{sj},$$

estimated on the sample, will have intercept and slope coefficients that differ from

$$y_j = \alpha + \beta x_j + \varepsilon_j,$$

as estimated on the full group, containing those truncated out of the observations (see Chapter 3, Section 3.3.3). That is, $\beta_s \neq \beta$.

CHAPTER 2

AN INTRODUCTION TO SURVEY COSTS

In the planning of a survey, effort should be directed toward the reduction of all of the errors that it is possible to reduce, but the effort should be apportioned with a view to producing the greatest possible usefulness with the funds available.

W. Edwards Deming, "On Errors in Surveys,"
American Sociological Review, *1944, p. 359*

2.1 RATIONALE FOR A JOINT CONCERN ABOUT COSTS AND ERRORS

Costs are rarely treated seriously in the texts used to teach students about survey design. They are even less often viewed as a serious topic, worthy of their own attention, by survey statisticians or survey methodologists, who prefer to concentrate on error. From one perspective, however, survey costs and errors are reflections of each other; increasing one reduces the other. In a limited number of *sample* design problems this has been acknowledged since Neyman's (1934) time. In those sampling problems, the design option that minimizes sampling error for a fixed total survey cost is identified. For example, what allocation of the sample to strata should be chosen to minimize the sampling variance of the overall sample mean, if there is $100,000 available for the survey?

In practice many survey designs are fitted within cost constraints. That is, the funding agency limits the amount of money allocated, so that any research accomplished must cost less than or equal to that amount. Unfortunately, these ceilings are often not set with any design options in mind but instead are informed scientific and political hunches about what areas should be given priority funding. Clearly, the harmful potential of this practice is that opportunities for great increase in quality sometimes cost only a fraction more but are ruled out, given the ceiling. For the survey designer the potential harm is that design options may be limited to those affordable without consideration of a slightly more expensive option which could greatly improve quality.

Surely, the opposite extreme is neither tolerable nor desirable—the minimization of error without cost constraint (usually the answer to the command, "I want the best survey you can do!"). It is a rare case that such a demand is wisely given, and probably even rarer that a survey design could be mounted to achieve that end.

2.2 USE OF COST AND ERROR MODELS IN SAMPLE DESIGN

Introductory sampling texts provide a simple design problem solved through the use of cost and error models. The designer is given a certain amount of money, C, to do the survey. Furthermore, the strata for the design have already been identified, and the decision has been made to draw simple random samples in each stratum. The question is what sample size should be drawn from each stratum in order to minimize the sampling variance of the sample mean, given fixed cost. Defined out of the designer's concern are the other sources of survey error affecting the accuracy of the sample mean (e.g., nonresponse error, measurement error).

The sampling variance of the sample mean, $\bar{y} = \sum W_h \bar{y}_h$, in a stratified random design has a simple expression:

$$\mathrm{Var}(\bar{y}) = \sum_1^H \frac{(1-f_h)\,W_h^2 S_h^2}{n_h}$$

$$= \sum_1^H \frac{W_h^2 S_h^2}{n_h} - \sum_1^H \frac{W_h^2 S_h^2}{N_h},$$

where $\bar{y}_h$ = sample mean for the hth stratum;

$\quad f_h$ = sampling fraction in the hth stratum ($h = 1, 2, ..., H$);

$\quad W_h$ = proportion of the frame population in the hth stratum;

$\quad S_h^2$ = element variance for the variable y in the hth stratum;

$$\sum_i (Y_{hi} - \bar{Y}_h)^2 / (N_h - 1);$$

$\quad n_h$ = sample size in the hth stratum;

N_h = total number of elements in the hth stratum frame population.

Note that the values in question, the n_h's, are components in the error model; hence, the sampling error is affected by the decision faced by the designer. The allocation of the sample cannot be allowed to cost more than C, and thus costs become a constraint within which solutions must be identified. To solve the problem a cost model needs to be developed which contains terms that are also present in the error model. In other words, we need to determine the costs of each of the design units in the error model (the n_h's). As is typical with this approach, each of the units which acts to improve the quality of the survey statistics (in this case numbers of sample elements) also brings with it a cost. The larger the sample size in any one stratum, the fewer the resources available to the researcher for other strata. In this simple problem there is only one set of cost components that varies across sample designs—the n_h terms. All other terms are fixed as part of the essential survey conditions.

The traditional cost model is

Total cost = Fixed costs + Variable Strata Costs,

$$C = C_o + \sum_1^H C_h n_h \,,$$

where C_o = fixed cost, to be incurred regardless of what sample size is chosen;

C_h = cost of selecting, measuring, and processing each of the n_h sample cases in the hth stratum.

Note that the cost model is parameterized in terms that are shared by the error model, the n_h's. The cost model could have been presented as a function of supervisors and interviewers' salaries, materials costs, computer time, and so on, but that would have failed to represent clearly the fact that some costs rise as sampling variance decreases. Those are the only parts of the cost equation that will determine the optimal allocation, the others will determine the overall sample size that can be purchased but not what proportion of the units should be allocated to each stratum.

Minimizing Var($\bar{y}$) subject to the constraint that the total resources available are C is the same as minimizing

$$\sum_1^H \frac{W_h^2 S_h^2}{n_h} - \sum_1^H \frac{W_h^2 S_h^2}{N_h} + \lambda \left(\sum_1^H n_h C_h + C_o - C \right),$$

where λ is the some constant.

This minimum is achieved when the partial derivatives of the expression with respect to n_h are set to zero. That is,

$$\frac{-W_h^2 S_h^2}{n_h^2} + \lambda C_h = 0,$$

so that

$$n_h \sqrt{\lambda} = \frac{W_h S_h}{\sqrt{C_h}} \quad \text{and} \quad f_h = \frac{n_h}{N_h} = \frac{\sqrt{\lambda} N S_h}{\sqrt{C_h}}.$$

Thus, the sampling variance of the overall mean is minimized given the H strata if the sampling fractions are set equal to the expression above. This is often presented as

$$f_h = \frac{n_h}{N_h} = \frac{K S_h}{\sqrt{C_h}},$$

where K is some constant chosen to produce the desired sample size, costing C in total. The solution is thus to set the sampling fractions in each stratum proportional to the element standard deviation and inversely proportional to the square root of the cost per interview in the stratum.

This solution is limited to one source of error (sampling error), one statistic (the overall sample mean), and one variable (y). It conditions on the fact that strata have been predefined. It assumes that the cost model is well specified; that is, the functional form of the model is correct (all units in the hth stratum have identical cost, C_h, and no additional costs other than C_o are incurred), and the parameter values used are correct for the survey to be done (C_o and C_h values are those to be incurred in the upcoming survey).

Because of its practical limitations this model is little used in day to day survey work (most often because each statistic in the survey might yield a different optimal allocation across strata). It *is* used routinely, however, to assess whether departures from proportionate allocation seem to be warranted for all major survey statistics. Thus, it is used as a standard not for optimizing but "proximizing" (see Kish, 1976). Unfortunately, in our view, the solution's applicability to the problem at hand is not always critically examined. While the sampling variance formula in the problem has its justification in statistical theory, there is no such justification for the cost model. However, the "optimal" allocation solution is conditioned on the cost model being appropriate. In the strict sense, the designer must justify that the cost model applies to his/her problem separately for each problem encountered.

Of special interest is the fact that the cost model proposed

1.   is linear in form (a simple increasing function of sample size),

2.   is continuous in nature (except for the integer constraints of sample size),

3.   is deterministic (there are no stochastic properties),

4.   has an implicit domain of applicability with all integer numbered sample sizes, $n_h = 1, 2, ..., N_h$.

The issues addressed below include the following:

1. How such a cost model might be made of more practical use to the survey designer, especially when cost-efficient designs are sought in terms of sampling and nonsampling errors. (The next section reviews some criticisms of using cost-error models in search of optimality).

2. What alterations to traditional survey cost models might make them more practical. (It attempts to extract some principles of cost model construction.)

3. Whether analytic or empirical approaches are most likely to offer useful guides to the designer. (It proposes simulation models for checking the implications and exploring the weaknesses of cost and error models.)

2.3 CRITICISMS OF COST-ERROR MODELING TO GUIDE SURVEY DECISIONS

Legitimate questions have been raised about the practicality of cost and error modeling at the survey design stage. Are those efforts most useful as idealized protocols for the designer (i.e., guidelines from which the practicing designer must depart)? Is the approach another example of theory being ahead of day to day practice? In a thoughtful commentary, Fellegi and Sunter (1974) offer a set of criticisms of attempts to address multiple sources of error and cost to guide design decisions. They deserve discussion:

1. *There are practical constraints.* Not all courses of action are open to the researcher; even if a design can be identified with desirable error properties, if it is infeasible for other reasons, it cannot be considered seriously.

Comment: The task of identifying feasible alternative designs takes place prior to the selection of that single feasible design that is most efficient. Instead of choosing the optimal design without any conditions, we always choose the optimal design, conditioning on fixed resources and feasibility. This must be explicitly acknowledged.

2. *Major alternative survey designs do not present themselves within a fixed budget.*

Comment: This is true. With an increase of 50 percent in a survey budget, it is sometimes possible to achieve an 80 percent reduction in error: for example, the choice of a telephone interview option, when considering a mailed questionnaire survey may entail fixed costs sufficiently high that even the smallest sample size could not be purchased by the allocated budget. It is the designer's obligation to assess whether the budget could be expanded to permit a threshold jump in quality, but lacking that ability, the feasible alternatives again define the limiting conditions of the design choice.

3. *Components of the loss function may not be continuous, let alone differentiable, over the whole range of possible designs.* Loss functions in this context describe the increase in error associated with departures from an optimal design. Differentiable loss functions permit simple identification of optimal design parameters. Loss functions may reflect sudden increases in error that correspond to the passage of an important threshold.

Comment: This is obviously true. For example, results by Fowler and Mangione (1985) imply that tape recording of interviewers may offer important (discontinuous) reductions in interviewer errors in delivery of the questionnaire. However, the fact that the loss function is not differentiable is important only if one is seeking a mathematical expression for the optimum. Instead the investigator should be seeking to locate the discontinuities in the loss function because they typically offer great reductions in error with minimal cost impact.

4. *In a complex design the error reduction functions will be complex and may depend in some way, often impossible to see in advance, on the ingenuity of the designer in using a single technique to reduce simultaneously a number of error components.* For example, since personal interviewers make multiple visits to sample segments seeking those previously uncontacted, they can also at very low marginal cost attempt persuasion efforts for others in the same segment.

Comment: The cost-error approach is not foolproof; indeed, it should offer a discipline in which creative solutions can be evaluated, not necessarily in which they will be identified automatically. As we noted above, the approach offers a systematic way to evaluate alternatives that have been identified. The hope that it automatically yields optimality is naive and dangerous.

5. *Terms in the error function may interact in some unknown way.* Examples of this abound. There is evidence that response errors on survey questions are a function of the respondent's true value on the item (e.g., rich persons tend to underestimate income; poor persons tend to overestimate income).

Comment: This is true; it offers a challenge to the designer but also offers the promise of ways of understanding the error source, if the different groups can be identified. For the designer, sensitivity analyses need to be performed on such interaction terms in order to check how important they might be to the efficiency of a particular design.

6. *Important interactions may exist between different surveys.* Although developing a national field force of personal interviewers may not be cost efficient for a single survey, relative to using a centralized staff of telephone interviewers, it might become so in the context of several surveys being conducted by the same organization. The cost-error modeling approach in and of itself does not recognize this.

Comment: This is true and must be part of the consideration of the specification of the cost model. This forms the practical part of defining the "essential survey conditions."

7. *Major surveys are seldom designed to collect only one item of information.* This means that a single "optimal" design for a multipurpose survey cannot be identified.

Comment: Again, this is an important weakness in the approach only if the sole purpose is to solve for the optimal design, instead of using cost and error models to evaluate alternatives. The investigator who knows when faced with the fact that different statistics are best estimated with different designs must *explicitly* determine the relative importance of the statistics. This can only benefit the design relative to the blind acceptance of some standard solution.

8. *A survey is seldom designed to measure variables at a single level of aggregation; subclass statistics are also important.*

Comment: Same as point 7.

9. *The time constraint of the survey may inject another set of considerations very much related to the balance between different sources of error.* For example, although field follow-up of nonrespondents to a mailed questionnaire survey may be an efficient alternative, the time required to conduct the follow-up may eliminate it as a feasible design.

Comment: This again is a source of limitation to the design alternatives that need to be considered.

10. *The method spends part of the budget to obtain data on costs and errors of components of the design, yet it offers no guidance on how much money should be spent on those evaluation activities.*

Comment: This is true. However, the system can provide a retrospective evaluation of the loss of efficiency or wastage of resources attached to a design more naively chosen.

In short, building cost and error models is a vacuous exercise if its sole purpose is the identification of optimal designs. That is an elusive goal, for the reasons reviewed above. The other extreme, however, is the use of standard designs, ones used for earlier surveys with very different goals, because of their convenience.

Rather than either of these extremes it seems preferable to explore, within a given range of resource allocation, survey design options that produce different error levels for the survey statistics desired. The value in the cost/error approach is, however, in the discipline it forces on the designer, the seeking of cost and error parameters, of specification of analytic purposes, and of administrative options. In the absence of good estimates of costs and errors, simulations of design options, exploring the likely range of error and cost parameter values, can alert the designer to risks posed by various design alternatives.

2.4 NONLINEAR COST MODELS OFTEN APPLY TO PRACTICAL SURVEY ADMINISTRATION

The example of sample allocation in a stratified element design above conditions on the specification of a cost model. If the cost model is not correctly specified, the optimal allocation changes. Thus, our first principle is that the specification of the cost model does affect results.

The traditional cost model for the stratified random sample allocation problem becomes less attractive when one gets nearer to the features that actually have costs attached to them and that correspond to errors in the statistics. For example, sometimes some strata produce unusual problems of access of sample persons or reluctance of sample persons to cooperate with an interview request. The first cases from the strata may require novel, time-consuming, expensive trial and error to determine the best technique for obtaining the survey measures. In general, there is evidence that the efficiency of interviewers increases with the number of interviews completed. As interviewers encounter different reactions from sample persons, they learn what tactics are successful in obtaining the interview. Thus, the relationship between sample size in a stratum and costs may not be linear, but curvilinear, so that the cost per unit is a decreasing function of sample size in the stratum. This model would be most appropriate for designs in which different interviewers were employed to conduct the data collection in different strata; for example, the strata may be in different geographical regions and interviewers may be hired from those areas to do the work.

To illustrate this principle in a simple way, consider the same stratified random element sample problem as above. The survey will be performed by a new administrative staff, new supervisors, and new interviewers. The target population consists of households in a metropolitan area. Two strata are defined: the central city and the suburbs. The central city has high crime rates, pervasive sample household suspicions about the intent of the survey, and unusual patterns of persons being at home which present difficulties for the interviewers. The suburbs present an easier task, one conforming to the training guidelines of a typical survey organization.

As above, the sampling variance formula for the overall sample mean is

$$\text{Var}(\bar{y}) = \sum_1^H \frac{(1 - f_h) W_h^2 S_h^2}{n_h},$$

where in this case

$H = 2$, $h = 1$ for the suburbs, $h = 2$ for the central city;

$W_1 = .5$, $W_2 = .5$ for this illustration;

$S_1 = 3$, $S_2 = 1$;

and the f_h's are so small that $(1 - f_h) \doteq 1$ for $h = 1, 2$.

The cost model is specified in a way to reflect the effects of experience with difficult cases in the central city. With experience the costs of gaining the cooperation of the inner city sample households declines; one possible cost model is a quadratic:

$$C = C_o + C_1 n_1 + C_2 n_2$$

$$= C_o + C_1 n_1 + (A_2 + B_2 n_2) n_2$$

where, for illustrative purposes,

C = total resources available for the survey, \$55,000;

C_o = fixed cost of doing a survey, \$10,000;

C_1 = cost per sample case in the suburbs, \$11.04;

A_2 = base cost per sample case in the inner city, \$33.12;

B_2 = change in cost per sample case with increasing sample size in the inner city, \$−0.0125.

The coefficients reflect a decline in cost somewhat larger than that found empirically for interviewers with increasing experience. Figure 2.1 presents the total interview costs of the two strata under the quadratic cost model. The curve of the quadratic model illustrates that it becomes an unacceptable cost model when the total sample size in the central city

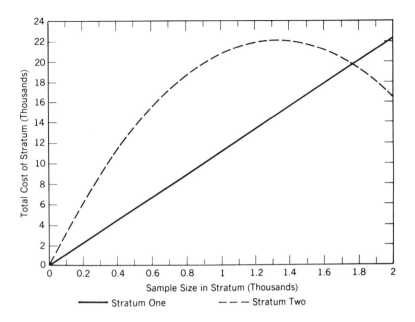

Figure 2.1 Total cost of stratum sample, by sample size, by stratum.

becomes larger than about 1100. Thus, the cost model might be stated with the limitation that $n_2 < 1100$ or so.

This model is compared to the simple linear cost model in which

$$C = C_o + C_1' n_1 + C_2' n_2 ,$$

where for our purposes,

$C_1' = \$11.04;$

$C_2' = \$33.00$, the cost of interviewing after at least 10 cases have been completed, if the quadratic model is true.

Figure 2.2 presents the sampling variance of the overall mean by different allocations of the sample, as indicated by the proportion of the total sample allocated to the suburbs (between 66 and 95 percent of the sample to the suburbs). In those figures the sampling variance for a design given the simple cost model is compared to the sampling variance, given the altered cost model. This illustrates that the optimum allocation for the simple cost model would be approximately 84 percent to the suburbs, but for the quadratic about 78 percent to the suburbs. If the

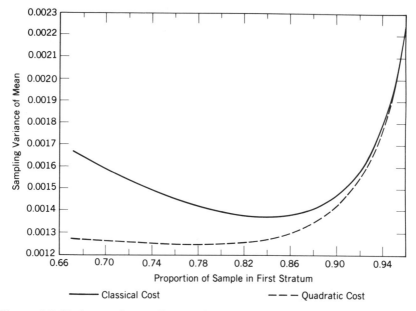

Figure 2.2 Variance of overall mean by allocation to first stratum, under the classical and quadratic cost models.

quadratic cost model applies and we erroneously chose the 84 percent allocation, a (.001272/.001246) − 1 = 2 percent loss of sampling variance would be incurred. This is a minor deviation and is probably of no importance to most researchers. Of more interest is the fact that the shape of the variance curve changes with the cost model. The quadratic model diminishes the differences in costs between strata as the sample size in the central city decreases (the left portion of the graph) and thus the quadratic model yields a statistic whose sampling variance is less sensitive to the strata allocation. The curve is much flatter for the quadratic model, demonstrating to the designer a wider range of equally desirable designs than would be implied by the simple cost model.

This illustrates that cost model specification does affect conclusions about desirable sample design features. Different conclusions would have been reached if there had been no attempt to tailor the cost model to the situation faced by the designer. With this illustration, the differences might be large enough to dictate a different solution. It is obvious that this would not always be the case. As the B_2 coefficient approaches zero in the quadratic equation, the simple cost model is approached. The observation made here is that some attention needs to be paid to the specification of the cost model to determine whether the simple form is sufficiently appropriate to the survey administration being proposed.

2.5 SURVEY COST MODELS ARE INHERENTLY DISCONTINUOUS

Fellegi and Sunter (1974) noted that loss functions are often discontinuous. This sometimes reflects the fact that a survey often involves the use of large and expensive discrete administrative units. For example, as the size of the data collection effort in a national personal interview survey grows, the number of field supervisors (or regional offices) must be increased. These additions involve large sums of money and thus greatly affect the cost per interview in the sample range in which the addition is made. Implicit in much survey administration is the notion of capacities of each administrative unit—for example, each regional office can oversee 1000 interviews per month, each supervisor can be responsible for 30 interviewers in the field, and each shift supervisor can supervise 9 telephone interviewers in a centralized facility.

These are averages about which there might be great variation, some variation due to costs of any change within an existing organization, and other variation due to the problem of small numbers. The source of the variation associated with natural inertia in survey organizations is most easily illustrated with an example of larger than normal surveys. Imagine an existing survey organization accustomed to 10 regional supervisors, yielding a supervisor to interviewer ratio of 1:20 (for the approximately 200 interviewers used). If a large survey is introduced into this organization, say one requiring 300 interviewers, it is likely that no new supervisors would be hired, and the supervisory ratio would be increased to 1:30 for that study. Thus, the supervisory ratio may vary over the range of numbers of interviewers and jump in a discontinuous way only after some threshold ratio has been passed. In this example, a new supervisor might be hired only when a 1:40 ratio has been passed.

The other source of variation in averages relates to the problem of small numbers. The best example here is in centralized telephone facilities. Morning shifts for telephone interviewing of household samples have traditionally been found to be less productive of interviews per hour than evening hours, and smaller numbers of interviewers are often scheduled to work those hours. For example, in a facility with 20 carrels, filled during evening hours, perhaps only three to five interviewers might be used in the weekday morning hours. While there might be two supervisors working on the fully staffed shifts, one might be used on the morning shift. This person is in a sense underemployed, relative to the night shift supervisors.

The discontinuities in cost models imply that partial derivatives, as in the stratified random sample above, do not exist. Hence, no single optimal design exists. Instead, the goal of the designer must be to seek to identify the boundaries of plateaus in costs. The discontinuities offer the researcher large benefits if they can be identified. For example, Figure 2.3 shows a hypothetical administrative model for the relationship between number of interviewers in a centralized facility and the number of shift supervisors to be used. One supervisor would be used if from one to six interviewers were employed; two, if seven to 19 interviewers were used; three, if 20 to 30 interviewers were used. If supervisors are paid $10.00 per hour, interviewers $5.00 per hour, and two interviewer hours are required to obtain one 30 minute interview, then the cost of each interview can be described by [$10 (number of supervisors)/(number of interviewers) + $5×2. This shows that similar costs per interview can be achieved with very different numbers of interviewers employed. However, jumps in costs per interview arise when additional supervisors are hired (i.e., between 6 and 7 interviewers and 19 and 20 interviewers). This cost model is plotted in Figure 2.4.

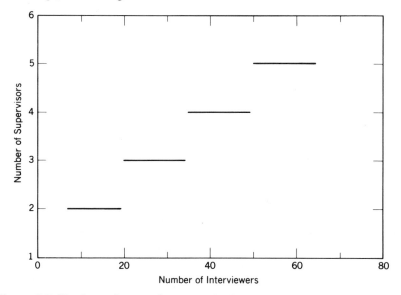

Figure 2.3 Number of supervisors required by number of interviewers in a centralized telephone interviewing facility under relatively fixed ratio of supervisors to interviewers.

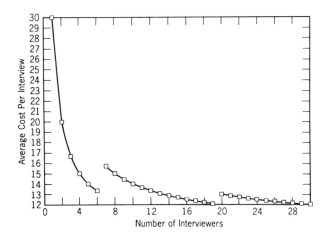

Figure 2.4 Cost per interview by number of interviewers resulting from decision to fix supervisor ratio.

2.6 COST MODELS OFTEN HAVE STOCHASTIC FEATURES[1]

Cost data have generally been obtained from documents prepared by accountants or field personnel, with little serious study of the sources of the data, data collection errors, or the variability ignored when a single fixed value is used for a particular cost parameter. The usual method of optimization under fixed constraints used in survey design has the advantage of straightforward results for the design features in a survey. But the straightforward nature of the optimization has had the detrimental effect that statisticians have ignored stochastic variation in costs and its impact on the subsequent solution to the allocation problem. Stochastic variation can greatly complicate the solutions for optimal survey design. Nonetheless, there is substantial experience which indicates that costs do vary across interviewers and other units in the design:

---

[1] Much of this section appears in Groves and Lepkowski (1985).

1. Some interviewers are consistently able to complete assignments more quickly than others, suggesting that there is variation in interviewing costs associated with the interviewer.

2. Costs of interviewing a particular type of unit have been observed to vary by day and time of contact.

3. The proportion of a sample that is eligible for a survey may vary from one sample to the next. Lower eligibility rates for a particular sample will lead to higher costs of interviewing eligible sample units because more sample units must be screened to identify the sample of eligible units.

4. Costs of contacting and interviewing cases in large metropolitan areas are frequently larger than those in nonmetropolitan areas. Within urban areas, costs of contacting and interviewing units in locked apartment buildings are higher than for single family structures. Samples that have different proportions of metropolitan and nonmetropolitan units or locked apartment buildings and single family structures will have different per unit costs for interviewing; that is, there is a covariation between the sampling units and the per unit costs.

These and other sources of variability in the costs of data collection create surprises in the administration of surveys. The potential for error in the design is influenced both by the choice of an expected value for the cost parameters (which may be incorrect for a particular survey) and by the shape of the distribution of the costs across units. If the distribution of costs is highly skewed, for instance, the choice of a mean cost in an allocation problem could lead to substantial problems in the administration of the survey.

Two examples will illustrate the principle. First, consider the allocation of a sample to two equal sized strata of size $N_1 = N_2 = N/2$. Let S_h denote the standard deviation of a variable in the hth stratum, and let $S_1 = 16$ and $S_2 = 29$. The objective is to find the allocation of the sample size, n, to the two strata such that for a fixed total budget, C, the variance of the sample mean is minimized.

A cost model for this two stratum design is given by the expression

$$C = C_o + \sum_1^2 C_h n_h ,$$

where C_o denotes the fixed costs of conducting the survey (regardless of the sample size), C_h denotes the per unit costs of a completed interview in the hth stratum, and n_h denotes the sample size in the hth stratum. Suppose that the per unit costs are not fixed values but random variables such that C_1 is distributed as a $N(25, 9)$ random variable and C_2 is distributed as a $N(64, 36)$ random variable. That is, the per unit cost of a completed interview in the first stratum has an expected value of $25, but across identical surveys 95 percent of the per unit costs would be between $19 and $31. Similarly, in the second stratum the expected per unit cost is $64, but the cost will vary between $52 and $76 in 95 percent of the surveys.

Under this stochastic cost model, the allocation of sample to the two strata which achieves the smallest variance for the sample mean will depend on the values of C_1 and C_2 observed for a particular survey. Given fixed values of the costs, $\tilde{C}_1$ and $\tilde{C}_2$, the standard solution to the allocation problem is to choose the same sizes in each stratum as

$$f_h = K \left(\frac{S_h^2}{C_h} \right)^{1/2} ,$$

where K is a proportionality constant chosen to achieve the fixed overall budget for the survey. For this example, fixing C_1 and C_2 at their expected values (i.e., $C_1 = \$25$ and $C_2 = \$64$), the conditional optimal solution allocates 47 percent of the sample to stratum 1 and 53 percent to stratum 2.

The distribution of optimal allocations conditional on particular combinations of observed C_1 and C_2 is shown in Figure 2.5. (The allocation is expressed as the percentage of the total sample that should be allocated to the first stratum. In the figure, the dots represent the relative frequency of a range of equal sized intervals of the allocations, while the line is a smoothed function fit to these points.) The expected value of the distribution is 47 percent, but approximately 68 percent of the optimal allocations are between 45 and 49 percent allocation to the first stratum, and only about 5 percent of the allocations are outside the interval from 43 to 51 percent.

The decision about the optimal allocation is now conditional on the unknown values C_1 and C_2 which will be observed in the survey. One choice for the allocation is to select the allocation corresponding to the

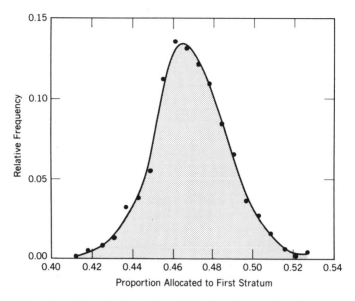

Figure 2.5 Empirical distribution of conditional optimal allocations to stratum 1 under stochastic variation in costs.

deterministic solution where $\hat{C}_1 = E(C_1) = \$25$ and $\hat{C}_2 = E(C_2) = \$64$, that is, the expected value of the optimal allocation distribution. Figure 2.5 can be used to inform the designer about the probability that another optimal allocation should have been used for this problem. For example, if the variance of the distribution in Figure 2.5 is very large, then a given allocation is unlikely to achieve true optimality, given lack of knowledge of the costs of interviewing to be achieved. In that sense, the allocation decision has less influence over the cost efficiency of the design. Reduction of cost variation will improve the researcher's control over the achieved precision of the survey, given fixed resources.

Stochastic cost models may also be used for the problem of determining an optimal interviewer workload. Let k denote the number of interviewers, b denote the number of interviews conducted by an interviewer, C_k the cost of recruiting, hiring, and training an interviewer, and C_b the interviewer cost for completing a single interview. A cost model for data collection can then be expressed as

$$C = kC_k + nC_b \ ,$$

where $n = k \cdot b$ is the total number of completed interviews. Under a simple response error model the variance of a sample mean is given by

$$\text{Var}(\bar{y}) = \frac{\sigma_r^2}{n} \left(1 + (b-1)\,\rho \right),$$

where σ_r^2 denotes the simple response variance and ρ denotes intra-interviewer correlation for responses obtained by the same interviewer (Kish, 1962). The optimal workload for an interviewer which minimizes the variance of the sample mean is given by

$$b_o = \left(\frac{(1-\rho)\,C_k}{\rho\,C_b} \right)^{1/2}.$$

Suppose that both C_k and C_b are not fixed constants but random variables with known distributions across the sample of interviewers selected for a survey. For the per interviewer cost of a completed interview, let $\mu_b = E[C_b]$ and $\text{Var}(C_b) = E[C_b^2] - \mu_b^2$, and for the recruiting, hiring, and training cost, let $\mu_k = E[C_k]$ and $\text{Var}(C_k) = E[C_k^2] - \mu_k^2$. For each set of fixed C_k and C_b, the optimal interviewer workload can be determined using the expression for b_o. Since b_o is a ratio of random variables, the variance of b_o can be approximated by

$$\text{Var}(b_o) \approx \frac{1}{4} \frac{(1-\rho)\,\mu_k}{\rho\mu_b} \left(\frac{\text{Var}(C_k)}{\mu_k^2} + \frac{\text{Var}(C_b)}{\mu_b^2} - \frac{2\,\text{Cov}(C_k, C_b)}{\mu_k\,\mu_b} \right),$$

where $\text{Cov}(C_k, C_b)$ denotes the covariance between C_k and C_b. The variance of b_o can then be estimated by substituting sample estimates for the parameters in the approximate variance.

As an illustration, the costs of recruiting, hiring, training, and completing an interview were obtained from the Survey Research Center's monthly Survey of Consumer Attitudes. The national random digit dialing sample survey is based on interviews lasting about 25 to 30 minutes each. During the period 1978 to 1984, interviewer assignments for the survey were assigned randomly within shifts, and data were collected for each interviewer-month on numbers of completed interviews and time per completed interview, as well as for other aspects of the monthly surveys.

Data on recruiting, hiring, and training costs are available from several large-scale recruiting and training sessions conducted during the period. These sessions cost approximately $300 for each interviewer completing training. About 14 percent of interviewers completing the training require additional training to improve their skills to a level required by the supervisory staff. The additional training costs an average of $70 per interviewer receiving the additional training. Thus, the cost of recruiting, hiring, and training, C_k, has a binomial distribution with $C_k = \$300$ for 86 percent of the interviewers and $C_k = \$370$ for the remainder. The stochastic nature of the variability in training costs occurs because the training staff selects the recruits without the ability to predict which ones will absorb the training quickly and which ones will require extra effort. For the purposes of this illustration, the 14 percent of the interviewers with the lowest response rates on their first month of interviewing in the Survey of Consumer Attitudes were assigned the value of $C_k = \$370$. The mean recruiting, hiring, and training cost per interviewer was therefore $309.83 with a variance $\text{Var}(C_k) = 594.14$.[2]

A cost monitoring system which is part of the centralized facility produces records for each interviewer's performance on each month of the Survey of Consumer Attitudes. Each entry in the dataset contained data on a given interviewer for a given month (i.e., each record is based on an interviewer-month). For the purposes of this illustration, interviewer-month data from 1978 through 1983 were used from interviewers who completed at least 10 interviews in the month. Since monthly interviews vary in length, an adjustment was made in both response rates and costs to eliminate this monthly variation. This was accomplished by subtracting the monthly means from each interviewer-month observation and then adding the grand mean calculated over 6 years of the entire study. Figures 2.6 through 2.10 provide distributions of the adjusted data and various statistics derived from them. As for Figure 2.5, both the frequency for selected ranges of values (i.e., the dots on the figures) and a smoothed curve fit to those frequencies are presented.

Figure 2.6 shows the distribution frequency of interviewer-months by the costs of completing an interview, C_b, during the period 1978 to 1984. That is, Figure 2.6 presents the frequency distribution of C_b where each observation is the average cost of completing an interview for a specific interviewer in a given month. These costs range from $8.00 to $35.00 per interview, are roughly symmetric in distribution around the mean cost of $\hat{\mu}_b = \$17.55$, and have a variance $\text{Var}(C_b) = 9.064$.

The covariance between C_k and C_b is $\text{Cov}(C_k, C_b) = 4.911$. This reflects the fact that interviewers who required more training initially are

---

[2] Using the interviewer-month as the observational unit, the mean training cost is $306.10 with a $\text{Var}(C_k) = \$389.14$. The lower mean reflects the fact that interviewers not requiring the extra training sessions completed more months of interviewing during the period examined.

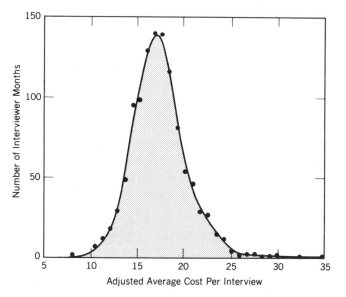

Figure 2.6 Distribution of adjusted average cost per interview of 1115 interviewer–months from the Survey of Consumer Attitudes, 1978–83.

also less efficient at producing completed interviews during production. Finally, the within interviewer correlation was estimated for key items on the survey and had an average value of approximately $\hat{\rho} = 0.01$ (Groves and Magilavy, 1986).

Using the average values for C_k and C_b from the Survey of Consumer Attitudes, the optimal monthly number of completed interviews for an interviewer is $\hat{b}_o = 41.553$ cases per month. This would be the result obtained if the traditional fixed cost optimization procedure were used.

The value of b_o was also estimated for each interviewer-month observation, and the distribution of these b_o values is shown in Figure 2.7. The mean of the distribution of these b_o values (42.265) is approximately the same as that estimated from the average C_k and C_b. The distribution is slightly positively skewed (skewness = 0.47) and slightly more peaked than a normal random variable (kurtosis = 1.11). About 95 percent of the optimal workload varies between 36 and 50 cases per interviewer. In a survey requiring 2000 completed interviews between 40 and 56 interviewers would be hired.

After examining the variability in optimal workloads associated with particular combinations of cost values, it is apparent that losses in the variance of the mean might be experienced if the workload chosen were not the optimum. Suppose that the workload chosen were b_o', based on assumed per unit costs C_b' and actual training costs C_k, but the actual

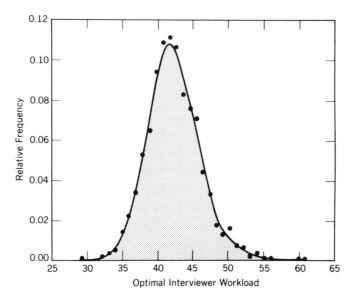

Figure 2.7 Empirical distribution of conditional optimal interviewer workload under stochastic variation in costs, from the Survey of Consumer Attitudes, 1978–83.

costs C_b and C_k indicate that the optimal workload should be b_o. Then there will be a loss in variance when b_o' is used instead of the optimum b_o. The proportionate loss in variance can be expressed as

$$
L = \frac{\left(1 - \left(\dfrac{C_b}{C_b'} \right)^{1/2} \right)^2}{1 + \dfrac{C_b}{C_b'} + b_o' \left(\dfrac{C_b}{C_k} + \dfrac{\rho}{(1-\rho)} \right)}.
$$

From this expression, the distribution of proportionate losses conditional on different actual optimum workloads b_o can be constructed.

For example, using the distribution of per unit costs in the Survey of Consumer Attitudes, the proportionate loss distribution when a workload of $b_o' = 42.3$ cases per interviewer is used instead of the optimum workload is shown in Figure 2.8. The workload 42.3 is the expected value of the optimal workload distribution in Figure 2.7. Figure 2.8 shows that, given

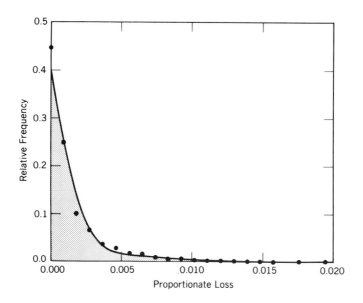

Figure 2.8 Empirical distribution of proportionate loss in variance for a fixed total budget and an interviewer workload of 42.3 under stochastic variation in costs.

the cost distributions, a loss of 1 percent (or even 0.5 percent) in the variance of the mean would be very unusual.

The distribution of losses can be used to inform design decisions about workloads in a centralized facility. For example, what losses might be possible if total survey costs were reduced by hiring fewer interviewers than that suggested by the expected optimal workload? That is, if only one-half the number of interviewers indicated from a study of optimal workload were hired and given a workload of twice the optimum, the loss distribution could indicate the probability of various types of loss in response variance. In particular, the data in Figure 2.7 suggested an optimal interviewer workload would be 42.3 cases per interviewer, but suppose only one-half the number of interviewers needed to achieve this workload are hired and each is assigned twice the optimal workload, 84.5 cases. The proportionate loss distribution would be that of Figure 2.9. The expected loss in variance is about 10 percent, and losses of 10 percent or more are experienced in 50 percent of the cases. Figure 2.10 presents a similar figure when a workload of 21.1 cases per interviewer is used instead of the optimum of 42.3.

One may view Figures 2.8, 2.9, and 2.10 as two-dimensional planes cutting through the three-dimensional surface of proportionate loss frequencies for values of the actual interviewer workload and the optimal interviewer workload. A presentation of the full surface rather than the three marginal distributions in Figures 2.8, 2.9, and 2.10 would provide more complete information about the chances of different losses, given various workload decisions being experienced.

Both the example of sample allocation to strata and the example of interviewer workload illustrate the contribution that stochastic cost models can make to survey design. With these the designer is reminded that the cost structure implied by average cost parameters will not necessarily be experienced in the survey and that changes in costs imply changes in optimal design features. Indeed, an empirical distribution of possible values of the optimal design parameter is produced. The investigator can assess the dispersion in the distribution to assess how variable the optimal parameter value might be and can assess the risks of possible losses in precision due to departures from the optimum. This information serves to refine the understanding of potential losses in nonstochastic models when nonoptimum solutions are chosen. The stochastic cost model reminds one that losses can be larger or smaller than that expected with average costs.

2.7 DOMAINS OF APPLICABILITY OF COST MODELS MUST BE SPECIFIED

Error models often are applicable to a very broad range of options. For example, the simple random sample variance for the mean $\bar{y} = \sum y_i/n$ is

$$\text{Var}(\bar{y}) = \frac{(1-f)S^2}{n}$$

where $S^2 =$ population element variance, $\sum_i (Y_i - \bar{Y})^2/(N-1)$ and applies to $n = 1, 2, ..., N$. This wide range of applicability does not usually apply to cost models. For example, it is unlikely that the administrative organization and costs of conducting a survey of a sample of size 10 would

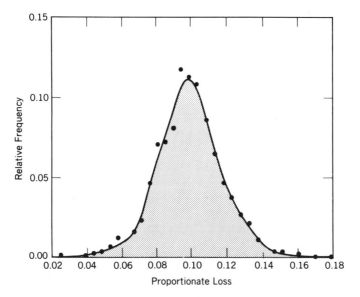

Figure 2.9 Empirical distribution of proportionate loss in variance for a fixed total budget and an interviewer workload of 84.5 under stochastic variation in costs.

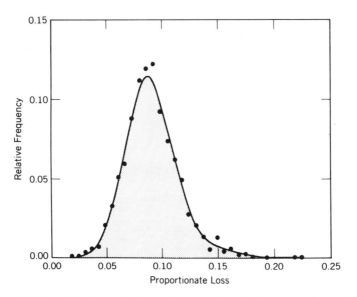

Figure 2.10 Empirical distribution of proportionate loss in variance for a fixed total budget and an interviewer workload of 21.1 under stochastic variation in costs.

even faintly resemble one of a sample of size 100,000 from the same population.

Some of the cost models described in the previous sections are based on real data collected at the Survey Research Center. Whenever such data are used they are dependent on the administrative structure and staffing decisions used at SRC. Furthermore, they are restricted to the variation in scope of designs used. This produces some practical limitations to the use of the cost models, illustrated by the following description.

SRC normally conducts national telephone surveys of size 500 to 5000 interviews, with survey periods that last from 3 weeks to several months. Interviewers work from 18 to 35 hours per week in a facility with one room with 18 carrels and one with 10 carrels. The length of interviews that are normally attempted range from 5 minutes to 55 minutes, using a respondent rule that is typically a random selection from among those household members 18 years or older.

Therefore, when a cost function relating interview length with interviewer effort is estimated, it must be conditioned on the domain of interview lengths that have been experienced by the facility. For example, Figure 2.11 plots an "efficiency ratio" used at the Survey Research Center by the average length of the interview for several years of telephone survey experience. The efficiency ratio (the y-axis in Figure 2.11) is defined as the ratio of total interviewer hours spent on a survey to the total hours spent actually engaged in interviews with respondents. The average interview lengths range from 10 to 50 minutes in the years examined. A linear model, Efficiency Ratio = $6.67 - 0.082$ (Interview Length in Minutes), is fit to the data. It is doubtful, however, that the same model would apply to interview lengths of 60 minutes, 90 minutes, or 2 hours or more.

The limitation of data to a set of typical design features implies that any extension of the model outside that range must be done without the support of experience. Rarely in the survey literature is this acknowledged, and instead homogeneity of functional form is assumed. In this case, it is assumed that the model applies for all interview lengths, $L = 1, 2, ..., \infty$.

It is clearly dangerous to apply a model in a context for which no data have been assembled. In the absence of data based on actual experience, an independent set of estimates might be assembled to describe the functional relationship between the costs and number of design units for the domain being examined. For example, in the case of the 2.11 example, one might assert that as the length of the interview increases:

1.   the initial refusal rate will increase;

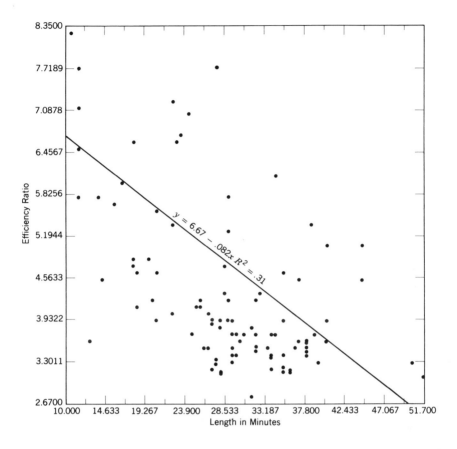

Figure 2.11 Ratio of total interviewer hours required for project to total interviewer hours spent questioning respondents by average length of interview in minutes.

2.   the success at refusal conversion efforts will decline;

3.   the number of interviews taken in multiple sessions will increase; and

4.   the number of broken off or partial interviews will increase.

These ingredients might be estimated by several knowledgeable administrators in the facility and a functional form specified. That function should then be backfitted to the data collected for shorter interviews. The purpose of this step would be to see if the difference between the predicted value under the model and the actual values are of a sign and magnitude that are sensible.

From one perspective such an effort attempts to correct a misspecification of the functional form of the cost model; that is, instead of

$$C = C_o + C_n n \,,$$

where n is the number of interviews, the appropriate cost model is

$$C = C_o + C_n' n' + C_r rn + C_2 tn \,,$$

where C_n' is the cost of interviewing the n' easily cooperating cases, where C_r is the refusal conversion cost for the the rn cases requiring such efforts and C_2 is the extra cost of completing partial interviews for the tn cases requiring such efforts. The parameters rn and tn are functions of interviewer length.

2.8 SIMULATION STUDIES MIGHT BEST BE SUITED TO DESIGN DECISIONS

Cost and error models for survey optimization probably have remained simple in general practice because (1) the survey statistician has not had extensive data on cost and error components, (2) closed form solutions to optimization problems are not available when complex cost and error models are used, and (3) they add complexity to the already difficult task of designing a complete survey. Gathering better cost data needs to be given some priority (this is discussed in various sections below). Furthermore, the cost models built should reflect accurately the complexities of the administrative procedures used in the survey. Design solutions need to be found given the appropriate cost models; cost models should not be built simply because they offer easy mathematical solutions to the design problem.

With complex models, various optimization problems can be approached with large-scale computer simulation. These simulations can be used to address the common design decisions of the survey statistician—optimal allocation to strata in the sample selection, optimal workloads for interviewers, and optimal number of waves in a panel design. The solutions will be found within the constraints implied by the total budget for a survey. Since it is likely that closed form-solutions to such problems will not exist with complex cost and error models, simulation approaches will be useful to measure the sensitivity of results to changes in various design, cost, or error parameters.

2.9 IS TIME MONEY?

Some researchers work in environments in which timeliness is a valued attribute of a survey estimate. Overwhelmingly, these researchers tend to be describers. This book does not discuss design alternatives to maximize timeliness, nor the loss of utility of a survey estimate that is not timely. In one sense, it assumes that any improvement in timeliness can be obtained through the greater expenditure of resources. That is, time *is* money. Despite this assumption, the issue deserves comment.

First of all, there appear to be different conceptions of timeliness. Sometimes the value of a piece of survey-based information decays rapidly over time. For example, if y_t represents the proportion of Democrats who voted for the Democratic presidential candidate at time t, knowing y_t only at $y_{t+6 \text{ mos.}}$ would be almost valueless for a newspaper or television news program sponsoring a survey. (It may be of great value for a political scientist studying trends in the importance of party identification. It may achieve its value only much later for historians.) Thus, one conception of timeliness concerns the decline over time in the value of information about past population characteristics.

Another conception of timeliness is the value of current information about a characteristic that is changing rapidly over time. For example, if one is interested in the proportion of persons in the United States suffering from an influenza epidemic now, knowing the proportion affected 2 months ago is only slightly better than having no information. The rapidly changing values of the proportion over time imply that the proportion 2 months ago has little predictive value for the current proportion. In this perspective, timeliness has less value for statistics that are relatively invariant over time (e.g., the sex ratio of the U.S. population).

A case that illustrates both conceptions of timeliness is the use of political polling to predict the outcome of a future election. The earlier an accurate prediction can be made about an election, the most valuable is the prediction. Conversely, knowing voter preferences 4 weeks earlier has little value on the day before the election. Both campaign strategists and members of the news media need current data collected over a brief period to minimize the risk of intervening events affecting the survey outcome. Untimely estimates both prevent the use of the information to change campaign strategy *and* give little useful information about the current situation.

In light of this book's focus on costs and errors, one interesting question is whether issues of timeliness are conceptually distinct from those of costs and errors. Can specified levels of error be obtained under any time constraint merely by changing the budget of a survey? Can specified costs of a survey be obtained under any time constraint merely by changing the desired error levels? Does a combination of error and cost constraints completely determine the time at which survey estimates are available? Can the time at which estimates with specified error levels are available be hastened merely by spending more money?

The answers to these questions depend on how extreme the demand for timeliness is. There are situations when increased timeliness can be obtained merely by spending more money. If, for example, a health survey studying dietary habits is planned for a 2 month survey period, doubling the number of interviewer hours (perhaps with more interviewers hired) and making other related adjustments could provide estimates with a single month's interviewing. Little impact on coverage, nonresponse, or respondent errors might result. The survey would cost more but would provide the estimates more quickly.

There are other situations whereby greater timeliness can be obtained only if greater error can be tolerated. The use of surveys to obtain immediate reaction to world events is a good example: for example, a household survey conducted by a television network over a 24 hour period between a political candidate's withdrawal from a campaign and the next evening news broadcast. Timely data are needed because attitudes are truly changing quickly. Truncating the interviewing period from weeks to hours increases the nonresponse rate among the sample drawn. It is unlikely that any amount of money could overcome the errors of nonresponse with such a survey period. That is, timeliness is obtained at the expense of greater error in the estimate.

This last example illustrates the fact that surveys are inherent compromises. No survey estimate exists without error; no amount of money spent on a survey will eliminate all error. Similarly, maximum timeliness is impossible in some circumstances, given a desired error level. Instantaneous estimates from surveys cannot be obtained. However, relatively timely estimates are possible with surveys but depend on decisions about relative costs and error tolerances.

Thus, most demands for timeliness have direct impacts on costs and errors. Elucidating the interrelationships of different costs and errors will provide insights into some timeliness issues. Some extreme time constraints make survey methods impossible, but most are related to the topics discussed in later chapters.

2.10 SUMMARY: COST MODELS AND SURVEY ERRORS

Several survey cost models exist to guide the identification of optimal sample designs. They are, in general, simple functions of sample design features (e.g., number of clusters selected, number of persons interviewed). They gain their applicability from the fact that most surveys are conducted with limited resources. When paired with functions describing the relationship between sample design features and sampling error, they permit efficient allocation of those limited resources.

The systems of cost and error modeling work well in that context because the error function is well specified. That is, under probability sampling theory the sampling error of large sets of statistics can be expressed as functions of design features (e.g., stratification used, number of units selected). Sampling error is a *measurable* property of the research design.

In the absence of well grounded estimates of coverage, nonresponse, and various measurement errors, the use of cost and error modeling to guide design decisions has not often been used. Clearly, the payoff of cost modeling is diminished without the prospect of determining a design feature that minimizes an error source under a cost constraint. However, ignoring cost implications of alternative designs affecting nonsampling errors is imprudent.

Survey researchers have given much less attention to survey cost models than to survey error models. Most existing models are linear functions of some survey parameter (e.g., number of interviews), yet many nonlinearities are found to exist in practice. Most cost models are continuous in those parameters, but discontinuities in costs often arise when administrative changes accompany certain changes in design. Most cost models are deterministic, assuming applicability over any replication of the survey, yet costs can vary greatly because of chance occurrences in the selection of the sample or choice of interviewers. Finally, most cost models are not presented as applicable to only a restricted set of surveys (e.g., ones having a particular administrative design, using a certain number of interviewers), yet they are clearly inapplicable under some ways of conducting the data collection. Finally, survey staffs seem to give too little attention to studying those components of survey design which produce large costs or to examining how changes in procedures can affect costs.

Thus, survey designers could improve designs if more attention were paid to cost model construction and parameter estimation. The extension of the concept of "measurability" to include ability to estimate changes in costs under limited design changes might spur designers to take cost models seriously. At the present time only sample design changes are routinely evaluated in terms of costs and errors. Designers most often claim that since the nonsampling errors tend to be unmeasurable no guidance can be given on alternative design features affecting them. This stance ignores the measurability of costs and the ability to bound some of the errors. The incentive to obtain some measures of nonsampling errors can increase if the costs of activities devoted to reducing them are routinely presented to the designer.

Ideally in a book with this title discussion of each error source would be accompanied by examination of the cost implications of designs that offer different error properties. When possible, such discussions do appear. This is true for sampling error and nonresponse error, but for only a few of the measurement errors. This is unfortunate and limits the illustration of how cost considerations might best inform design decisions. In Chapter 4 on nonresponse and Chapter 11 on measurement errors associated with mode of data collection, cost and error models used in simulation studies for survey design options are described. Similar approaches could be used in other errors when efforts to reduce the error produce different resource requirements. Hopefully, the future will bring other such examples.

CHAPTER 3

COSTS AND ERRORS OF COVERING THE POPULATION

Bu Fu, the sorcerer to the boy Chi Po, to whom he is teaching painting: "No, no! You have merely painted what is! Anyone can paint what is; the real secret is to paint what isn't."

Chi Po: "But what is there that isn't?"

O. *Mandel,* Chi Po and the Sorcerer: A Chinese Tale for Children and
Philosophers, *1964, as cited in R. Smullyan,* 5000 B.C. and
Other Philosophical Fantasies, *1983, p. 112*

3.1 DEFINITIONS OF POPULATIONS RELEVANT TO THE SURVEY

Although survey data sets describe a fixed set of respondents, they are generally used to learn about a much larger group of persons, the full population that is the focus of the research. The examples of survey samples given in most elementary texts first describe a well–defined population of persons (e.g., the household population of a city, students in a school) from which a probability sample is drawn.[1] Measurements are taken on this sample in order to learn of the characteristics of the full population. Unfortunately, several complications are encountered during this process in real research. This chapter discusses one arising from failure to include some elements of the population in the materials used to identify persons to measure, resulting in "coverage error."

Any discussion of coverage error in surveys must be presented in conjunction with a discussion about the populations of interest to the researcher. Because of the traditional concerns in survey research with

---

[1] A "probability sample" is one giving every member of the population a known, nonzero chance of selection.

inference to groups external to the sample cases themselves (in contrast to the concerns with internal control of experimental research), there have been some distinctions among different types of population (some of them only abstract concepts) relevant to the survey. Kish's (1979) enumeration of different populations of importance to the survey researcher is useful to understanding the problems.

The **population of inference** (or inferential population), loosely stated, is the set of persons to be studied, with explicit delimitations with regard to the time interval over which research interest is focused. This set may not be of finite size. For example, assume a researcher is interested in attitudes of the U.S. society as it existed between July 1, 1984 and November 7, 1984. There are an infinite number of moments at which attitudes might be measured among the people who were alive during that period. In other cases, the population is more closely finite, for example, in a survey estimating the proportion of schoolchildren with single parents in a school district as of the first day of the school term.

The **target population** is the set of persons of finite size which will be studied. The discrepancy between the population of inference and the target population can generally be a well-defined one. The distinction is made by the researcher, whether implicitly by treating the target population as if it were the population of inference or explicitly by noting eliminations of the population of inference that result in the target population. For example, although the Current Population Survey which measures unemployment and employment in the United States produces monthly unemployment rates, the questions in the survey address only employment of the sample person during the week containing the 19th of the month. This was an arbitrary restriction of the research to produce a target population of greater accessibility, those persons in households during the week of the 19th. Similarly, U.S. household samples conducted by nongovernmental agencies often exclude households in Alaska and Hawaii, and persons who are institutionalized (e.g., inmates in prisons, patients in hospitals), although their population of inference may be the full United States.

The **frame population** is a set of persons for whom some enumeration can be made prior to the selection of the survey sample. In a simple case the "sampling frame", the listing of units used to draw the sample, may really be a list of persons, in other cases it may be a list of area units (e.g., counties or city blocks) to which persons can be linked in a simple way (e.g., their residences), while in still other cases it is a list of households (e.g., by addresses, telephone numbers). A more formal definition of the frame population, from Wright and Tsao (1983, p. 26), is "the materials or devices which delimit, identify, and allow access to the elements of the target population. In a sample survey, the units of the frame are the units to which the sampling scheme is applied. The frame also includes any auxiliary information (measures of size, demographic information) that is used for (1) special sampling techniques, such as, stratification and probability proportional to size sample selections; or for (2) special estimation techniques, such as ratio or regression estimation." The frame population is typically the "universe" referred to in many statistics books. These texts thus ignore the distinctions made above among the frame population, the target population, and the population of inference. Examples of frames used to cover the human population include sets of area units (e.g., counties, enumeration districts) that divide the population geographically, directories of residents of the country made by private companies, and telephone numbers for the residential population.

The **survey population** is the set of people who, if they were selected for the survey, would be respondents. They are accessible to the survey interviewers. They have the physical and mental capabilities to provide information about themselves. They are willing to respond to the survey request. This concept assumes that each person has some propensity to respond or not to respond to the survey, and that these tendencies exist prior to the sampling step of the survey. This is an abstraction that can typically not be observed prior to the survey requests themselves. We shall refer to discrepancies between the survey population and the frame population as nonresponse to the survey.

Despite this litany of different populations relevant to a survey researcher, we have failed to define the term "coverage error." Coverage error generally refers to the discrepancy between statistics calculated on the frame population and the same statistics calculated on the target population. Coverage error arises from the failure to give some units in the target population any chance of being included in the survey, from including ineligible units in the survey, or from having some target population units appear several times in the frame population. To the extent that these units have distinctive characteristics the survey statistics will be biased. "Bias" here means that all possible surveys using the same survey design and the same frame population will be subject to the same error, due to omission of the noncovered units. "All possible surveys" is an abstract concept, never operationalized in practice. It is used here to note that although other features of the survey may have effects on errors that vary over possible replications of the survey design, the absence of target elements from the frame population will have effects that are constant over possible replications.

The reader will note also that coverage error does not arise merely because the survey attempts to study a *sample* of the persons in the target population. Even if a census of the target population were attempted, using the frame materials, those omitted from the frame would not contribute to the statistics.

3.2 COVERAGE ERROR IN DESCRIPTIVE STATISTICS

As will be true with other survey errors, the effect on linear survey statistics of coverage error can often be expressed as a function of two components.[2] If Y is any linear statistic, the effect of target population elements being omitted from the frame population can be illustrated by expressing Y as

$$Y = \frac{N_c}{N} Y_c + \frac{N_{nc}}{N} Y_{nc},$$

where Y = value of the statistic on the full target population;

N_c = number in the target population covered by the frame population;

N_{nc} = number in the target population not covered by the frame population;

N = total number in the target population;

Y_c = value of the statistic for those covered by the frame population;

Y_{nc} = value of the statistic for those not covered by the frame population.

The above expression can be altered to illustrate the nature of the error due to noncoverage. Adding the term $(N_{nc}/N)Y_c$, to both sides yields

$$Y + \frac{N_{nc}}{N}Y_c = \frac{N_c}{N}Y_c + \frac{N_{nc}}{N}Y_c + \frac{N_{nc}}{N}Y_{nc},$$

but $N_{nc}/N + N_c/N = 1$, by definition. Thus, that expression reduces to

---

[2] A linear statistic is one that is an additive combination of observations taken in the survey, a weighted sum of random variables.

$$Y + \frac{N_{nc}}{N}Y_c = Y_c + \frac{N_{nc}}{N}Y_{nc}$$

or

Frame Population Value = Target Population Value + (Proportion not Covered) × (Frame Value − Value for Those not Covered)

$$Y_c = Y + \frac{N_{nc}}{N}(Y_c - Y_{nc}).$$

Coverage error is a function of both the proportion of the target population that is not covered by the frame and the difference on the survey statistic between those covered and those not covered. Thus, even with large portions of the target population missing on the frame, if those missing resemble the full population (in the extreme, if their value on the survey statistic is the same as for those covered by the frame), then there will be no bias due to noncoverage. In contrast, if a small portion of the target population is missing from the frame, but they have very distinctive values on the survey statistic, then large noncoverage bias may result.

Most of the time a survey researcher has no way of knowing how large a coverage error applies to a particular statistic. The proportion of the target population that is not covered by the frame , N_{nc}/N, is generally unknown, and the differences between the noncovered and covered population are unknown. Furthermore, although N_{nc}/N is constant over all statistics calculated on the full sample, it will vary over different subclasses on which estimates might be calculated (i.e., coverage rates of different subgroups in the target population will vary). For example, later discussion in this chapter describes lower coverage of young black males in U.S. household sample surveys. In addition, the $(Y_c - Y_{nc})$ term can vary over all statistics. For example, the effect of omission of young black males may be larger for estimates of the proportion of persons in the labor force than for estimates of the proportion of persons who have vision problems that could be corrected through eyeglasses. *Coverage error is a property of a statistic, not a survey.* In a few specific cases, some reviewed below, special studies can be mounted to describe coverage error properties of some statistics.

3.3 AN APPROACH TO COVERAGE ERROR IN ANALYTIC STATISTICS

While most researchers acknowledge the potential damage of coverage error to descriptive statistics, many believe that it is only a small problem for analytic statistics.[3] The fact that some persons in the target population are excluded, they argue, has no effects on whether one variable is related to another in the frame population. There is less burden on the researcher to demonstrate that his target population is well covered by his frame population. Instead, there is the burden of justifying the proper identification of variables that affect the survey measure of interest, the proper specification of the analytic model. There are thus differences between the practice of researchers using survey data to identify *causes* of phenomena (modelers) and those who are interested in describing the characteristics of a known, fixed population (describers).

3.3.1 Inferential Populations in Analytic Statistics

One argument advanced regarding analytic models estimated from survey data asserts that analysts are rarely interested in fixed, finite populations when they build models to describe social phenomena. Rather, they consider the attitudes or behavior under study to be the result of a stochastic (random) process. The observations that they collect in the survey are outcomes of experiencing this stochastic process. Indeed, any population at a moment in time is only one of an infinite number of possible sets of outcomes of this process. Their interest in collecting data is in understanding the parameters of the model, or stated in a more extreme form, in discerning the physical or social laws that produce the outcomes they observe. They are interested in the model, not necessarily in the particular fixed population which provides the observations. In some sense, they are interested in all possible sets of observations that could be produced by the stochastic process, a *superpopulation* of outcomes. Notions of coverage error as described above seem irrelevant to this perspective.

Malinvaud (1966, p. 65) notes that, in most econometric analysis, "the phenomenon to be studied may be considered to resemble a process entailing random determination of certain quantities which are in this case assumed to be random in the universe as in the observation sample. "

---

[3] "Analytic statistics" refers to measures of relationships among variables with the intent of obtaining information about causal relationships among phenomena.

Klevmarken (1983), in citing this, notes that in most cases no finite universe (in our terms, population of inference) is ever defined by the researcher. Rather, the focus of the investigations is on the theoretical model, stochastic in form, operationalized and tested with the data. Haas (1982, p. 103) presents the argument that knowledge is accumulated in sociology "not by deriving hypotheses from more general ones as in the hypothetico–deductive model, but by sampling new populations." He notes that a scientific law must be true under specified conditions irrespective of time and place: "...part of the faith of science is that nature is uniform throughout and that therefore a hypothesis found to be true of one subset of a populations' will be true of any other subset" (p. 104). Failure to cover the target population well in the investigation can harm conclusions from the analysis only if predictor variables are omitted from the analytic models being promoted by the study or the functional form of the models is misspecified in some other way.

In contrast, sample surveys implicitly assume lack of uniformity. It is that assumption which creates a need to study several cases, to sample from the frame population. Since we cannot fully specify the conditions under which a hypothesis (a model) is true, randomization is introduced to assure lack of confounding of results with some unmeasured variable. (Note that this confounding is possible only if the response model affecting the dependent variable is not fully known.) In practice, as Blalock (1982. p. 112) notes, "often 'populations' are selected primarily because of their convenience for research, as for example in the case of a social psychological study of prejudice conducted within the confines of a nearby community. Here, the 'population' has no particular relevance to the limits imposed by a theoretical class (say that of white Americans)." Theories based on these procedures can be general only in the sense that they are true of very large, historically specific populations. As Anderson and Mantel (1983) note, replication over different target populations is one method to increase the confidence in the universality of results. Usually such replication is combined with enhancement of the model. Hence, the second applications of the model are sought not by statistical concerns of failure to cover all types of persons but by the assertion that the model is misspecified and should be altered in functional form or in the set of predictors included in the model.

3.3.2 Treatment of Inferential Populations in Survey Analysis

Many times the real population of inference is poorly described by the researcher; other times, researchers make distinctions between descriptive and analytic goals of the survey. For example, Coleman in *The Adolescent*

Society (1961) notes that the purposes of the study were "to inquire into the nature and consequences of adolescent social climates for the purposes within them, and to learn what factors in the school and community tend to generate one or another adolescent climate" (p. 332). The survey selected ten schools that varied in size and "urban–suburban–rural location, but with enough preliminary observation of the schools to insure that they encompass the desired variations in adolescent status systems" (pp. 336–337). Formal definitions of the population of inference are not presented. Is Coleman's interest in the "adolescent cultures" of U.S. high schools in the Fall of 1957? Is he interested in high schools that no longer existed at that time, or is he interested in those that were yet to be built? Although he does not explicitly say so, it is likely that he is interested in identification of causes or factors that influence the characteristics of adolescent schools. (These are called "analytic" versus "descriptive" goals of a survey.) He is therefore interested, as he says, in achieving "desired variations in adolescent status systems" within the sample. The phrase "desired variations" probably refers to the need on the part of the researcher to obtain variation within the sample on those characteristics that are thought to influence the character of the adolescent social systems. Only with such variation can the "effect" of those variables on the social system be assessed. This is a problem in statistical analysis, but where do notions of a population of inference come into play here?

Coleman's work might be considered the analysis of the construction of social systems; he desires to understand the parameters of that process, the causes and the effects. He is interested only in identifying those causes. If he learns that there is a subset of the population for which a different set of causal principles (different variables) apply, his reaction would most likely be to incorporate the differences in a more complex explanation of social systems. Some would term this an elaboration of the causal model to include the new causal factors. The burden on the researcher in this perspective is the identification of the causal mechanisms, not in the delimitation of the population of inference or of the target population. Unfortunately, such attention begs important questions about the circumstances under which the causal principles are asserted to apply. Do the results apply to adolescent social systems in high schools in the 1930s in the United States? Do the results apply to 3 year and 4 year high schools? Do they apply to private and public high schools, to boarding schools and those where students live at home? There is no discussion of such issues.

Another example illustrates the same problems more clearly. Duncan et al. (1972) combine analyses of national household populations with those of other populations. The theme of their work is the understanding of the process of "status transmission between generations" (p. xxi). The authors explicitly define the population to be studied: "adult male population of the contemporary United States in the central ages of working force participation" (p. 31). But they note that

> On occasion, well-defined sub-populations within this population are considered for separate study. It will be noted that the vagueness of the terms "contemporary" and "central ages" allows some considerable latitude in making specific decisions. Even more to the point, all sorts of approximations to the target population were accepted when there seemed to be reasonable grounds for doing so.

With these guidelines the authors analyze in one monograph a national household sample: one studying all civilian noninstitutionalized men 16–24 years old; one studying men 21-64 years old in the household population of the three county area around Detroit, Michigan; one studying couples who recently had their second child, based on birth records of "several major metropolitan areas"; one studying graduating high school seniors in 1957 in Wisconsin; and one studying males living in Lewanee County, Michigan, who were 17 years old in 1957.

The various data sets are attractive to the analysts because they contain different sets of measures, offering unique tests of various parts of their model of status transmission. They test similar models in the various data sets, expecting replication of findings, because, in their words, "a minimum requisite for the orderly accumulation of scientific knowledge is that findings of one investigation must recur in other investigations, supposing that sufficient precautions to assure comparability have been taken, as they obviously should be if such accumulation is to occur" (p. 47).[4]

The lack of concern with strict identification of the population of inference, in one case, or with the discrepancy between various frame populations and the population of inference is common to much analytic use of survey data. Underlying the perspective is an assumption that the factors identified as causal predecessors are sufficient to understand the phenomenon under study. This assumption is related to one of homogeneity between the small frame population studied and the full target population. The same causal model applies in both parts of the

---

[4] The criteria set by Duncan et al. for "comparability" appear to include relative equality of pairwise correlations among the variables in their models.

target population. If the model is perfectly specified this will be true, and hence, the critical focus on the nature of model specification follows.

3.3.3 A Link Between Coverage Error and Analytic Modeling

The works of Tobin (1958), Heckman (1979), and Goldberger (1981), summarized by Berk and Ray (1982) and others, can be used to illustrate a link between the statement of coverage error on a descriptive statistic given above and its effects on estimates of parameters in a causal or analytic model. Suppose there is an interest in estimating a simple regression model,

$$y_i = \alpha + \beta x_i + \varepsilon_i \,,$$

where y_i = value for the ith element of the target population on a dependent (endogenous) variable;

α = intercept;

β = regression coefficient for the independent (exogenous) variable;

x_i = value of the independent variable for the ith person in the target population;

ε_i = deviation of the y_i value from the model based estimate, specific to the ith person in the target population.

Heckman and others use the phrase "original population" to signify the target population and "selected population" to signify that group eligible for the survey measurement. In their terms a "selection bias" exists when some portion of the original population is omitted from the selected population. Note that the survey researcher's use of "coverage error" is indistinguishable from this concept.[5] Estimates of β above can be damaged by failure of the frame to cover parts of the target population. They would note that the process by which a portion of the original population was screened out produces a selectivity bias in our

---

[5] The notion of "selection bias" is also applicable to failures to give a chance of observation to some target population units because of nonresponse or deliberate exclusion from the survey population. Thus, the results described in this section are also applicable to the treatment of nonresponse error, discussed in Chapter 3.

observations. The researcher is permitted to obtain measures on only a subset of the original population.

Figure 3.1 illustrates a specific type of noncoverage error that has clear effects on estimates of β in the model above. Here the frame population systematically omits target population elements that have values of y_i less than some threshold; in the figure the threshold is $y_i < z$. The line described by the model in the full target population is denoted by the solid line. In the target population as a whole, y_i is found to be an increasing function of x_i. When the linear function is estimated by the elements that are contained in the frame population, denoted by the dashed line, the line has a smaller slope, and the estimated coefficient, β^*, has a negative bias. Under an assumption of normality for the x's, Goldberger (1981) shows that the relationship between the β in the target population and the estimated coefficient, β^*, is

$$\beta^* = \beta \, \frac{V^*(y)/V(y)}{1 - \rho^2[1 - V^*(y)/V(y)]} \, ,$$

where β^* = estimated coefficient in the covered population;

β = true coefficient in the target population;

$V^*(y)$ = variance of the endogenous variable in the frame population;

$V(y)$ = variance of the endogenous variable in the target population;

ρ^2 = coefficient of determination for the regression model in the target population.

The bias term for a simple linear descriptive statistic was a function of the proportion of cases missing from the frame population and the difference in their values on the statistic versus that of the covered population. For a regression coefficient, however, the bias is a function of the relative variance of the dependent variable in the target and frame populations, and the fit of the model in the target population. If the variance of the dependent variable in the selected population is reduced, as in the case in Figure 3.1, by truncating the distribution from below, the measured regression coefficient in the frame population is smaller than the true value in the target population. This effect is not as pronounced when the

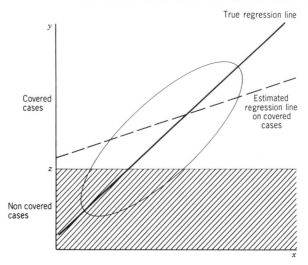

Figure 3.1 Effect on estimated regression line under truncation of the sample on the dependent variable (y); all $y_i < z$ not covered.

model has a good fit in the target population, but for many social science models this is not the case.

In the language of this approach, Figure 3.1 describes an example of "truncation," whereby some elements of the target population are never covered and others are always covered. A further restriction is that the exclusion is based on a value of the endogenous variable (y) in the model. No such bias in the regression coefficients occurs when such truncation yielding noncoverage is based on values of the independent variables.

Most cases of noncoverage, however, do not fit these conditions of truncation. More likely is the case in which some target population elements are omitted, not because of their values on a dependent variable but as a result of some other process. In our case, the process is the construction of the frame materials, the listing of the target population. The "selection" process is that which identifies elements of the target population to be members of the frame population. It is rarely the case that the dependent variable in the survey analysis is a measure of the probability of listing on the frame. Rather, the process that omits elements from the frame is separate from the process creating values on the dependent variable. Such cases are termed the result of "incidental selection" in the language of Heckman (1980).

Figures 3.2 and 3.3, taken from Berk and Ray (1982), can be used to illustrate Heckman's approach to incidental selection bias. We describe the figures using terms compatible with the survey coverage error literature. Figure 3.2 describes the process of constructing the sampling

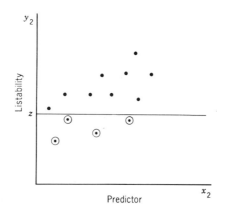

Figure 3.2 Illustration of exclusion from measurement of elements below a threshold value of the "listability" variable (y_2), as a function of a predictor (x_2); circled points, with $y_2 < z$, are omitted from frame population.

frame. The Y axis, for variable y_2, might correspond to a measure of "listability," the ease with which a target population element is placed on the list. Below a certain threshold of this "listability" characteristic, the element is missed, omitted from the frame population. Listability is presented as a function of an exogenous variable, x_2. (As an example, we describe below characteristics of households that are related to their omission from area and telephone frames.) Corresponding to Figure 3.2, therefore, is a coverage model, an analytic specification of the process by which elements are covered by the frame.

Figure 3.3 presents the effects of this selection process on the relationships between variables of substantive interest to the investigator. Y_1 is the endogenous variable in the substantive model, x_1 is the regressor. Note that the process of frame construction leads to the omission of observations throughout the scatterplot. Circled elements are omitted because they fell below the threshold of listability. Depending on how these observations fall on the scatterplot, the regression coefficients will depart from those in the full target population. If Figure 3.2 can be described by

$$y_2 = \alpha_2 + \beta_2 x_2 + \varepsilon_2$$

and Figure 3.3 by

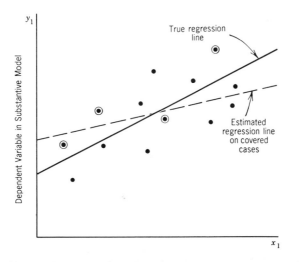

Figure 3.3 Illustration of "incidental" exclusion from measurement of elements, with various values on the dependent variable (y_1); circled points are those in Figure 3.2 which fell below the threshold of listability.

$$y_1 = \alpha_1 + \beta_1 x_1 + \varepsilon_1,$$

then the relationship between the β_1 in the target population and its estimate, β_1^*, in the frame population, under assumptions of multivariate normality, is

$$\beta_1^* = \beta_1 - \Phi\left(\frac{\omega_{12}}{V(y_2)}\right)\beta_2$$

where $\Phi = \dfrac{1 - V^*(y_2)/V(y_2)}{1 - \rho_2^2(1 - V^*(y_2)/V(y_2))}$

$V^*(y_2)$ = variance of the endogenous variable, y_2, in the frame population;

$V(y_2)$ = variance of the endogenous variable, y_2, in the target population;

ρ_2^2 = coefficient of determination for substantive model, $y_2 = \alpha_2 + \beta_2 x_2$, in the target population;

ω_{12} = covariance between error terms, ε_1 and ε_2, for the coverage process equation and the substantive equation.

Since both ω_{12} and β_2 can be either positive or negative, the bias in β_2^* can either be positive or negative. It is not a simple proportional effect, as was the case for truncation based on the endogenous variable. Note also that if the errors in the coverage equation (for y_2) and the substantive equation (for y_1) are uncorrelated, there is no bias in the estimates for β_1. This corresponds to the case where, after consideration of x_1 (the exogenous variable on the substantive equation), there is no relationship between the likelihood of being covered by the frame population and y_1, the endogenous variable in the substantive equation. It is clear, however, lacking that situation, that inferences to the target population based on the frame population will be in error when the frame misses elements according to the equation for y_2.

Thus, it is clear that either with a frame that excludes elements with certain values on the dependent variable or a frame that excludes elements that pose unusual problems, the estimates of coefficients in analytic models can be in error. Coverage error can affect both simple and complex statistics calculated on sample surveys. We should also note that the same approach as above can be used to describe the effect of nonresponse errors on estimates of analytic statistics. Finally, the specification of the selection bias model can be viewed a statement of a theory of coverage error. It should describe those factors that produce coverage error, both characteristics of those missed and characteristics of the frame construction process that missed them.

3.4 COMPONENTS OF COVERAGE ERROR

The above discussion introduces the reader to the conceptual complexity of coverage issues and the relevance of different populations to a survey researcher's work. A more traditional treatment of noncoverage can be given if we restrict our attention to surveys of finite human populations, those with inference to a population fixed in eligibility rules for persons, geographical extents, and time. This section describes the nature of coverage problems of such target populations. Attention to the expression describing the nature of coverage error on a linear statistic,

$$Y_c = Y + \frac{N_{nc}}{N}(Y_c - Y_{nc}),$$

may imply that the only discrepancy of import between the target population and the frame population is that of missing elements on the frame. This *is* the one source of coverage problems that cannot be eliminated by alteration of the survey design, but there are also practical problems related to other coverage issues.

Yates (1948) and Kish (1965) present a clear typology of coverage problems. In the figures below the "F's" represent elements on the frame population, cases that are listed in that set. The "T's" are members of the target population, elements that should be studied by the survey. A perfect frame is signified by the first case, where there is a one-to-one correspondence between the frame population cases and the target population cases (Case I).

Case I

Coverage errors arise when such a correspondence does not exist. One of the most frequently occurring coverage errors, undercoverage or noncoverage, is illustrated by Case II, in which some elements of the target population do not appear in the frame population.

Case II

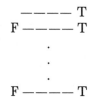

An example of this case is discussed in detail below, the use of a frame population defined by telephone numbers to sample the U.S. household population (target population). Approximately 7 percent of U.S.

households are not linked to elements of the frame population in any simple way.

Case III is the existence in the frame population of elements that are not members of the target population.

Case III

These are "foreign" elements that are not of interest to the researcher. Almost all survey work encounters examples of this. Lists of addresses are often used for personal visit interviews of the household population, but they often contain addresses of businesses or institutions, units not part of the targeted household population. If sufficient information can be obtained on these units prior to the survey, they can be purged from the frame population. There *is* a mismatch between the frame and target populations; that is, a statistic calculated on the entire set of frame population elements will not have the same value as that calculated on the target population.[6]

Case IV illustrates the existence of multiple elements on the frame population attached to one element in the target population.

Case IV

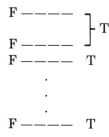

---

[6] In some cases, a statistic may not be defined for the full frame population. For example, if measures of rental value are desired, some housing units, occupied by owners and not rented, are inapplicable to the survey measurement. They are not members of the target population.

This is an example of a coverage problem sometimes labeled "overcoverage." The impact of this case on survey errors depends entirely on whether the sample design or analysis of the survey data takes this into account. However, a careful inspection of the frame population could locate such duplicates and purge them from the population. Alternatively, as is described more fully in Chapter 6, weighting of the target population element that has multiple appearances could correct for its "overcovered" state, if that were known at the time of analysis.

Case V is the situation in which one frame element corresponds to several target populations elements.

Case V

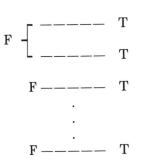

Personal visit samples of adults based on lists of addresses can be thought of as examples of this case. Each housing unit on the list corresponds to one or more adults. In this case there is not a one–to–one correspondence between frame elements and target elements, but a one–to–many correspondence. Each target population element, however, is attached to one and only one frame element, and thus, in some sense, there is "even" coverage of the target population. In such cases, as noted in Chapter 6, if a sample survey from the frame households interviews all persons in each sample household, all adults in the population have equal chances of falling in the survey. Alternatively, one adult can be selected at random from each household and their survey data weighted to reflect varying chances of selection by size of household.

The reader will note that most overcoverage problems can either be eliminated at the frame construction phase or cause estimation problems after the survey. Many times frame problems prompt the use of questions in the survey questionnaires about the nature of the frame element or the correspondence between the frame element and target elements (e.g., "Is this the only telephone number assigned to this household?"). In one sense, the frame population is the researcher's window into the target population and all efforts necessary to measure the purity of that window must be executed prior or during the survey itself.

3.5 COVERAGE PROBLEMS WITH THE TARGET POPULATION OF HOUSEHOLDS

As noted earlier, this book concentrates on surveys of the household population. Using such surveys as examples of our concepts above, the *population of inference* (as opposed to the target population) is often all persons within certain geographical extents of the country being studied (as in the study by Duncan et al., 1972, of occupational mobility in the United States), sometimes all persons in Western society (and the discrepancy between the target population and the population of inference is eliminated through arguments that the United States resembles other Western nations on the topic of the survey), and sometimes, as in epidemiological surveys, all humans. The *target population* consists of all persons who live in housing units in the territory studied at a particular point in time. Eliminated in the move from the inferential population to the target population, therefore, are persons in institutions (as in the Current Population Survey coverage for unemployment statistics), in remote portions of the country (as in many of the surveys of the World Fertility Survey, as a cost-saving measure), on military bases (as in many national surveys by academic organizations), and persons in a more-or-less permanent state of movement (transients who live on the street, nomadic peoples in developing countries).

At some point in the survey process the frame population contains a list of housing units in which the persons live. That is, for most household surveys, the target population and frame population are based on different element definitions. The target population consists of persons, while the frame population consists of living quarters. Therefore, a rule of correspondence between the housing unit and person populations is needed before the survey can proceed to contact persons. There are two alternative rules that are used. One, the *de jure* rule, attaches each person to one and only one housing unit based on where they "live." The words "housing unit" and "live" are the key terms in the execution of the frame (as will be discussed below). Another, the *de facto* rule, attaches each person to the residence at which they are staying at a fixed point in time (e.g., where they were on June 23, 1986). Finally, for household surveys, the *survey population* consists of those members of the frame population who would consent to the interview request. It is therefore a theoretical concept for *sample* surveys, because requests for interviews are never made of all members of the population, only sampled persons.

Given the identifications of the various relevant populations for household surveys, it can easily be seen that some issues of coverage error will be related to correspondence between persons and units on the frame. There will be issues of the linkage of housing units to listings on the frame, and the linkage of persons to various housing units. These can be discussed as issues of coverage of housing units and of persons within housing units.

3.5.1 Coverage of Housing Units

Coverage of housing units should be discussed separately for frames of housing units based on lists of geographical units (area frames) and frames based directly on lists of addresses, telephone numbers, or other identifiers of households (list frames). Area frames are most often used for personal visit household surveys in the United States. The frame consists of lists of mapped area units (counties, cities, census tracts, enumeration districts, blocks). The actual list of housing units is made after the sampling of area units in which the survey will be conducted. (Thus, the sampling and frame construction steps are combined.)

Coverage error can result from conceptual differences between units in the frame and units in the target population (as we shall see later in using the telephone numbers as a frame for all households), but it can also arise with operational weaknesses in constructing a frame based on a well-defined conceptual unit. Thus, for household surveys, coverage error cannot be discussed usefully without an operational definition of "housing units." As with most concepts it seems simple, almost obvious, until it has to be applied in uncontrolled settings.

Most survey definitions of housing units resemble that of the U.S. Bureau of the Census, used in decennial censuses. The following is a shortened version of the definition given to listers for the Survey Research Center.

> "Housing units" are living quarters used by their occupants separately from other persons. The "living quarters" of a person mean the usual place of residence, where a person lives and sleeps most of the time. (Housing units may be occupied by a single family or by two or more family units.)

> The application of the "housing unit" definition becomes complex in buildings with multiple living quarters. If a single

building is occupied by persons who live and eat apart from each other, the living quarters of each person are to be considered distinct from those of others if the living quarters have direct access from the outside. Persons are considered to be "living apart" from others when they own or rent different living quarters. Persons are considered "eating apart" when they provide and prepare their own food *or* have complete freedom to choose when they eat, and they do not have to pay a fee for the meals whether or not they are eaten. Living quarters have "direct access" when (1) they have an entrance directly from outside the building, (2) they have an entrance from a common hall, lobby, or vestibule, used by the occupants of more than one unit or by the general public. Access is not direct if the only way to get to the living quarters is through another person's living quarters.

Some of the common examples of housing units are single family houses, row houses, town houses, small multi-unit structures (e.g., duplexes), flats, garden-type apartments, apartments over or behind commercial structures, or high-rise apartments. More unusual examples are apartments of resident staff members in hospitals, hotels, or prisons. For mobile home parks and boat marinas the spaces are generally the unit counted for housing units (an empty space viewed as a vacant unit).

There are many living quarters that are not considered housing units. These include:

(a) institutional quarters, occupied by persons under the care or custody of others (e.g., children in an orphanage, patients in hospitals, nursing homes or mental institutions, inmates in a prison);

(b) some special noninstitutional quarters [e.g., convents and monasteries, student dormitories, fraternities and sororities, nonstaff rooms in shelters and flophouses, boarding and rooming houses (for persons who pay for their meals as part of their fee)];

(c) transient or seasonal living quarters. These must have at least five units operated by the same management. At least half of the units must be occupied by people who stay less

than 30 days or who pay a daily rate, to qualify as
"transient" living quarters. To qualify as "seasonal" the
facility must be closed at least one season of the year.

Only to those persons who have worked for some time with this
operationalization of the concept of "housing unit" does the above
definition seem simple. Even then, most instructions for listing housing
units contain a safety clause which acknowledges that some cases are not
classifiable using the definition. Hours of discussion can be spent by those
with years of experience in household surveys, trying to determine the
appropriate status of particular living situations. In fact, however, most
applications of the definition in the field are made by interviewers or
part-time staff for whom the listing operation is a small part of their full
activities and who may have little experience in the task.

Fortunately for coverage rates, most units present no problem.
However, those that do present problems tend to be unusual living
situations, ones that have occupants distinctive on a host of
socioeconomic data [keeping in mind $(Y_c - Y_{nc})$]. For example, in
violation of zoning ordinances, some owners of small businesses in
crowded ethnic areas use their business as their living quarters, despite no
evidence of this fact during normal business hours. How are these to be
handled? A second structure sits behind a house in a suburban area and is
used to house visitors and guests of the family. Should this be listed as a
housing unit? A trailer sits behind a house and is occupied by 6 of the 12
members of the family in the house. Do the house and trailer count as one
unit or two? In an inner city area a house is encountered with most of its
windows broken, the grass uncut, apparently unoccupied. Are these
possible living quarters fitting the housing unit definition or should this
structure be omitted?

A further practical problem of constructing a frame of housing units
concerns the temporal definition of the target population. For example, a
lister encounters a house under construction, a basement dug and the
concrete poured. Is this part of the target population? It may be "born"
into the population within a few months, but it would not meet most
conceptions of "living quarters" in its current state. If the target
population were to be defined to be the set of housing units existing at the
time of the visit of the lister, then the site would clearly not be included.
If the listing were made in preparation for a survey of the housing unit
population at a later point in time (and this is most typical), the site
might be included.

Survey organizations vary in the length of time in advance of use of
the frame that listings are made and hence vary in their inclusion rules for
units under construction. Most rules encourage listers to err on the side

of overinclusion of units under construction, avoiding noncoverage of a future target population, risking inclusion of units not appropriately labeled a housing unit. (Again, problems of overcoverage can be more easily detected than problems of undercoverage.) The application of any rule defining the temporal extents of the target population, however, is complicated by the fact that rarely are listings made at the same time and day throughout the entire frame population. Listers are asked to do their work within weeks (sometimes months) and hence observe their part of the population at different times than when other listers are working. This leads to variable coverage errors for different parts of the target population.[7]

In extreme cases coverage issues reduce to interpretations of words as perceived by the lister or by persons contacted at the address. These words include "providing and preparing their own food," "common hall," "renting," and others. These critical observations do not imply that this definition of "housing unit" is particularly poor. Indeed, the years of experience in its use suggest that large improvements are unlikely. The almost infinite number of situations that people have for their day-to-day living arrangements challenges any definition of a target population based on residences. Unavoidable loss of coverage arises with any operationalization. To counteract this, most instructions guide listers to include units that are on the margins of eligibility, rather than excluding them. Overcoverage is a problem that reveals itself during data collection or analysis and thus is a measurable error; noncoverage error yields itself to no simple measurement procedures.

Information from the Census of Population and Housing. Some of the literature on the coverage problems of housing units in area frames concerns failures of the decennial U.S. Census of Population and Housing to cover housing units. The census uses very different methods to cover the population of housing units than most sample surveys using area frame methods. It begins with address lists prepared by the U.S. Postal Service and commercial firms, updates them near the census date, and mails census questionnaires to the addresses. About 83 percent of the mailed questionnaires were returned in the 1980 decennial census effort (Bailar, 1983). Personal visit enumerators visit the remainder.

Estimates of noncoverage of the census can be made using demographic methods, using administrative records on births and deaths, immigration and emigration data, and prior census data to estimate the

---

[7] It also leads to the use of frame updating procedures at the time of selection of the sample from the frame. One of these is the half-open interval technique, described below.

current population. These methods are generally restricted to population estimates and to counts of the full population not subgroups. Other estimates are made by matching studies. As with all modern censuses, the 1980 U.S. Census conducted a Post-Enumeration Survey, using more expensive methods than could be used on the entire population (better trained personnel, more intensive search methods). Checks are made to determine which of the persons located in the post-enumeration survey were counted in the census.

For the 1980 U.S. Census, assessment of coverage based on demographic analysis suggests better coverage of the population than the 1970 U.S. Census achieved. Table 3.1 shows estimated undercounts (positive numbers) or overcounts (negative numbers) for different demographic groups (Passel and Robinson, 1984). The overall estimated undercount is 1.0 percent of the population in 1980, versus 2.6 percent in 1970. This overall rate, however, masks the lower coverage rate of black males (8.8 percent undercount) and the apparent accurate count of nonblack females (0.0 percent overcount).

Data from the 1950 U.S. Census Post-Enumeration Survey show that half of the omissions are completely missed buildings; 20 percent were missed units in enumerated buildings; others were changes in vacancy status. Misclassification of the occupancy of units appeared to be a problem in 1970 and 1980; in 1970, 16.5 percent of the units originally classified as vacant were occupied in a reinterview evaluation survey (Bailar, 1983). An important cause of missed units is their presence on the boundary of an enumeration district. Mobile home parks appear to pose unusual problems for listers, these related to the transient nature of such structures. In a 1980 U.S. Census pretest operation, Harahush and Fernandez (1978), estimate that 3 to 4 percent of units in such areas are missed by the listing operations. Multiunit structures also challenge census enumerators, especially those that are single unit structures altered to have multiple units.

Coverage Errors in Area Frame Surveys. Area frames for household surveys have traditionally been attractive because they offer theoretically complete coverage of the household population. With area frames all housing units are assigned to a single point in space. All spaces in the geographically defined population are included in the area frame population. Therefore, all housing units can be covered by the area frame. These principles, however, are not achieved in the *practice* of area frame sampling.

In practice, after the selection of area units, for example, census enumeration districts or blocks (see Chapter 6), field staff members visit the areas to list the housing units one by one. It is thus the combination

Table 3.1 Population Estimates and the Percent Net Underenumeration, by Race and Sex, 1980 (Legal Residents Only) and 1970

Race, Sex, Year	Estimated Population (000's)	Net Percent[a] Undercount (%)
1980		
All classes	226,717	1.0
Male	111,132	1.5
Female	115,585	−0.4
Black	28,064	2.0
Male	13,604	8.0
Female	14,350	2.7
White and other races	198,652	−0.2
Male	96,996	0.6
Female	100,704	−0.9
1970		
All classes	207,891	2.2
Male	102,062	3.1
Female	105,829	1.4
Black	24,444	7.6
Male	11,952	10.1
Female	12,492	5.3
White and other races	183,447	1.5
Male	90,109	2.1
Female	93,338	0.9

[a]Base of percentage is estimated population.
Source: R.E. Fay, J.S. Passel, and J.G. Robinson, "The Coverage of Population in the 1980 Census," *1980 Census of Population and Housing*, U.S. Department of Commerce, 1988, p. 26, Table 3.1.

of the list of area units and the listing of housing units that forms the frame population. Some housing units pose more problems of detection that do others. Some of these problems concern application of the definition of "housing unit," just as was the case for listing for the decennial census. It is expected that the same kind of units tend to be missed in area frame listing operations as are missed in census operations.

Kish and Hess (1958) describe one of the few studies of area frame coverage by noncensus operation. They estimate coverage errors by comparison of the results of Survey Research Center (SRC) listers in sample blocks and counts of housing units from the 1950 U.S. Census for the same blocks. The listing operations mostly took place between 1951 and 1952, and no later than 1954. Thus, the 1950 U.S. Census counts, even if they were perfect for April 1, 1950, might be poor representations of the number of housing units at the time of listing. To adjust for this, Kish and Hess used as their indicator of coverage rates

$$\frac{\text{(Addresses Listed on Block by SRC)} \times \text{(Dwelling Units per Address in Region Group)}}{\text{(Dwelling Units on Block in 1950 Census)} \times \text{(Growth factor for City)}}$$

Using this ratio to estimate coverage rates requires two assumptions:

1. Homogeneity of the ratio of dwelling units to addresses in the regions chosen as adjustment groups. (These groups were based on city size and region). If the particular block in question had an unusual ratio of dwellings to addresses, the ratio would overestimate or underestimate the coverage rate.

2. Homogeneity across blocks within a city in growth of number of housing units. The growth rate of the city was itself estimated by the growth rate between 1940 and 1950 and number of building permits in the city since 1950.

Because of these assumptions, Kish and Hess recommend the use of the ratios as indicators of *relative* coverage rates across blocks, not absolute coverage. Kish and Hess use a self-weighting sample of housing units from a sample of 629 blocks. They examine values of the ratios for different subsets of the blocks, by characteristics of the blocks and the cities in which they are located.

Table 3.2 presents the values of these ratios by city size and various block characteristics. The table shows that the lowest ratios are not found in the largest cities, but in the group with 50,000 to 99,000 population (0.92 coverage ratio versus 0.96 for the entire group of sample blocks). There is consistent evidence of largest noncoverage among housing units in poorer areas, using both census data and ratings of economic status by the SRC listers. In conflict with expectations, blocks with single unit dwellings had the lowest coverage ratios (0.93) versus those with units of two to seven dwellings (0.95), apartments above stores (0.94), apartment houses (1.04), or mixed blocks (0.99). The authors were also surprised to find that unusually shaped blocks were not most susceptible to low coverage ratios. Instead, simple rectangular blocks had the lower coverage ratios (0.95 versus 0.96 for combinations of blocks, and 1.00 for irregular shaped blocks).

Table 3.2 Coverage Rates of U.S. National Area Frame Household Samples, for 1958, at Survey Research Center

Characteristics of Block	City Size			
	50,000–99,000	100,000–249,000	250,000 and over	Total
Census average property value or rental price				
High	1.04	0.94	1.13	1.07
Medium	0.96	0.92	1.03	1.00
Low	0.88	0.94	0.94	0.93
Average of economic ratings assigned by SRC lister				
High	1.18	1.05	1.00	1.05
Medium	0.90	0.91	0.99	0.96
Low	0.89	0.92	0.99	0.94
Prevalent type of structures in block				
Single dwellings	0.88	0.90	1.00	0.93
2 to 7 dwellings	0.94[a]		0.97	0.95
Apartments above stores	0.91[a]		0.96	0.94
Apartment houses	—	—	—	1.04
Mixture of several classes	1.06	0.94	1.00	0.99
Nature of block boundaries				
Simple rectangle or polygon	0.92	0.91	0.98	0.95
Combination of 2 or more blocks	0.94	0.98	0.96	0.96
Irregular shape or large area	0.89	0.94	1.16	1.00

[a]Rates for combined city size 50,000–249,000.

Source: L. Kish and I. Hess, "On Noncoverage of Sample Dwellings," *Journal of the American Statistical Association*, Vol. 53, 1958, pp. 509–524, adapted from Table 513.

There is an unfortunate dearth of studies on errors in listing housing units in area frame surveys. The only routine checks that are made on such listings are based on comparisons with census counts in the sample area or searches for unusual patterns of addresses among listings within a block. There *is* often a built-in check of listings for some designs that use one staff to list the areas initially and another to check the listings and then interview at selected units. In contrast to errors attached to nonresponse and mistakes in administering the questionnaire, however, failure to list houses is not the subject of routine monitoring procedures in area frame designs. Hence, the common magnitude of omitted units is not known even for ongoing surveys.

3.5.2 Coverage Errors for Persons Within Housing Units

Since the real target population for household surveys is often persons not housing units, another step in frame construction concerns the listing of persons within sample housing units. As with listings of housing units, an operational definition is required to guide the listers' work. A *de jure* operational rule, used by the Survey Research Center, is to count as members of the household:

> 1. All persons staying at the housing unit at the time of contact if:
>
>> (a) the unit is their usual or only place of residence, or
>>
>> (b) a place of residence is maintained for them both at the unit and somewhere else.
>
> 2. Persons not staying at the unit at the time of contact, if a place of residence is held for them at the unit and no other place of residence is held for them.

A chart, like that appearing in Figure 3.4, is given to interviewers to aid them in implementing the definition.

As with listings of housing units, some of the difficulty of applying the *de jure* residency rule concerns transiency. For example, if an interviewer arrives at a housing unit to list the household the day before it is to move out of the unit, should it be listed? What if she arrives on the day of the move?

As with developing the frame of housing units, some time limits must be set on the application of the inclusion rules. To the extent that the definition includes blocks of time, it becomes possible for a person moving during that block of time to be included in more than one household. However, in practical survey work it is infeasible to insist that all household listings for the survey be done at the same moment. The appeal of the *de facto* residency rule is its attempt to fix a moment in time to define residency, but to the extent that interviewer visits to housing units are spread over time the burden on the respondent to recall who was staying in the unit at the designated time becomes larger. Often survey organizations use the rule that the application of the *de jure* definition is made at the first contact of the interviewer with the household. Following

Have a place of residence here?	Have a place of residence elsewhere?	Include in household	Examples
		1. PERSONS STAYING IN SAMPLE UNIT AT TIME OF CONTACT	
Yes	No	Yes	(a) Just "lives here" (b) Lodger (c) Servant
Yes	Yes	Yes	(a) Has country home or town house (b) Has summer home or winter home (c) Student living here while at school, or soldier while in service (d) Home on military leave or school recess
No	No	Yes	(a) Waiting completion of new home (b) Takes turns staying with children or parents
No	Yes	No	(a) Helping out with new baby, or during illness (b) Visiting friends or relatives (c) Works or eats here, sleeps elsewhere
		2. PERSONS ABSENT FROM SAMPLE UNIT AT TIME OF CONTACT	
Yes	No	Yes	(a) Traveling salesperson on the road (b) Railroad worker on a run (c) In general hospital (d) On vacation or visiting (e) Absent on business
Yes	Yes	No	(a) Has country home or town house (b) Has summer home or winter home (c) Away at school or in service (d) In prison or nursing home or special hospital

3. IF "DON'T KNOW" ON ANY OF THESE CRITERIA, INCLUDE IN HOUSEHOLD

Figure 3.4 Chart for determining members of the household.

this rule, a unit that is vacant on the first visit but occupied on the second is generally listed at the second visit. Thus, overcoverage of mover households is dependent on the time interval during which interviewers make their first successful calls on housing units.

In addition to the problems of attempting to fix the population in time, however, new sources of coverage problems that result from the listing process arise. Listing of persons is not based solely on the interviewer's observational activities but usually requires the questioning of a household member. The level of coverage of persons within households is thus dependent on the behavior of the interviewer *and* the household member supplying the information (a source of variability in coverage error over replications of a survey procedure). Brooks and Bailar (1978) state that the majority of noncoverage in the Current Population Survey arises not from missed housing units but from missed persons within households. Korns (1977) estimates that such coverage problems affect poor persons and males more than other groups.

One of the most interesting treatments of coverage error uses a nonsurvey technique to assess coverage problems of surveys. Using techniques of participant observation common to ethnography and cultural anthropology, Valentine and Valentine (1971) lived in a black and Hispanic inner city community for several years, examining social, economic, and nutritional activities of the area's residents. A double-blind comparison of household composition using survey and participant-observation techniques was conducted. Census Bureau interviewers listed 33 households in a block within the area studied by the anthropologists. Independently, they had identified the household compositions.

The comparison reveals the strengths and weaknesses of the two methods. One woman living alone in an apartment unit, and listed by the interviewers, was viewed by Valentine and Valentine as a lodger with a family in another unit. That is, the anthropologists viewed the person as a member of another household. (Whether this person would have been listed as a household member by that household could not be addressed with the data because the interviewers were not asked to list in that area.) Another household listed by the interviewers was omitted by the anthropologists because the family lived in the area only for a short time, was not known by their neighbors, and thus there were no ethnographic data available. Three other units, when listed by the survey interviewers, contained persons who were assigned to other units by the ethnographers. These included visitors to the units, persons living in institutions at the time of the interviewer's visit, and family members not present who live elsewhere. The cause of some of these problems is the use for the phrase "all persons staying here" in the listing instructions and the misunderstanding on the part of the informant that families instead of households were being listed.

The preponderance of evidence from the study suggests undercoverage problems with the survey technique. The comparison involves 153 persons living in 25 households and thus is severely limited in its inferential power. However, the nature of the omissions from the survey listing offers insights into weaknesses of the methodology. Table

Table 3.3 Comparisons of Counts of Persons Identified by Ethnographic and Survey Methods by Demographic Group

Group	Percentage of Listings		
	Ethnography	Census Bureau	Undercount
Male, 0–18 years	32%	35%	6%
Female, 0–18 years	31	34	9
Male, 19 or older	18	9	61
Female, 19 or older	19	22	7
Total	100	100	17
n	(153)	(127)	(26)
Male–headed households	88%	28%	68%
Female–headed households	12	72	—
Total	100	100	—
n	(25)	(25)	—

Source: C.A. Valentine and B.L. Valentine, "Missing Men," report submitted to the U.S. Bureau of the Census, 1971, p. 40, Table 1.

3.3 shows that the largest noncoverage problems arise within the group of males aged 19 or older, where 61 percent of the group was missed by the survey interviewers! Associated with this grievous noncoverage is the misclassification of households as female-headed, following the definition of "head of household," used by the census.

What lies behind these gross undercoverage problems? One is the problem of communicating to the informant the extents and limits of the concept of "household." The second involves benefits to the informant of misreporting the existence of adult males. Almost all households were found by ethnographers to subsist on multiple sources of financial income, involving conventional employment, welfare, and extra- or illegal activities. Welfare restrictions reward households with no wage-earning adults. Disclosure of illegal activities is also minimized by failure to report household members who engaged in them. Such pressures lead in the neighborhood to "a tacit agreement that each individual's and each family's affairs are their own business." The ethnographers believed that this explanation would apply to all the units in which male heads were not reported.

The type of subpopulation studied in this ethnographic research does not form the majority of the population but does contribute no doubt to large portions of the coverage problems of sample surveys. When there is evidence that response error exists that is well justified by the perceived threat of potential harm to a household's livelihood, strategies of more careful training of listers or more carefully designed inclusion rules appear fruitless. What appears extant here is the breakdown of trust between the survey organization and the sample household regarding confidentiality and utility of the survey data. The goals of the survey are not believed or shared by the target population, and they choose to omit themselves from the frame.

Although there appears to be no experimental efforts to increase the trust in survey methods as a way to improve coverage, there is some literature in survey methodology regarding questioning techniques for listing household members. These have the goal of probing the informant more carefully about members of the household. Tenebaum (1971) reports on a study within a Census Employment Survey to improve within-household coverage by adding six probe questions.

The stock questions were:

What is the name of the head of this household?
What are the names of all other persons who are living or staying here?

Then,

    1. Have I missed
        — any babies or small children?
        — any lodgers or boarders who live here?
        — anyone who usually lives here but is away at present traveling, at school, or in a hospital?
        — anyone else staying here?

To these stock questions were added:

    2. How many other people stay here some of the time?

    _____ Number

3. Are any of these people staying here now?
  __ YES (List any staying here if not already listed)
  __ NO

4. Does anyone who is not staying here now consider this place home? (Exclude Armed Forces members.)

  __ YES (Ask 5)
  __ NO   (Skip to end)

5. Does he have some one place he usually stays?

  __ YES ELSEWHERE (Skip to end)
  __ YES HERE (List)
  __ NO   (Ask 6)

6. Do you expect him to return here within two months?

  __ YES (List)
  __ NO

From a sample of 60 urban and seven rural sampling areas, over 6000 sample households were used in an analysis of how many additional household members were included because of the six new questions. A total of 19,861 persons were listed prior to question 1, only 83 were added at question 1, and only 69 more at questions 2 to 6. Even among that small number not all the added persons fulfilled the requirements of household membership. Of the 83 added from question 1, 42 or 50 percent were treated as household members; of the 69 added by questions 2 to 6, only 14 or 20 percent were treated as household members. This means that the added questions increased coverage by less than 0.3 percent. The persons added to the household because of such questions tend to be white, relatives of the household head, and less than 30 years of age. In short, this kind of probing does not by itself appear to constitute a solution to the within-household coverage problem.

Measurement of household composition has also received some attention in the literature on telephone survey methods. In that mode of data collection it has been found that refusals often occur before a full household listing can be taken. In reaction to this, streamlined methods of counting adults in the household have been used. Much of this has been examined in the context of selecting one adult at random among all adults in the household. One strategy for selecting one adult from the household merely inquires about the number of male and female adults and then selects one adult from among all by identifying the sex and relative age of the selected individual (e.g., "I would like to interview the second oldest female in the household"). Thus, instead of the lengthy questioning of the respondent to list the adults one-by-one, the interviewer merely asks:

First, could you tell me how many people 18 or older live there?

How many of these are female?

To compare this method of enumerating the adults in the household with the traditional listing by relationship and age of all the adults in the household, Groves and Kahn (1979) used the two-question method to select a respondent and then asked the respondent for a complete listing of the household at the end of the interview. They viewed the answers given at the end of the interview as more accurate, both because they involved more probing questions and because more rapport had typically been established with the respondent than was the case with the phone answerer.

Table 3.4 presents the comparison of number of adults in the two methods. A total of 9 percent of the sample households had discrepancies between the two methods. Groves and Kahn note that households with young adults had a relative larger frequency of errors. There is a tendency for the smaller households to be described as larger and the larger households as smaller by the simple two-question approach. This probably relates to differing understandings of "adult" and "live here" in the two-question scheme. Because of the overestimation and underestimation, however, the total numbers of adults in the sample households are very similar between the two approaches.

Within-household coverage of persons is an area which could be studied cheaply by reinterviewing samples of listed households and enumerating them again. This could be done in conjunction with verification interviews that are sometimes conducted as checks that the interviewers did indeed visit the sample unit. Verification of the household listing could also be made at the end of the interview with the selected respondent, like the Groves and Kahn experiment above. However, such a procedure might be threatened by interviewer tendencies to seek confirmation of the original listing and thus underestimate actual errors.

Table 3.4 Percentage of Households Reporting Different Numbers of Adults in Two–Question Enumeration Within Categories of Numbers Recorded in Full Household Listing

Number in Full Household Listing	Number Using Two–Question Scheme								Total	n
	One	Two	Three	Four	Five	Six	Seven	Eight or More		
One	**93.8**	4.6	1.6	0	0	0	0	0	100.0%	368
Two	1.9	**94.0**	2.4	1.4	0.3	0	0	0	100.0	934
Three	1.3	10.9	**78.6**	7.5	1.3	0	0.4	0	100.0	239
Four	0	11.1	11.1	**74.0**	1.9	1.9	0	0	100.0	54
Five	7.1	21.4	7.1	7.1	**57.3**	0	0	0	100.0	13
Six	0	0	0	0	0	**0**	0	0	—	0
Seven	0	0	100.0	0	0	0	**0**	0	100.0	1
Eight or more	0	0	0	0	50.0	50.0	0	**0**	100.0	2

Source: R. Groves and R. Kahn, *Surveys by Telephone: A National Comparison with Personal Interviews*, Academic Press, New York, 1979, p. 62, Table 3.6.

115

3.5.3 Telephone Number Frames for the Household Population

As the cost of conducting surveys with personal visit methods has increased over time, survey researchers have used telephone survey methods more frequently. This is an example of lower cost increasing the attraction of one frame over another, despite differences in coverage rates. With this method each person is linked to a housing unit using the same definitions as above. Housing units are then linked to a telephone number, based on the location of telephone instruments reached by that number. The full telephone frame population consists of all possible 10-digit numbers within the telephone system of the United States.:

```
- - -  ⁻  - - -  ⁻  - - - -
```

    Area   Prefix   Suffix
    code

There are about 110 area codes in the United States in active use. These are geographical units that, for the most part, do not cross state boundaries. Small states have only one area code; large states can have several. There are currently over 35,000 prefixes that are active in these area codes. These contain either business or household numbers. Finally, each prefix contains at most 10,000 numbers (0000 to 9999). This frame therefore contains over $35,000 \times 10,000 = 350,000,000$. This is obviously many more than needed to cover the full set of housing units in the United States, less than 90,000,000. Indeed, only about 22 to 25 percent of the numbers on the telephone frame are working household numbers (Groves, 1978). Thus, the frame contains many foreign elements, including business numbers, telephone company service numbers, and nonworking numbers.

In addition to coverage problems involving nonhousehold numbers, the telephone frame also suffers from "overcoverage" of the household population. This problem arises from the fact that some households have more than one telephone number. About 3 to 4 percent of U.S. households have more than one telephone number, the vast majority of which have two (Groves and Kahn, 1979). In uses of the telephone frame, sample persons are asked directly how many telephone numbers they have at home, and survey statistics use weights to adjust for the overcoverage.

By far the largest concern with telephone frames for the household population is undercoverage, the failure of all households in the country to subscribe to telephone service. The nontelephone household population has diminished in size steadily throughout the last decades, but universal service does not exist.[8] The best information on coverage rates of the telephone frame comes from personal visit surveys of the U.S. Census Bureau. The National Health Interview Survey (NHIS), a monthly cross-section sample of all U.S. households, includes a series of questions about telephone subscription by the sample households. Table 3.5 presents data from the 1985-86 NHIS on the percentage of persons without telephones.[9] The results here are similar to others (Groves and Kahn, 1979; McGowan, 1982) and show that the most powerful correlate of telephone subscription is family income.

Over 29 percent of those families with less than $5000 in annual income do not have telephones, but less than 1 percent of those with incomes greater than $30,000 do not have telephones. This finding is not surprising when one realizes that membership in the telephone frame population entails a cost to the household. Monthly subscription costs for telephone service vary greatly over the 1500 operating companies in the United States, higher in rural and remote areas than in urban areas. The costs of subscriptions form larger portions of poor persons' income and thus are more significant influences in the decision to have telephone service. Auxiliary support for this reasoning comes from lower coverage rates in rural areas. It is also useful to note that older persons, despite their lower average incomes, tend to be relatively well covered by the telephone frame. (This may result from a substitution of the telephone for personal visits to businesses, friends, and relatives that would disproportionately affect the elderly.) Finally, we must caution that since the estimates in Table 3.5 are based on a personal visit survey, the estimated percentage of *total* persons missed by telephones may be underestimated (if those not covered by the frames used in the personal visit survey are disproportionately nontelephone households).

The figures in Table 3.5 address only noncoverage *rates*, not the characteristics of the noncovered persons on survey variables. McGowan (1982) presents some data on noncoverage *bias* of the telephone frame for the National Crime Survey (NCS) victimization rates. The NCS is a national household survey of persons aged 12 years old and older, based on census address frames supplemented by various frames of units missed

---

[8] Indeed, the divestiture of the American Telephone and Telegraph system in the early 1980s led to speculations that coverage would decline because of higher base prices for service. No data are yet available to speak to that speculation.

[9] Note that these estimates of noncoverage by telephone are subject to errors from the coverage of housing units by the census address frame and auxiliary frames used in the NHIS. It is possible that the set of households covered by telephone frame methods is not merely a subset of those covered by the NHIS methods. Instead, some households reached by telephone may be missed with NHIS methods.

Table 3.5 Percentage of Persons Not Covered by Telephone Frame Based on the National Health Interview Survey, 1985–86

Characteristic	Percentage in Nontelephone Households
Race	
White	5.8
Black	15.6
Other	10.9
Age	
Under 6 years	12.3
6–16 years	8.5
17–24	11.2
25–34	8.0
35–44	5.0
45–54	4.2
55–64	3.7
65–74	3.2
75 and over	3.3
Family income	
Less than $5,000	29.2
$5,000–$6,999	20.0
$7,000–$9,999	18.1
$10,000–$14,999	12.9
$15,000–$19,999	7.7
$20,000–$24,999	3.6
$25,000–$34,999	1.8
$35,000–$49,999	0.9
$50,000 or more	0.4
Unknown	8.0
Marital Status	
Married, spouse in household	4.8
Married, spouse not in household	16.2
Widowed	4.5
Divorced	9.8
Separated	18.8
Never married	8.1
Under 14 years	10.6

Source: O. Thornberry and J. Massey, "Trends in United States Telephone Coverage Across Time and Subgroups," in R.M. Groves et al. (Eds.) *Telephone Survey Methodology*, 1988, Table 4.

by the last census. All households, both those with and without a telephone, are eligible for the survey. A set of questions in the personal visit interview ask whether the household has a telephone. Table 3.6

shows that the nontelephone population is much more heavily victimized than the telephone population. The differences between the nontelephone and telephone victimization rates are estimates of $(Y_c - Y_{nc})$, in the equation above. The differences between the telephone population rates and the total population rates are estimates of $(N_{nc}/N)(Y_c - Y_{nc})$, the bias in the victimization rates from noncoverage if only the telephone frame were use to estimate the rate of the full household population.

The nontelephone household population has victimization rates that are sometimes two to three times those of the telephone household population. For example, the violent crime rate per 1000 persons is about 73 for persons without telephones, but only 31 for those with telephones. However, because only 6 to 8 percent of the persons do not have telephones, the overall error due to noncoverage if a telephone frame were used is only $(33.69 - 30.71) = 2.98$ per 1000 persons. The victimization rate of the total population would be underestimated by the telephone frame population alone (and the difference is beyond that expected by sampling error at the 0.95 confidence level), but the difference is only about 10 percent of the rate itself. Is that an important difference? The answer to that question depends entirely on the substantive purposes of the statistic.

The coverage bias due to the telephone frame when describing the full household population will be sensitive to the nature of the survey statistic. For statistics highly variable by income groups, the error could be quite large, despite the fact that only small portions of the target population are missed by the frame. Noncoverage error associated with the telephone frame can be especially large when subpopulations of the full household populations are studied. For example, surveys of welfare recipients among the household population would likely suffer from much larger noncoverage error because the *rate* of telephone subscription in the welfare recipient subpopulation is much lower than that in the total population.

3.6 MEASUREMENT OF AND ADJUSTMENTS FOR NONCOVERAGE ERROR

3.6.1 Noncoverage Error Estimation

Even if we restrict our attention to simple means and totals, it is very difficult to estimate coverage error in practical survey work. The survey data themselves provide neither an estimate of (N_{nc}/N), the rate of

Table 3.6 Criminal Victimization Rates for Telephone and Nontelephone Persons and Households, National Crime Survey, 1976–1979 (Rate per 1000 Persons or Households)

Crime Statistic	Rate per 1000 Persons or Households		
	Nontelephone	Telephone	Total
Total personal crimes	181.42	125.15	129.16
Crimes of violence	72.75	30.71	33.69
Rape	2.72	0.82	0.95
Robbery	16.45	5.42	6.21
Assault	53.58	24.48	26.54
Crimes of theft	108.67	94.44	95.47
Total household crimes	319.43	220.71	229.27
Household burglary	151.47	80.78	86.82
Household larceny	149.17	122.96	125.32
Motor vehicle theft (per 1000 households)	18.79	16.97	17.12

Source: H. McGowan, "Telephone Ownership in the National Crime Survey," U.S. Census Bureau memorandum, 1982, Table 6. All differences between same statistic for different groups are statistically different from zero except those involving motor vehicle theft, and the differences involving the total population for rape, crimes of theft, and household larceny.

noncoverage, nor of $(Y_c - Y_{nc})$ the difference between the frame population and the noncovered group on the survey statistic, Y.

Estimates of noncoverage error require the creation or access of data external to the survey data themselves (as in the use of area frame data to measure telephone frame noncoverage). For area frame surveys this means the appeal to census data (*and* the assumption of no coverage error in the census data) or the implementation of special coverage check studies. Special coverage check studies generally consist of using more experienced personnel and/or more detailed methods to do the same listing (or equivalent frame construction) work, with subsequent comparisons of listings. The assumption made in the estimation is that the more expensive method has no coverage error.

One example of a special coverage check study comes from the National Survey of Black Americans, an area frame household survey of blacks conducted by the Survey Research Center. Since black households form roughly 10 percent of all households in the United States, the surveying of that subgroup entails large screening costs, especially in areas where few blacks were found to reside by the last census. To reduce the costs of frame development in those areas, instead of screening the residents at each listed unit to learn their races, screening was done only at a subset of houses. In addition to inquiring about the race of the residents of the house visited, however, the interviewer asked whether any blacks lived in the area described by a map of the sample block (or other sample area chosen). If none of the visited houses identified a black household in the sample area, no further visits were made in the area. If a black family were identified, interviews were attempted at those units. This method relies on sample area informants to provide frame information on sets of potential sample housing units and thus saves large amounts of screening costs.

As a check on this method, interviewers were asked to visit all households in a subset of the blocks, both to ask the screening questions of each visited household and to use each as an informant about the residence of black persons in the sample area. Since the subset of blocks on which both methods were used was a probability sample of all areas, the check yielded estimates both of the rate of noncoverage of the cheaper method, and also of the characteristics of the black persons who would not be covered by the cheaper method. This is an example of using an expensive method in a subsample to evaluate a cheaper frame construction method used throughout the survey.

Finally, although this discussion has presented coverage error as a result that would have consistent effects over replications of the survey using the same frame construction techniques, it is useful to note that coverage error may vary over replications of the survey. This variation would arise if the frame construction procedures themselves might vary in their adequacy over time. This is a practical problem in all survey work because the decision to include a household or person on the frame is ultimately made by a lister or interviewer. In area frame designs, human listers assemble the frame at later levels of sampling. They vary in their ability to do this well. In addition, the difficulties of listing vary over selection units, and thus the proportion of the population missed in any one survey is a function of which units happened to be selected into the sample. Thus, noncoverage error is both a bias and a source of variable error. The coverage check study above does not measure the variation in error likely to be experienced in replications of the survey.

3.6.2 Multiplicity Sampling and Measurement

Multiplicity (or network) sampling is a method by which elements of the target population are conceptually linked to others in the target population (Sirken, 1970). Although it is generally referred to as multiplicity *sampling*, it is based on a technique of frame construction and needs no sampling to be used. Often in uses of multiplicity sampling for the household population each person is viewed as linked to other members of his nuclear family (parents and siblings). Each occurrence of a person on the frame population could therefore be considered a listing for a family cluster of persons. Using the notation that was shown earlier, this could be represented as

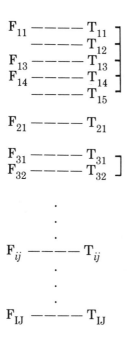

where F_{ij} = listing on the frame population of the jth person in ith family;

   T_{ij} = the jth person in the ith family in the target population.

This example shows that family $i=1$ in the target population contains five members. Three of the five, $j=1$, 3, and 4, appear on the frame; two do not, $j=2$, 5 (e.g., they are moving between residences at the time of the survey). Family $i=2$ contains only one person, and that person is listed on the frame. Family $i=3$ contains two people and they are both listed. Even though two people in family $i=1$ are not listed themselves, they are considered as connected to three different elements on the frame through their family network (indicated by the lines to the right of the target elements). In this view each person in the target population is associated with potentially several members of the frame population. Each person "appears" on the frame a number of times equal to the number of persons in his family that appear (termed his "multiplicity"). Note that a person can be covered by the frame even when he himself does not appear on the frame. It is this property which allows the multiplicity approach to improve the coverage properties of any frame.

Multiplicity approaches were first used only as a measurement tool. Nathan (1976) used a multiplicity measurement to estimate births in a population. Using a tax list frame he drew a sample of households and enumerated all births in the sample households. He also asked each sample woman to identify her mother, sisters, and daughters (if any) living in the country and report any births for them. To estimate the total number of births in the population, reports of each birth were weighted inversely proportional to the number of women in the family of the mother, who appeared in the area frame. This weighting adjusts for the fact that the area frame supplemented by the multiplicity measurement "overcovers" women in larger families.

Multiplicity survey techniques were also used to measure census undercoverage. In a 1980 U.S. Census pretest in Oakland, California, Sirken et al. (1978) supplemented the usual census enumeration of the household with the following questions:

A1. Did you live in Oakland on the census date?
A2. How many of your parents lived in Oakland then?
A3. How many of your siblings lived in Oakland then?
A4. How many of your children lived in Oakland then?

For each person reported in A1 to A4:

B1. Where in Oakland did (...) live on census date?

C1. Does (...) live in Oakland now?
C2. How many of (...)'s parents now live in Oakland?
C3. How many of (...)'s siblings now live in Oakland?
C4. How many of (...)'s children now live in Oakland?

Each of the questions above is needed to assemble estimates of undercoverage in the census. Questions A1 to A4 gather data on the family network of the person. Note that the network is restricted to the family living in Oakland for purposes of this test because only data from Oakland could be used to check on the quality of reporting and coverage. Question B1 is asked to obtain information on the reported network member useful for matching to other census information. Questions C1 to C4 determine how many other persons are linked to the reported network person.

A check of census enumerations can be performed for those relatives reported in the multiplicity questions. If they do not appear in the listings they have been missed by the normal census procedures. Each person should be reported in the multiplicity questions in numbers equal to the number of relatives living in Oakland at the time of the census.

However, practical problems plague the implementation of the multiplicity technique. People do not always know the addresses of their relatives; almost half of the addresses reported in the Oakland pretest were incomplete or not locatable. Some respondents refuse to give the interviewer information about their relatives. Although in the Oakland pretest less than 1 percent of the households presented this problem, in tests of the method for the Hispanic population and the elderly populations by the Survey Research Center, intolerably high rates of refusals have obtained. A successful use of such questions occurred in a study of black family relationships, where the rationale for the inquiry about other family members might have been more clearly understood by respondents.

3.6.3 Multiple Frame Methods

Often the problem facing the survey designer is not an absence of a tight conceptual definition of the target population (or, for that matter, the population of inference) but rather choosing a frame population that is adequately similar to the target population. Most often the problem of identifying a desirable frame population involves an explicit balancing of costs and errors, and most often the problem involves the sampling of subpopulations. For example, Lazerwitz (1964) describes the problem of sampling the Jewish household population (the target population), with a choice of a list of members of Jewish organizations or an area probability sample (frame populations). The area frame would offer complete coverage of the Jewish household population but would force the screening of scores of sample addresses to locate one eligible family. The list frame offers cheap access to eligible persons, but large coverage errors (both from a small portion of the population being on the list and likely distinctive characteristics of those who are members of Jewish organizations). Similar frame problems arise when sampling farmers (an occupational group) or farms (a kind of use of the land). Traditional area frames cover these subpopulations, but because the target populations are so small relative to the frame populations, screening costs would be very high.

In such situations dual or multiple frame designs are often considered. A multiple frame design combines two or more inadequate frames in the same survey to offer coverage rates that exceed those of any single frame. The examples above are special cases, where one frame (the area frame) offers theoretical complete coverage but has very high costs, relative to a cheap frame with poor coverage. Multiple frame designs draw samples from each of the frames simultaneously, interview the sample persons, identify on which of the different frames each sample person is listed, and combine the interviews from all sample persons, using weights to account for memberships on different sets of frames. The Current Population Survey of the U.S. Census Bureau is a multiple frame design. The majority of the persons are sampled from a list of addresses from the last decennial census of population and housing; some come from a list of building permits issued since the census (U.S. Bureau of the Census, 1977). Kish and Hess (1958) describe a personal visit survey design that combines city directory listings of addresses prepared by commercial firms with area frame listings. Steinberg (1965) describes the use of list and area frames for surveys of farmers in the United States.

A dual frame household survey design involving a telephone and area frame was investigated by Groves and Lepkowski (1985). Here the dual frame is attractive because although the telephone frame offers cheaper access to sample households, it fails to cover 6 to 7 percent of the U.S. households. In the dual frame approach, a random digit dialed sample is drawn and screened for household numbers, and interviews are taken from sample households on the telephone. Simultaneously, an area frame survey is conducted, sample addresses visited, and interviews taken from each. In each personal visit interview the respondent is asked whether the housing unit has a telephone (this to determine whether the unit was also covered by the telephone frame). The final survey estimates have the form, for means,

$$\bar{y} = p\bar{y}_{a1} + (1-p)[\theta\bar{y}_{a2} + (1-\theta)\bar{y}_b] \, ,$$

where p = proportion of sample households without telephones;

$\bar{y}_{a1}$ = sample mean for households without telephones;

$\bar{y}_{a2}$ = sample mean for households with telephones, selected from the area frame;

$\bar{y}_b$ = sample mean for households with telephones, selected from the RDD frame;

θ = mixing parameter.

The overall mean is thus estimated by a weighted average of means from nontelephone and telephone households. The mean of the telephone households is itself a weighted average of means from the telephone and personal visit samples (using the mixing parameter, θ). Hartley (1962, 1974) investigated properties of this estimator for the simple random sample case; Casady et al. (1981) extended it to the case of dual frame telephone and area frame surveys. One of the chief problems in dual frame sample surveys is deciding what portion of the sample should come from each frame. This decision requires a balancing of costs of sampling and interviewing persons from the two frames, as well as error properties of the two. This problem is discussed in some detail in Chapter 6.

Although multiple frame designs can offer coverage error reduction, they face administrative problems that single frame designs avoid. These include:

1. Some persons in the target population will have multiple appearances in the various frame populations. If a purge of these multiple appearances is attempted before data collection, large costs and errors of mismatching might result. If no purge is attempted beforehand, for each sample person, the researcher must determine the membership status for each of the frames. Sometimes the persons can report accurately about their listing status for each frame (as in the case of the telephone–area dual frame survey); in others, they cannot (as in the telephone survey design using directory listings and methods of random generation of telephone numbers).

2. Often the use of a multiple frame approach entails the maintenance of several different data collection procedures. A list frame (as in the case of farms) often contains more information about persons than an area frame. This sometimes permits cheaper methods of data collection (telephone or mail methods). Many times this is accomplished

with a centralized staff. The area frame often requires the maintenance of a staff for on-site inspection or listing of addresses, with personal visits typically following listing. The creation of two or more staffs using different data collection methods is an overhead that is avoided by single frame methods.

3. Purging of multiple listings prior to data collection is rarely done in multiple frame designs. Hence, the estimators for the survey involve the use of weights attached to the various sets of persons covered by different combinations of frames. These estimators are thus more complex than those typically needed for survey analysis.

3.6.4 The Half–Open Interval

The multiplicity approach described above is one form of linking of frame elements to others in the target population not covered by the frame. Another linking mechanism is often used in frames that have a spatial or other ordered property to them. This method, the half-open interval, links to every frame unit all eligible target population members that lie after it in the order of the frame, but before the next listed unit in the frame (Yates, 1948). It is called the "half-open" interval because elements that follow the listed element are linked (not those preceding it) to the listed element. The result of this procedure is that each frame element corresponds to a potential cluster of target population elements, one from the frame and others not listed on the frame but following the listed element. Therefore this process, if successful, leads to each element in the target population being associated with one and only one element on the frame population, but some elements on the frame population associated with more than one element in the target population.

The half-open interval is often used to correct the listings made for area frame samples. At some point in the process of using area frames, errors of omission are made in listing addresses in sample areas. Listers are instructed to start at one corner of the area and list sequentially around the area; thus, the lists are ordered. Usually this listing takes place somewhat before the survey data collection. At the time of the survey, interviewers are instructed to implement the half-open interval technique, inspecting the area between a listed unit and the next listed unit for omitted housing units. If they discover a missed unit after a selected unit, they are to add it to the frame and to the sample.

There are various practical problems in this method that impede its success:

1. The technique works when spaces of land can unambiguously be assigned to each listed unit. The land starting at one listed unit, up to but not including that of the next unit, is to be included. Property lines, fences, or other delimiters are often used to define the beginnings and ends of listed units in simple single unit dwelling situations. The space associated with a listed unit becomes less clear, however, when apartment units on different floors are listed. (Here, some practices call for links between floors for the last listed unit on one floor and the first on the next higher floor).

2. The technique does not handle well errors in listing that involve overcoverage, the listing of units outside the boundaries of the selected area. It is a remedy for undercoverage.

3. The technique is often implemented by interviewers, who are evaluated more closely on their success at obtaining cooperation from sample persons and their interviewing technique than their ability to locate missed units.

Despite the problems facing the technique, it continues to be used in area frame samples because of its relatively low cost of implementation, judged to be negligible by most survey organizations.

Half-open interval techniques have been considered for telephone frame samples of the household population, but, to our knowledge, these have not been implemented. They would call for a telephone household contacted by an interviewer to identify whether the household "next" to theirs has a telephone. If it did not, the nontelephone household would be linked to the sample unit. Efforts at personal visit contact would then be made. The ambiguity of various notions of the "next household" and ignorance of one household about another's telephone status are large challenges to this method in the telephone frame.

3.6.5 Use of Postsurvey Adjustments for Coverage Error

Another approach for dealing with coverage error avoids any attempt at correction of the frame prior to the survey but attempts to correct the survey data for coverage error *after* the data collection is completed. This technique is akin to that of poststratification to reduce sampling error, discussed in Chapter 6. It can offer elimination of any errors due to coverage problems, but only under limiting assumptions. It is discussed most often in the context of a simple linear statistic like a mean or population total estimate.

The approach requires the identification of subgroups of the population (1) for which the value of the survey statistic (say, Y_h) is the same for those covered by the frame and those missed by the frame, (2) that experience different coverage rates, and (3) for which the proportion the subgroup forms in the *target* population is known (say, W_h). The survey statistic adjusted for noncoverage is

$$\sum W_h Y_h.$$

The first criterion for groups generally cannot be assessed in practice. It is most often believed that the smaller the group in absolute size, the more likely it is that those not covered are more like those covered by the frame. Thus, some groupings used in practice are as small as 1 percent subclasses of the population. The existence of different coverage rates can be argued from studies like those discussed above. Race, gender, and age are found consistently to be correlates of noncoverage in area frames; income is a strong correlate of noncoverage in telephone frames. The third criterion, the knowledge of proportions of the target population represented by the subgroups, is achieved generally only with census data adjusted for any known coverage problems. There are generally no better estimates available, but there is also no assurance of their accuracy.

CPS data are adjusted by demographic techniques for undercoverage. These adjustments are meant to correct for both failure to cover full households and failure to cover persons within listed households. One way to estimate undercoverage is to examine how estimates might change when adjustments for undercoverage are made. Table 3.7 shows that estimated percentages of persons employed change by less than 0.5 percent with two different adjustments for noncoverage—one based on independent estimates of the population, the other on demographic and administrative records. Both sets of weights make less than one percentage point change in the estimates. The two adjustment schemes, however, give slightly different results. With the use of administrative records, the largest changes are found among nonwhite males (moving the estimate from 92.1 to 91.2 percent). The same estimate incurs minor effects with the adjustment from demographic sources alone (92.1 to 92.0 percent). The source of data for estimates of proportions of the population in subgroups (whether it is based on demographic methods alone or demographic and administrative information) can affect the nature of the adjustment.

Since the Current Population Survey is subject to nonresponse as well as coverage error, the application of weights to counteract noncoverage

Table 3.7 Persons Employed as Percentage of Labor Force for Two Stages of March 1973 Estimation Compared to Percentage Adjusted for Census–CPS Undercount, Persons Aged 16 and Over

Group	Before Adjustment for Coverage	Adjusted by Demographic Methods	Adjusted by Demographic and Administrative Records
White males	95.6	95.6	95.3
White females	94.5	94.8	94.4
Nonwhite males	92.1	92.0	91.2
Nonwhite females	89.5	89.8	89.3
Total	94.7	94.8	94.4

Source: C. Brooks and B. Bailar, *An Error Profile: Employment as Measured by the Current Population Survey,* Statistical Policy Working Paper 3, U.S. Department of Commerce, September 1978, p. 6, Table 3.

simultaneously makes adjustments for nonresponse. This means that the reader cannot infer from Table 3.7 alone that nonwhite males are most susceptible to coverage error. They might also be disproportionately subject to nonresponse. Furthermore, such weighting procedures can sometimes decrease the sampling variance of estimates (under the label of "poststratification"). Hence, use of population weights for adjustment of errors of nonobservation is a multipurpose endeavor, performed in hopes of decreasing noncoverage, nonresponse, and sampling error.

3.7 SURVEY COST ISSUES INVOLVING COVERAGE ERROR

The last chapter noted the rule of thumb that survey errors are inversely proportional to survey costs. This is most clearly exhibited by coverage errors. Coverage errors arise in the development of a sampling frame. In some cases (the listing of household addresses in an area frame) these developments resemble full scale surveys, requiring detailed observations on population units. In other cases (use of telephone directories) they are built from materials created for purposes unrelated to survey research. These two cases demand very different cost considerations.

The latter case is the simpler. As an example, consider the option of using a telephone directory frame or random digit dialing (a full area code/prefix frame) in a sample of telephone households. In this case the coverage error properties of each frame can be taken as fixed, given the selection of the frame. The appropriate cost considerations involve the cost of acquisition, sampling, and screening for eligible units. For the directory frame, costs of acquisition depend on whether large numbers of directories are required (as in a national sample), locally available (and free) directories exist, or a useful cost model for coverage error would incorporate costs of activities that are related to the success at covering different parts of the target population. It would therefore pertain to frames which were under the control of the researcher, not those acquired from some other source for purposes of the survey researcher. In the latter case, the only possible design decision is the choice of alternative frames.

3.8 SUMMARY

Coverage error in a survey statistic is most often conceptualized as a bias, a deviation from the full population parameter which is relatively constant over possible replications of the survey, given the same design. The "population" in this usage is the target population, consisting of all units that are deemed eligible for the survey measurement. Discrepancies between the target population and the frame population produce coverage error. The frame provides identification of some or all target population members and may include units outside the target population. The "same design" in the usage above includes the choice of sampling frame or procedures use to build a sampling frame.

In simple descriptive statistics, coverage error is a function of the coverage rate and the difference in values between the covered and noncovered population statistics. In analytic statistics, the error remains a function of differences between the covered and noncovered groups but involves differences of variances and covariances of variables involved in the analysis.

Most coverage errors are best viewed as fixed properties of survey designs, constant over replications of the survey. They are biases that resist measurement within the resources of the survey itself. Instead, survey designs must incorporate some internal variation to measure components of coverage biases. For example, use of two different methods to enumerate the household (for within-household coverage issues) might be applied to probability subsamples of a survey and coverage error differences estimated through their comparison.

Reinterviews using stricter frame development procedures serve the same purpose. Alternatively, some source of information outside the survey must be used to evaluate the coverage of the design. Census-based data are often used for such purposes.

Attempts to build frames of the household population are never completely successful. The U.S. decennial census disproportionately misses minority groups, especially males. Some of the problems of covering the household population appear to stem from difficulties in applying the traditional survey definition of "household" to some situations. Others arise because of difficulties in discriminating different housing units in unusual housing situations. Finally, based on ethnographic methods, there appear to be potential financial harm or social embarrassment that can arise from accurately enumerating household composition.

The use of telephone number frames to study the U.S. household population fails to cover poorer households and those in rural areas, for whom the marginal cost of telephone subscription is higher. The elderly population is disproportionately well covered by telephones. Telephone number frames also include nonworking numbers and nonresidential numbers, which must be eliminated prior to interviewing for household surveys.

Some coverage errors, especially those flowing from work of individual interviewers, have variable components over replications of the design. If different interviewers were used, different coverage errors might apply to the survey. Thus, the notion of coverage error variance has some utility.

There are methods of reducing coverage errors within household surveys. Use of multiplicity and multiple frame methods attempt to enrich the frame materials used in the design. The half-open interval attempts to correct during its use a frame already constructed. Finally, various postsurvey adjustments are used to reduce the magnitude of coverage error after a survey is completed.

CHAPTER 4

NONRESPONSE IN SAMPLE SURVEYS

And they heard the sound of the Lord God walking in the garden in the cool of the day, and the man and his wife hid themselves from the presence of the Lord God among the trees of the garden.

But the Lord God called to the man, and said to him, "Where are you?" And he said, "I heard the sound of thee in the garden, and I was afraid, because I was naked; and I hid myself."

Genesis *3.8–3.10*

Chapter 3 described coverage errors, arising from the exclusion of persons from all samples of the population to be studied. Like noncoverage, nonresponse is an error of nonobservation. Nonresponse is the failure to obtain complete measurements on the survey sample. Among all nonsampling errors it has captured the attention of many practitioners because the rate of nonresponse, the percentage of the sample not measured, is easily documented on many surveys. Nonresponse rates indeed are often used mistakenly as a measure of quality of the survey statistics.

The nonresponse rate is one component of the error but does not by itself fully measure nonresponse error. As was true for linear statistics affected by coverage error, nonresponse error produces its effects through two components, the nonresponse rate and the difference between nonrespondents and respondents to the survey:

Respondent Value = Full Sample Value
+ (Nonresponse Rate)×(Respondent Value − Nonrespondent Value)

$$y_r = y_n + \left(\frac{nr}{n}\right)(y_r - y_{nr})$$

where y_r = statistic estimated from the r respondent cases;

y_n = statistic estimated for all n sample cases;

y_{nr} = statistic estimated from the nr nonrespondent cases.

Thus, the error introduced to the survey estimate is a function of the percentage of the sample not responding to the survey and the differences on the statistic between respondents and nonrespondents [i.e., nonresponse error = (nonresponse rate)×(difference between respondent and nonrespondent values)]. The group of nonrespondents can be conceptualized as a fixed group of persons in the frame population (i.e., nonrespondent on all replications of the survey design). Alternatively, the nonrespondent group can be viewed as varying over possible replications because each person has a probability of responding on each replication. Sample persons measured on one replication of the survey may fail to respond on the next.

Furthermore, the expression for nonresponse error above is deceptively simple. There are different kinds of nonresponse, each of which might be associated with the failure to measure different kinds of people. In household surveys sample persons are not measured because they cannot be reached, because they are physically or mentally unable to respond, and also because they refuse to cooperate with the request for the interview. Thus, a more appropriate expression for a survey statistic might be

$$y_r = y_n + \left(\frac{nc}{n}\right)(y_r - y_{nc}) + \left(\frac{ni}{n}\right)(y_r - y_{ni}) + \left(\frac{rf}{n}\right)(y_r - y_{rf}) \, ,$$

where y_{nc} = statistic for the nc noncontacted sample cases;

y_{ni} = statistic for the ni sample cases that are incompetent
to provide the interview;

y_{rf} = statistic for the rf sample cases that refused the
interview.

In this case, $nc + ni + rf$ equals the nr term. This more complex expression is more enlightening because there is little prior belief that the values of the statistics for these different kinds of nonrespondents are similar. Clearly, even this expression overlooks other influences on nonresponse (e.g., urban/rural status, age of person).

Although both the expressions above emphasize the effects of nonresponse on the estimated value of statistics, nonresponse can also affect the variability in estimates over replications of the survey. Just as listers can affect coverage error in a survey statistic, interviewers can affect the levels of nonresponse error in a statistic. Interviewers vary in their ability to pursue elusive respondents, in their patience in administering questionnaires to persons with limited language or cognitive abilities, and in their ability to persuade the reluctant that the survey interview is a valuable experience. Thus, the choice of different interviewers might lead to different levels of nonresponse. If the survey were replicated on the same sample with a different set of interviewers, lower or higher nonresponse rates and nonresponse error might result. Nonresponse errors thus can have variable components (over conceptual replications of the survey and across interviewers working on the survey) and fixed components: both nonresponse *variance* and nonresponse *bias* may exist.

Finally, nonresponse is often separated into unit and item nonresponse. "Unit nonresponse" is used to describe the failure to obtain any of the substantive measurements from the sample person or household. "Item nonresponse" is the failure to obtain information for one question within an interview conducted with a sample person. The expressions above can be applied to both types of nonresponse, but the literatures on the two sources reflect the fact that (1) different psychological influences may prompt the two types of missing data to exist and (2) more information is known about those persons with item missing data on only a portion of the survey measures. This chapter focuses on unit nonresponse but offers a brief commentary on approaches to item missing data.

4.1 NONRESPONSE RATES

4.1.1 Components of Unit Nonresponse

Response and nonresponse rates are often used to evaluate surveys, but there are so many different ways of calculating response rates that comparisons across surveys are fraught with misinterpretations. Before comparing the many different calculations labeled "response rate," we examine the various possible dispositions of sample cases in a survey.

There are at least four survey design aspects that affect the response rate calculations:

1. Whether all units on the sampling frame are eligible for the survey (i.e., whether some units on the frame are not members of the target population). For example, some addresses sampled in household personal interview surveys are businesses, and thus not part of the survey population. In area probability samples of minority groups, many sample households may be ineligible for the survey because no minority group members live there.

2. Whether each unit sampled contains one sample element or many. Some household surveys attempt interviews with all members of the household (or all adult members) about their individual attitudes or behaviors; others use one informant to describe conditions of the housing unit. When no contact or cooperation can be obtained from the entire household, in the first design, several sample elements (persons) are lost; in the second, only one element (a housing unit).

3. Whether all sample persons have the same probability of selection. Some surveys oversample certain groups in the population to permit special analyses (e.g., some National Election Studies in the 1970s oversampled the black population because of interests in black electoral behavior). A black respondent's data, because of the higher probability of selection, would affect appropriately weighted overall statistics in a smaller way than a nonblack respondent's.

4. Whether substitution at the sampling stage is permitted by the design. Some designs permit the survey administrators to substitute a nearby or similar household for sample, one that cannot be contacted or which refuses the interview request. In such cases, a decision must be made about how the initial noninterview case is documented.

We review the implications of these design aspects in the next sections.

4.1.2 Nonobservation of Sample Units

The attempt to obtain interviews from sample cases inevitably doubles as an evaluation of the sampling frame. Some of the categories of disposition thus must reflect the appearance of cases on the frame that are not population elements. See Figure 4.1 for a list of disposition codes used in many SRC personal interview surveys. These cases do not yield survey measurements but are excluded upon initial contact by the interviewer. In an area frame design, these include structures that are not occupied during the survey period.[1] These are sometimes labeled "vacant dwellings" or "vacant units." Other units on the list may be businesses, not containing housing units where people live. Other listings may have been made in error and pertain to valid housing units outside the boundary of the area that was sampled. SRC calls these cases "SLIP's" for *Sample Listing Isn't Proper*. Finally, in studies in which some units do not contain eligible persons (e.g., studies of racial subgroups, the elderly, parents of young children, teenagers), some housing units are labeled as "NER," having *No Eligible Respondent*.

The above outcomes result from cases not part of the target population being on the sampling frame. Complete unit nonresponse arises in three ways for eligible cases: (1) inability to contact the sample household or person, (2) inability of the sample person to provide responses to the survey, and (3) refusals to the interview request. The last two are generally viewed as requiring contact with the sample unit, although many times if a sample of persons is conducted, one household member will refuse for another.

The number of noncontact cases encountered in a survey is affected by the number of repeated calls that interviewers make on sample cases. If a survey is conducted with a maximum of one call attempted for each sample case, the vast majority of nonresponse cases are noncontacts. In household samples of persons (either one or multiple respondents per household) "noncontact" is generally used to mean that the interviewer did not speak with anyone in the household, not merely that the designated sample person was not reached.

The inability of the sample person to provide responses to the survey is often placed in a diverse category called "noninterview for other reasons." This category has varying definitions among and within survey organizations. The tradition in some organizations is to attempt no interviews with persons who cannot speak English (i.e., they are removed from the denominator of the response rate calculations and from the

---

[1] Sometimes survey organizations specify that if the housing unit is unoccupied at the time of the first visit of the interviewer, the structure is to be labeled as a "vacant unit." Such a rule then defines the target population as all occupied housing units at the time of the first visit (a variable day and hour, depending on different interviewers' schedules).

IW	. . . .	Completed Interview
P-IW	. . . .	Partial Interview
APPT	. . . . . .	Appointment Made for Interview
NOC-RD	. . . . . .	No Contact with Selected R (HU listed and R determined)
NOC-RU	. . . . . .	No Contact, R Undetermined (HU listing not obtained)
REF-F	. . . . . .	Refusal by Selected Respondent
REF-O	. . . . . .	Refusal by Someone Other Than Selected R
REF-U	. . . . . .	Refusal, Respondent Undetermined (HU listing not obtained)
NER	. . . . . .	No Eligible Respondent
NIP	. . . . . .	Noninterview, Permanent Condition (e.g., senility, language, death, severe illness)
NIO	. . . . . .	Noninterview, Other Condition Not Permanent (e.g., ill or out–of–town for duration of the study)
HV	. . . . . .	House Vacant
SV	. . . . . .	Seasonal Vacant
VTS	. . . . . .	Vacant Trailer Space
ORE	. . . . . .	Occupant Residing Elsewhere
SORE	. . . . . .	Seasonal Occupant Residing Elsewhere
SLIP	. . . . . .	Sample Listing Isn't Proper

Figure 4.1 Abbreviations used by Survey Research Center interviewers to denote disposition of sample units in face to face surveys.

target population).[2] Other organizations attempt to use translators or to hire interviewers with Spanish (or other language) skills in areas where

---

[2] The rationale for this stems, no doubt, from concerns both with nonresponse error and measurement error associated with this population. Even when this group might be willing to respond, the translation of survey questions into another language and of the respondent's answers to English might be productive of errors in the recorded data.

such problems exist. A similar problem arises with the deaf and, for some surveys requiring visual abilities, the blind.

Refusals can result from the reluctance of the entire household to answer any of the interviewer's questions, from the denial of the interview by proxy (e.g., a husband refusing access of the interviewer to the wife, who was chosen as the respondent), or from the direct denial given by the selected respondent. The first type of refusal produces more difficult problems than the latter two. When a household–level refusal occurs, the interviewer typically is ignorant of how many eligible persons are household members. In a survey in which all persons in the household are to be measured, the interviewer is unaware of whether the household refusal corresponds to one or many nonrespondent cases, at the person level. In a survey in which one respondent is to be selected among those eligible, the interviewer is ignorant of the probability of selection of the resulting nonrespondent.

In telephone surveys there is a category of refusal which occurs only rarely in personal interview surveys—the partial or broken–off interview. Rarely are interviewers asked to leave the house after they begin a personal interview, but respondents apparently feel more justified in terminating telephone conversations in the middle of an interview.

Finally, there are categories of outcomes that are specific to individual surveys. If the sampling frame is a list of individuals, there may have been deaths since the frame was created. Deaths are most often treated as nonsample cases, although if the respondent is to report about past events, they might best be treated as nonresponse cases. In some quota sampling schemes (with procedures specifying a certain number of interviews for different classes of persons) there are sample cases that are not interviewed because the quota class which they occupy has been filled.

4.1.3 Classification of Outcomes

In practice, the disposition of each sample number is determined by survey field personnel. The application of the classification system is, however, subject to error. These errors arise because of both the inherent difficulty in determining the appropriate outcome and influences of evaluation procedures to misclassify results.

Many of the results which identify cases that are not part of the target population can easily be confused with nonresponse among eligible sample cases. For example, many houses that are vacant for the entire survey period are indeed retained for use by an eligible household. Families take extended vacations; temporary job assignments force people to leave their principal residences for short periods. An interviewer can repeatedly call without contact on a housing unit with a broken doorbell, residents who are hard of hearing, or occupants afraid to open the door to a stranger. After repeated unsuccessful efforts, interviewers might feel justified in coding the case as a vacant unit. If, in addition, there are no clear signs of habitation, the pressures to avoid a nonresponse might be overwhelming. Since individual interviewers are evaluated on the ratio of interviews to total eligible sample units they are assigned, they have an incentive to reduce the denominator of the ratio. Miscoding noncontact cases as vacant units produces this result.

In telephone surveys a similar phenomenon occurs because some nonworking numbers are not connected to nonworking number recording machines. When these numbers are called, a normal ringing tone is obtained. The ringing will continue indefinitely. These numbers are therefore difficult to distinguish from numbers assigned to housing units where no one has answered the ringing telephone. One result implies a case that is not part of the target population; another a nonresponse case. Many survey organizations verify that the interviewer has correctly classified cases by recontacting a sample of the units.

4.2 RESPONSE RATE CALCULATION

Response rates can be calculated in a myriad of ways, each implying to the naive reader different levels of success in measuring the complete sample. The alternative forms of calculations have been the focus of attention of several investigators (Kviz, 1977; and Bailar and Lanphier, 1978; Wiseman and McDonald, 1979). There do exist recommendations for preferred estimates (CASRO, 1982), but there is no universal compliance with the guidelines. Rather than using a single preferred response rate calculation, different survey researchers use different rates, each for a different purpose and each yielding a different measure of the completeness of the data collection on the sample.

In surveys for which all members of the sample frame are eligible for measurement, the following outcomes are relevant:

I = completed interviews,
P = partial interviews,
NC = noncontacted but known eligible units,
R = refused eligible units,
NE = noneligible units,
NI = other noninterviewed units.

One rate of interest to some survey personnel is the *cooperation rate*, defined by

$$\frac{I}{I+P+R},$$

that is, the ratio of completed interviews to all contacted cases capable of being interviewed. This rate might be used to characterize how successful interviewers were in persuading those able to do the interview to comply with the request.

Another rate of interest adds to the denominator those cases not contacted,

$$\frac{I+P}{I+P+NC+R},$$

still excluding those cases which cannot supply an interview because of mental or physical impairment. For example, the practice of excluding non-English speakers from the response rate calculations could be viewed as use of such an estimate. This estimate might be used to characterize how completely those sample cases which could provide survey measures did indeed produce them. Note, however, even for these purposes the estimate is flawed by underestimating the success of the field procedures. Not all the NC noncontact cases would be physically or mentally able to provide the interview, but they are all included in the base.

The most universally endorsed response rate is

$$\frac{I}{I+P+NC+R+NI},$$

where the denominator includes all sample cases in which an interview could have been completed. This rate most clearly estimates the proportion of all eligible persons measured by the survey procedures.

It is important to note that for sample surveys these rates are *sample statistics*, subject to sampling variability. That is, a different response rate would have been obtained if a different set of persons had been drawn into the sample. The rates would vary depending on the variability in cooperation and ease of contact among target population members.

More complex response rate calculation problems exist in three types of surveys: those with unequal probabilities of selection given to sample units, those with less than 100 percent eligibility in the sampling frame, and those using an intermediate selection unit (e.g., households) to which a variable number of target population members (e.g., adults) might be attached. In all these situations, a single response rate fails to describe fully the results of the survey implementation.

Sample designs assign unequal probabilities of selection to persons for a variety of reasons (see Chapter 6). A common reason is the desire to have a large sample of a relatively rare subpopulation (e.g., elderly persons in a study of health care, black voters in a study of electoral behavior). One way to obtain such disproportionately large samples offers no problems of response rate calculation—the identification of a stratum of such persons prior to selection. For example, a survey of the Hispanic population in the United States might use larger sampling fractions in the Southwest than in the Midwest. With such a stratification, the strata that share sampling probabilities can be combined for purposes of response rate calculation.

Another method of introducing unequal probabilities of selection causes more difficulty in response rate estimation. Such surveys assign higher probabilities of selection to units that contain the rare subgroup by having interviewers first obtain a short screening interview to identify whether the sample persons belong to the special group and then, with assistance of some device based on random selections, eliminating at random some portion of the cases that are not members of the rare group and retaining all members of the rare group. Through this procedure, the overall probability of coming into the sample for each person is unknown until the screening interview data are obtained on the full household. For example, a survey of attitudes toward child care might oversample persons with young children but include others at a lower sample fraction in order to estimate the distribution of attitudes in the entire population. To that end, a short screening questionnaire might be administered for each sample household. Those having young children would be included with certainty in the larger sample, while others would have a chance of exclusion from the larger sample. With completion of the screening interview in all sample households, all components of the response rate can be estimated. If, however, there is a refusal prior to completion of all the screening interviews, it is unclear which households would be included in the final study with certainty and which would have lower probabilities of selection. This threatens the ability to present two response rates, one for the households with young children and one for the other households.

In cases in which unequal probabilities of selection are used, it is unlikely that a simple unweighted response rate, $I/(I+P+NC+R+NI)$, is very informative. Instead, separate response rates for each domain with different probabilities of selection, $I_h/(I_h+P_h+NC_h+R_h+NI_h)$, might be preferred. To the extent that these separate domains will be used for separate analytic goals, such response rates inform each of those analyses.

An alternative single response rate is weighted by reciprocals of probabilities of selection,

$$\sum w_h \left(\frac{I_h}{I_h + P_h + NC_h + R_h + NI_h} \right),$$

where w_h = reciprocal of the probability of selection for the hth domain.

This expression would estimate, perhaps, the likely response rate that would result from these survey procedures if an equal probability sample might have been used. Only rarely, however, would such an estimate be of use to the survey analysts.

A similar problem exists when only the subgroup members are to be sampled; for example, if the above attitudinal survey did not want to estimate the attitudinal distribution for the full adult population but only for those persons with small children. Then a household–level refusal, without knowledge of whether there were parents of small children in the household, presents the researcher with the same ambiguity regarding the proper base of the response rate.

In surveys with multiple respondents per household, some statistics may be calculated on the person level (e.g., an analysis of the types of person who visit the doctor often) and some on the family or household level (e.g., a comparison of medical expenditures for families with and without children under the age of 5 years old). There are response rates that apply to both of these populations. The following outcomes of cases can be used:

I_r = completed person interviews,

I_h = housing units with at least one complete person interview,

E_h = housing units with complete household enumerations but no completed interviews,

P_r = partially completed person interviews,

NC_r = persons not contacted,

NC_h = occupied housing units without contact with any household member,

R_h = refusal for full household prior to enumeration of household,

R_r = refusal for an individual by respondents themselves or another in household,

NI_h = other noninterviews without enumeration of the household,

NI_r = other noninterviews of individuals.

With these components we can define a household–level response rate,

$$\frac{I_h}{I_h + NC_h + R_h + NI_h + E_h},$$

and a person level response rate,

$$\frac{I_r + P_r}{I_r + P_r + NC_r + R_r + NI_r + AVE \times (NC_h + R_h + NI_h)},$$

where AVE is the average number of persons in those households, often estimated by $(I_r + P_r + NC_r + R_r + NI_r)/(I_h + E_h)$.

Perhaps the most important observation about response rates is that they are likely to have multiple purposes and that calculation procedures should be fitted to the purpose at hand. One major purpose is the evaluation of the field activities. For this purpose several related rates are of interest:

1. The contact rate, $(I+P+R+NI)/(I+P+R+NI+NC)$, to assess how fully the sample was alerted to the survey.

2. The cooperation rate, $I/(I+P+R)$, to assess how well the field staff persuaded those contacted and able to respond.

3. The refusal conversion rate, the proportion of cases which initially refused to provide an interview but later agree to do so.

For purposes of estimating quantities that are related to the nonresponse error, we might prefer,

4. The response rate, I/(I+P+R+NC+NI), the complement of the term used in the mean square error model for nonresponse bias, the portion of the total eligible sample which provided a completed interview.

Individual response rates are often computed for separate administrative units in order to evaluate their performance. In large-scale personal interview surveys, the response rates for interviewers in different regional offices or under different supervisors might be examined. Individual response rates of interviewers are calculated and often used for salary or promotion reviews. These comparisons of response rates are useful evaluative tools only if the populations assigned to the different administrative groups present the same problems of access and cooperation. Without changes in the design this is rarely the case. In a national sample some interviewers are assigned work only in the New York City area; others, in the Henley, North Dakota area, a small town where interview requests tend to be met with friendlier response. Only when there is a random assignment of cases is the researcher assured, without making restricting assumptions, that the challenge given to each interviewer or supervisor or regional office is the same. Such designs can be implemented more easily in centralized telephone interviewing settings or in geographically confined surveys, where the marginal increase in travel cost associated with the randomized assignment is small.

4.3 TEMPORAL CHANGE IN RESPONSE RATES

4.3.1 Ongoing Surveys of Government Survey Organizations

Since the proportion of eligible sample persons interviewed is a term in the nonresponse bias expression, it is useful to compare the proportion across different surveys around the world. Evaluations of absolute numbers cannot usually be made. For example, without any limit on the resources available to the survey, we would prefer a response rate of 100 percent, other things being equal. There are, however, cost constraints for all surveys, and response rates are maximized only by using resources that may alternatively improve question wording, and interviewer performance (and thus reduce response error) or increase sample size (and thus reduce sampling error).

Since we cannot evaluate absolute magnitudes of response rates, of secondary value is the examination of response rates over time on ongoing surveys that have retained the same design. Marquis, in a controversial paper (1979), claimed that there was little evidence for declining response rates in household surveys in the United States. He examined several government and academic surveys. The first was the National Health Interview Survey, a monthly cross-sectional personal interview survey in which all persons present are interviewed in a group and proxy responses are obtained for those absent. The interviews last between 30 minutes and 1 hour for most families. Marquis cites data that show a 95 to 96 percent response rate (using the calculation with all eligible units in the denominator) in the 1960s and a 97 percent rate in the 1975 year. Figure 4.2 presents the response rates for the NHIS between 1967 and 1985; as Marquis noted, the overall rates are quite stable, moving between 95 and 97 percent during the period. This overall rate, however, masks an important change in the *composition* of the nonresponse. Figure 4.2 shows that the refusal rate is increasing, but the other nonresponse categories are decreasing. Indeed, the proportion of the total nonresponse associated with refusals in the later 1960s is about 0.25, but it increases to the 0.60 range in the mid-1980s. It is likely that such a change affects both error and cost properties of the survey. The error structure of a rate could be affected if the difference between the respondent and the nonrespondent statistics varied across the categories of nonresponse. That is, in the expression

$$\left(\frac{nc}{n}\right)(y_r - y_{nc}) + \left(\frac{ni}{n}\right)(y_r - y_{ni}) + \left(\frac{rf}{n}\right)(y_r - y_{rf}) \, ,$$

$(y_r - y_{nc})$, $(y_r - y_{ni})$, and $(y_r - y_{rf})$ generally differ. This means that if the proportions of the sample falling in the various nonresponse categories change, then overall bias in the survey statistic can change.

For example, consider a case of a health survey in which the noninterview and refusal portion of nonresponse is disproportionate ill persons, unable or unwilling to provide the information about their health status. In the same survey the noncontact portion of nonresponse might be relatively more healthy than the full population, consisting disproportionately of young, active persons. Assume interest in a statistic, the proportion of persons visiting a doctor in the 6 months prior to the interview. Assume that initial efforts by interviewers yield a 70 percent response rate, a 20 percent noncontact rate, a 5 percent noninterview rate, and a 5 percent refusal rate. If a decision is made to increase the response rate converting all noncontacts to interviews, what would happen to nonresponse bias?

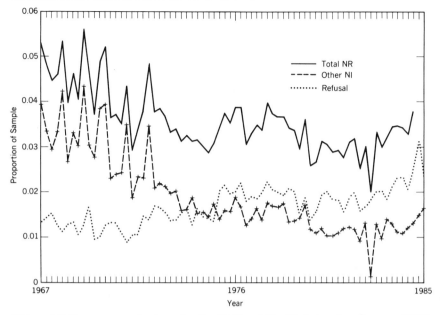

Figure 4.2 Nonresponse by type for the National Health Interview Survey, 1967–1985.

The answer to that question depends entirely on the attributes of the noncontacted portion of the sample, relative to those of the respondents and other nonrespondents. Assume that among the initial 70 percent who responded 0.10 visited a doctor in the last 6 months; that 0.05 of the noncontacted, 0.4 of the noninterviews, and 0.3 of the refusals visited a doctor. Table 4.1 demonstrates what would happen to nonresponse bias if nonresponse reduction efforts concentrated only on the noncontact portion. Among the full sample, 0.107 went to the doctor. If the survey stopped with the initial 70 percent response rate, the estimate would be 0.100, a bias of -0.007. If the rate were raised to 80 percent by interviewing the previously noncontacted, the bias would be -0.013. If it went to 90 percent the bias would be -0.027.

This example shows how higher response rates *can* lead to higher, not lower, nonresponse bias. This occurs when those initial nonrespondents who are converted to respondents are very atypical of the full set of initial nonrespondents. Similarly, stable response rates, achieved by reducing noncontacts in the face of more refusals, may merely acquire respondents very atypical of the full population, while losing to refusals those more typical.

In addition to unknown effect on nonresponse bias, for the NHIS the constancy of the overall response rate in the presence of increasing rate of

Table 4.1 Simulation of Effect on Total Nonresponse Bias of Increasing Response Rate Solely by Reducing the Noncontact Rate

(1) Response Rate	(2) Noncontact Rate	(3) Noninterview Rate	(4) Refusal Rate	(5) Estimate Based on Respondents	(6) Full Sample Estimate	(7) Nonresponse Bias (5)-(6)
70%	20%	5%	5%	0.1	0.107	−0.007
75	15	5	5	0.094	0.107	−0.013
80	10	5	5	0.08875	0.107	−0.01825
85	5	5	5	0.084117	0.107	−0.02288
90	0	5	5	0.08	0.107	−0.027

This assumes proportions visiting the doctor for the initial respondents (based on 70 percent response rates) of 0.1, the initial noncontacts of 0.01, the initial noninterviews of 0.4, and the initial refusals of 0.3.

refusal, no doubt, is achieved through increasing efforts at contacting those sample units that are difficult to find at home. Those callbacks on previously noncontacted units increase the cost per completed interview in the survey. Thus, the change in nonresponse composition was no doubt accompanied by a change in the cost of the NHIS.

The National Crime Survey, another U.S. Census Bureau survey, uses a different respondent rule, one requiring self-response for all persons in the household aged 14 and over. The survey is a rotating panel design for housing units; that is, if a housing unit is sampled it is visited every 6 months for a 3-1/2 year period. The interviews requires less than 15 minutes for most persons, but its length depends on how many victimizations are reported by the respondent. This survey is an example of a design which produces some difficulty in estimating person-level response rates because those households that are not contacted or for which household compositions are not known can contribute an unknown number of persons to the sample. The response rates in Figure 4.3 correspond to the proportion of all occupied housing units that provided at least one person interview. The refusal rate is the proportion of households that refused to provide any interviews. In contrast to the data from the NHIS, these figures show *both* a stable overall response rate between 1974 and 1985 *and* a stable ratio of refusals and other reasons for noninterviews *at the household level.* The overall response rate varies between 95 and 96 percent; the proportion of all noninterviewed households that refused tends to be about one third. The real changes that occurred over this 10 year period in NCS concern the proportion of eligible *persons* who provide interviews. This fraction (expressed as a ratio of persons refusing to total number of eligible households) increases from about 3 percent in the mid-1970s to over 6 percent in the mid-1980s (Figure 4.4). This implies that Census interviewers may not be finding it increasingly difficult to obtain a household roster from the sample units but may be experiencing more refusals after the roster is obtained. In the NCS case therefore, the person-level response rate is declining over the 10 year period.

Finally, the Current Population Survey, the U.S. labor force survey, provides monthly estimates of unemployment rates. It uses a respondent rule which allows any "responsible" member of a sample household to provide data for other members 14 years or older. This rule is thus different from both the NHIS and the NCS rules. Figure 4.5 presents the total noninterview rates for CPS households and the component of nonresponse from refusals between 1965 and 1985. The most recent overall noninterview rate is near 5 percent of the sample. The time series of rates is very erratic, reflecting the effects of various supplement questions (e.g., a March income supplement) on cooperation rates and other seasonal factors. As with the other surveys, the refusal rate has shown a steady increase over the 20 year period.

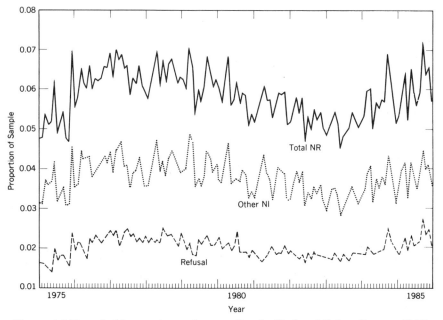

Figure 4.3 Household nonresponse by type for the National Crime Survey, 1974–1985.

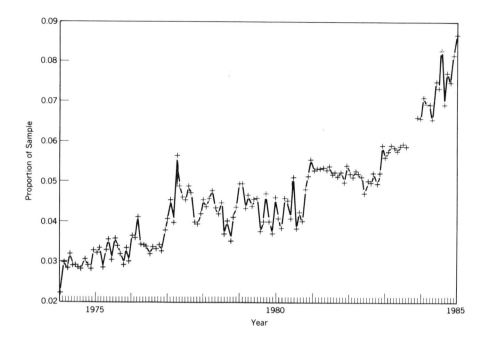

Figure 4.4 Person–level nonresponse rate for the National Crime Survey, 1974–1985.

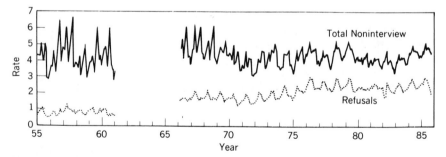

Figure 4.5 Total nonresponse and refusal rate for the Current Population Survey, 1955–1986. From T. DeMaio et al., "Cognitive and Motivational Bases of Census and Survey Response," *Proceedings of Second Annual Research Conference*, **U.S. Bureau of Census, 1986, Figure 5a.**

Response rate data from statistical agencies in other countries illustrate that the U.S. problem may not be uniformly shared. The

Table 4.2 Canadian Labor Force Survey Overall Nonresponse Rates by Year and Month

Month	Year					
	1973	1974	1975A	1975B	1976	1977
January	7.3	6.0	4.3	10.2	7.9	6.1
February	7.2	6.0	4.7	9.3	7.7	5.9
March	6.8	6.4	4.6	8.7	7.3	5.8
April	7.9	8.3	4.7	7.5	8.8	4.5
May	7.0	7.0	4.7	7.9	9.2	5.8
June	8.4	6.8	5.8	8.9	8.6	5.7
July	15.1	10.4	7.6	13.2	11.7	
August	10.9	8.8	6.3	11.0	9.2	
September	6.5	5.6	4.3	7.5	5.8	
October	5.7	5.5	4.5	7.5	5.9	
November	5.2	4.3	4.3	7.0	6.0	
December	6.6	4.6	5.3	7.6	5.3	

Source: R. Platek, "Some Factors Affecting Non-Response," *Survey Methodology*, Vol. 3, No. 2, 1977, pp. 191–214, Table I.

Canadian labor force survey, designed to measure unemployment and job-seeking behavior, uses a household respondent to provide information on all eligible persons. The nonresponse rates between 1973 and 1977 are presented in Table 4.2 but show no obvious trend. The table does show the effect of summer interviewing on nonresponse rates, with July rates being higher than other months. This effect is no doubt a function of the length of the survey period chosen for the labor force survey; if the survey purposes could be served by a month-long survey period, the nonresponse would likely be lower.

Nonresponse rates from the labor force survey conducted by The Netherlands Central Bureau of Statistics are higher than those in the United States and Canada (Bethlehem and Kersten, 1981), ranging from 13 to 20 percent in the 1970s. However, these rates are lower than those obtained in other surveys in the same country, for example, surveys on living conditions and surveys on travel (see Table 4.3).

Although the World Fertility Survey was not an ongoing survey project with replicated surveys within countries over time, it does provide evidence of the results of an attempt to implement surveys of similar designs in 42 different countries using local survey organizations to do the data collection. In all these countries women aged 15 to 49 years were selected to respond to a questionnaire about their fertility experiences. Table 4.4 presents the response rates for various countries. The refusal rates are exceptionally low for the WFS, almost always under 1 percent. The response rates appear to be uniformly high but vary between developed and less-developed countries. There is little evidence that the developing countries are facing similar problems with response rates; this appears to be a problem of the developed world.

Table 4.3 Total Nonresponse Rates for Various Surveys at the Netherlands Central Bureau of Statistics, 1973–1980

Year	Labor Force	Consumer Sentiment	Living Conditions	Travel	Holiday
			Survey		
1973	13.2				
1974			28.2		
1975	15.8	30.1			14.5
1976		28.1	23.0		12.9
1977	13.1	30.9	29.7		17.6
1978		36.1		33.0	21.9
1979	19.7	36.6	33.7	30.6	25.5
1980		36.8	35.6	32.1	

Source: J.G. Bethlehem and H.M.P. Kersten, "The Nonresponse Problem," *Survey Methodology*, 1981, Vol. 7, No. 2, pp. 130–156, Table 1.

Table 4.4 Response Rates for World Fertility Survey Countries on Household Interview

Country	Response Rate	Refused Among All Sample Units
Africa		
Cameroon	93.8	NA
Ghana	98.2	0.7
Ivory Coast	97.5	0.3
Kenya	92.8	0.3
Lesotho	99.7	0.1
Nigeria	93.4	1.1
Egypt	97.4	NA
Mauritania	98.6	NA
Morocco	96.7	0.1
Sudan	95.2	0.3
Tunisia	96.2	0.9
Asia and Pacific		
Jordan	96.2	NA
Syria	96.2	0.0
Turkey	85.2	NA
Yemen A. R.	98.2	0.4
Bangladesh	98.2	0.0
Iran	92.6	NA
Nepal	94.8	0.0
Pakistan	99.4	0.1
Sri Lanka	99.7	0.2
Fiji	94.2	0.2
Indonesia	96.7	NA
Republic of Korea	98.5	NA
Malaysia	99.8	0.1
Philippines	98.5	0.4
Thailand	98.7	NA
Americas		
Colombia	95.8	0.7
Ecuador	96.2	0.4
Paraguay	95.5	0.3
Peru	96.2	0.4
Venezuela	96.9	NA
Costa Rica	98.3	0.6
Dominican Republic	94.7	0.2
Mexico	96.0	0.6
Panama	99.3	0.1
Guyana	97.6	0.7
Haiti	89.6	0.1
Jamaica	92.9	3.2
Trinidad and Tobago	96.1	0.4
Europe		
Portugal	80.1	2.8

Source: A.M. Marckwardt, *Response Rates, Callbacks and Coverage: The WFS Experience*, WFS Scientific Reports, No. 55, April 1984, Table A1.

4.3.2 Ongoing Surveys of Academic Survey Organizations

In contrast to governmental surveys, surveys conducted by U.S. academic organizations exhibit a more dramatic temporal trend in response rates. Steeh (1981) notes increasing refusal rates between 1952 and 1979 for two long-run studies connected with the Survey Research Center, the Survey of Consumer Attitudes, and the National Election Studies. Figures 4.6 and 4.7, taken from that paper, show variability over consecutive editions of the surveys but a general trend toward higher rates of refusals. Steeh notes, as did Marquis earlier, that the increases in refusals are more dramatic in the large cities of the country. Table 4.5 shows that, if linear regression models are fit to the time series of refusal rates in the national samples, refusals are increasing more quickly over time in the large cities than in the other areas. Indeed, for the Survey of Consumer Attitudes, the rate of change per year in the percentage refusing (based on a regression estimator) is 0.676 for the large cities and 0.388 for the small towns.

4.3.3 Surveys Conducted by Commercial Agencies

It is more difficult to obtain response rate data from commercial survey organizations both because of the proprietary nature of some of the research they conduct and because the nature of the survey designs they use sometimes do not permit easy measurement of response rates.

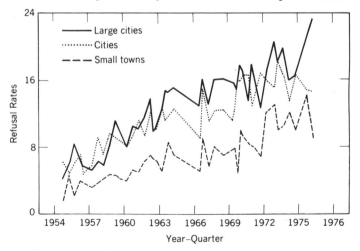

Figure 4.6 Refusal rates for the Survey Research Center, Surveys of Consumer Attitudes, 1954–1976.

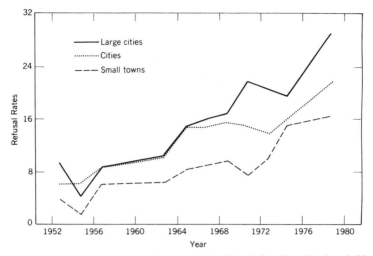

Figure 4.7 Refusals rates for the Center for Political Studies, National Election Studies, 1952–1978.

Table 4.5 Regression of Quarter–Year on Refusal Rates for National Election Studies, 1952 to 1978, and Surveys of Consumer Attitudes, 1953 to 1976, by Type of Place

Study and Place Type	Intercept	Regression Coefficient	R^2
National Election Studies, 1952 to 1978			
Large cities	5.0	.198	.89
Cities	5.9	.132	.89
Small towns	2.3	.118	.83
Total	4.3	.148	.92
Surveys of Consumer Attitudes, 1953 to 1976			
Large cities	4.3	.169	.84
Cities	5.1	.127	.80
Small towns	1.9	.097	.75
Total	3.9	.126	.90

Source: C.G. Steeh, "Trends in Nonresponse Rates, 1952-1979," *Public Opinion Quarterly*, Vol. 45, 1981, pp. 40-57, from Table 1.

Wiseman and McDonald (1979) present data showing that the median noncontact rate of 182 telephone surveys from 32 firms is 39.1 percent (denominator is respondents with known phone numbers). The median refusal rate is 28.0 percent (denominator is sample cases contacted).

4.4 ITEM MISSING DATA

In contrast to unit nonresponse, the definition of item missing data often varies across items within a single survey. Indeed, it would be more appropriate to refer to item missing data as a property of a statistic based on a survey measure. For some statistics on some questions (e.g., the percentage of likely voters favoring one candidate in an election) "don't know" may sometimes be taken as meaningful response, that is, one providing the researcher information about the population. On other questions and other statistics it has no informational content (e.g., to estimate median income of the population, a "don't know" answer provides no information).

Item missing data can occur in surveys because the interviewer fails to ask the question, the respondent is not able to provide an answer, the respondent refuses to provide an answer, or the interviewer fails to record the answer provided. Thus, item missing data rates have been found to be subject to effects of interviewer experience. In the first adaptation of the SRC Consumer Sentiment Survey to centralized telephone interviewing, the average missing data rate over 14 key items was 2.7 percent, for the second survey, 2.5 percent; for the third, 1.5 percent. Administrative experience and supervision of interviewers can reduce the item missing data rate.

Less success is likely when the cause of the missing data lies with the respondent. Table 4.6 presents missing data rates for income items in the Current Population Survey, the basis of unemployment statistics estimation in the United States. These rates far exceed the unit nonresponse rate of the survey. They probably stem both from inability of a household informant to provide household income data *and* from the reluctance of the respondent to reveal the value if it is known.

Missing data occur on individual items in the survey after the sample person has consented to be interviewed. It is much more a part of the measurement process than is unit nonresponse, and the cognitive and affective psychological influences to item missing data resemble those involving measurement error (treated in Chapters 7 to 11).

4.5 STATISTICAL TREATMENT OF NONRESPONSE IN SURVEYS

In recent years the attention of statisticians has focused on item missing data versus unit nonresponse. This work has constructed models of response probabilities, sometimes utilizing Bayesian analysis, examined alternative imputation procedures for missing data, and extended the

Table 4.6 Item Missing Data Rates for Income, Current Population Survey for 1980 by Age of Person, Males

	Missing Data Rates	
Ages of Males	One or More Income Items	All Income Items
14–19	17%	8%
20–24	16	7
25–34	14	6
35–44	20	8
45–54	25	11
55–64	27	11
65 and over	26	10

Source: W. Madow et al. (eds.), *Incomplete Data in Sample Surveys* Vol. 1, p. 24, Table 5.

analysis of missing data into the arena of analytic statistics. The three volume set prepared by a National Academy of Sciences panel on Missing Data in Surveys (Madow et al., 1983) is a thorough discussion of the issues as of that date. Kalton (1983a) offers a review of standard imputation procedures; Rubin (1986) describes use of multiple imputations which permit estimation of variance induced by the imputation process.

We focus our attention in this section not on item missing data but on unit nonresponse. It is here that the need for joint development of cost and error models comes clearly into play. Cost increases will normally accompany efforts to contact and interview larger portions of the sample. In surveys with low response rates in the absence of such efforts, the designer is faced with expensive decisions regarding what design features to use to reduce nonresponse error. Item missing data, although sometimes reduced through callbacks to sample cases, is often attacked by training procedures for interviewers and changes of the questionnaire, activities that are not as dramatically cost affecting.

A search of the statistical literature on unit nonresponse reveals only a few approaches that are unique to that discipline, double (or two–phase) sampling, optimization routines for callbacks, and postsurvey weighting schemes to reduce nonresponse effects. We begin, however, by discussing the components of the mean square error addressed by the procedures.

4.5.1 Properties of Components of Nonresponse Error

Whether or not this is done explicitly, the statistician forms a model of
nonresponse error effects on a given survey statistic. One, given earlier, is

$$y_r = y_n + \frac{nc}{n}(y_r - y_{nc}) + \frac{ni}{n}(y_r - y_{ni}) + \frac{rf}{n}(y_r - y_{rf}),$$

where y_{nc} = statistic for the nc noncontacted sample cases;

$\quad y_{ni}$ = statistic for the ni sample cases that are incompetent
$\qquad$ to provide the interview;

$\quad y_{rf}$ = statistic for the rf sample cases that refused the
$\qquad$ interview.

This assumes a linear statistic y, one formed by simple weighted
addition of individual observations. Such a formulation applies to a
sample mean, when a constant proportion of cases will be nonrespondent
in each replication of the survey. It does not apply for more complex
statistics, for example, a regression coefficient. There, as we saw with
noncoverage error through truncation of a dependent variable (see
Chapter 3), with assumptions of normal distributions for the regressor,
the relationship between the estimate based only on respondent cases, β^*,
and the coefficient on the target population, β, is

$$\beta^* = \beta \frac{\left(\dfrac{V^*(y)}{V(y)}\right)}{1 + \rho^2[1 - V^*(y)/V(y)]},$$

where $V^*(y)$ = variance of the endogenous variable among
$\qquad$ respondents;

$\quad V(y)$ = variance of the endogenous variable in the full target
$\qquad$ population;

$\quad \rho^2$ = coefficient of determination in the target population.

Both of these expressions present the effects of nonresponse error on
the point values of statistics, the sample mean and a sample-based

regression coefficient. The nonresponse error terms are biases, deviations between the sample-based estimates and the true value *expected* over replications of the survey. In considering the replications there is an implicit assumption that those aspects of the survey design that are related to the nonresponse error (e.g., refusal conversion efforts, callback rules) remain the same. The measurement of nonresponse bias, as is true of all survey biases, requires some data source external to the survey itself.

Variable Nonresponse Error. It is most often expected that nonresponse rates will vary because of sampling variability: the proportion of the sample cases that come from the group of potential nonrespondents will vary over samples selected with the same design. (That reason stimulates the calculation of standard errors about the response rates from single surveys.) There are, however, other sources of variability in nonresponse error that could be conceptualized. One of those that is well known by field managers but rarely reflected in statistical models of nonresponse error is that associated with interviewers. Some interviewers obtain high response rates; some obtain low response rates. Some have difficulty gaining the cooperation of certain types of respondents. These differences in performance remain despite the fact that all the interviewers might be given the same training, the same supervision, and the same workloads with similar mixes of cases. Thus, the amount of nonresponse bias in one implementation of a survey is a function of which interviewers are chosen to conduct the data collection. If a survey analyst is interested in estimating the variability in error in survey statistics over replications of the survey, then the interviewer as a source of variability in nonresponse error should be considered.

4.5.2 Statistical Models for Nonresponse

This section presents four different procedures which bring nonresponse errors into formal consideration during the choice of survey design. The first two options, using a technique called "double sampling" or "two-phase sampling," require the researcher to know various characteristics of groups in the population who differ in their likelihood of nonresponse. A relatively cheap method of data collection is used initially; then an expensive method (offering lower nonresponse rates) is used on the subsample of nonrespondents. Given the knowledge about nonrespondent groups, optimal sampling fractions for the second-phase sample can be determined.

The second two techniques do not require knowledge of the nonrespondent characteristics prior to the survey. The third technique obtains information directly from respondents about the likelihood that they might have been nonrespondents. It uses that information to construct weights used in analysis. Respondents with small likelihoods of providing an interview are given greater weight; those with high likelihoods, a smaller weight. The fourth technique is the opposite of a double sampling technique, in that the most expensive method is attempted first, on initial probability subsamples of the full survey sample. Given the cost and error properties of the initial subsamples, an optimal design is chosen for the remainder of the survey.

Callback Models with Double Sampling. Double sampling or two-phase sampling identifies a subsample of cases after the initial selection is made (Neyman, 1938). The first phase sample is that drawn and subjected to the initial data collection effort. The second phase sample is drawn typically using information obtained in the first phase. Such a design has appeal for the nonresponse problem because it offers a method of balancing costs and errors in deciding what efforts should be made to measure sample persons who elude casual efforts to interview them.

When applied to nonresponse, the approach identifies two kinds of sample cases: (1) those who provide the survey data in response to initial efforts and (2) those who do not. How "initial efforts" is defined is determined by how expensive callbacks and persuasion efforts are *and* how successful they measure the sample cases. After the initial efforts are finished, a second phase of the data collection operation is begun. A probability sample of the remaining nonrespondent cases is drawn, and expensive (and ideally completely successful) methods of obtaining measures on the sample nonrespondents are implemented. Since a probability sample of nonrespondents is drawn and then interviewed, it is used to estimate characteristics of all nonrespondents, and when it is combined with respondents from the first phase, survey statistics can be calculated that are free of nonresponse bias. To eliminate nonresponse error, a complete measurement of the subsampled nonrespondents is required.

Double sampling designs lend themselves to cost and error modeling approaches. The costs of contacting and persuading sample persons to cooperate with an interview vary between phases. The sampling and nonresponse error of resulting statistics depend on the second phase sample design and the success at measuring all sampled nonrespondents. One early work, that of Deming (1953), describes a double sampling scheme to minimize mean square error in a callback situation. Although other strategies (e.g., Politz and Simmons, 1949; Birnbaum and Sirken, 1950; Hansen and Hurwitz, 1958; Ward et al., 1985) exist, most share the characteristic of the Deming scheme of requiring knowledge of costs and errors prior to the initiation of the survey. Most also ignore distinctions between two categories of nonresponse with different cost and error characteristics—refusals and noncontacts.

Deming's approach assumes (1) groups that have different probabilities of responding at each call by an interviewer, (2) subsampling of nonrespondents remaining after the first call, and (3) full follow-up of those sampled nonrespondents (regardless of the number of calls required). The statistic of interest is the sample mean, and the design feature in question is the sampling fraction in the second phase. That is, the designer must decide how many of the nonrespondents after the first call should be followed up, and how many should be discarded. To answer the question the model requires knowledge of the differences in mean values on the survey variable among the groups (subclasses contacted on each callback) and of differences in element variances among the groups. The constraint is overall fixed amount of money to do the survey.

Deming assumes that each sample person has one of six response probabilities (0, 0.125, 0.250, 0.500, 0.750, or 1.000) on each of the calls. Note that these are constant across calls for a single person (i.e., there are some people who always have a 0.250 chance of giving an interview). There are other people who are always waiting to be interviewed (i.e., 1.000 chance). He also allows for some people never being interviewed; these could be those who repeatedly refuse the request.

The model examines the sample mean at each call, treating it as a weighted average of the means obtained at that call and at previous calls. The sample mean estimator after the third call is

$$\bar{y}_{1+2+3} = w_1\bar{y}_1 + w_2\bar{y}_2 + w_3\bar{y}_3,$$

where w_h = proportion of the respondents interviewed at the hth call, $h=1,2,3$;

$\bar{y}_h$ = the mean for cases interviewed on the hth call, $h=1,2,3$.

This is a simple linear combination of the means of interviews obtained on each call, weighted by the proportion of cases obtained on that call. As the number of calls increases the number of means combined will increase.

Deming provides an illustration of the technique using empirical data (see Tables 4.7 and 4.8). The costs of calls are set at \$3.00 for the first call and \$5.00 for each subsequent call. Thus, the total cost of calling on a case for the first time is \$3.00, but for two times it is \$8.00. These costs

Table 4.7 Illustrative Parameter Values For Deming Model

Response Probability Class	Proportion of Population in Class	Mean Value on Survey Variable	Element Variance on Survey Variable
0—Never interviewed	0.05	2.25	
0.125	0.10	2.00	2.00
0.250	0.10	1.75	1.75
0.500	0.20	1.50	1.50
0.750	0.25	1.25	1.25
1.00	0.30	1.00	1.00
Total	1.00	1.40	

Source: W.E. Deming, "On a Probability Mechanism to Attain an Economic Balance Between the Resultant Error of Response and the Bias of Nonresponse," *Journal of the American Statistical Association*, Vol. 48, No. 264, 1953, pp. 743–772, Table 1, with small changes.

are incurred for both interviewed cases (successes) and noninterviewed cases (failures).

The optimal fraction of nonrespondents to sample after the first call is conditional on whether one wants to follow the nonrespondent subsample for one more call, two more calls, three more calls, or what. To solve for the optimal subsampling fraction, a cost and error model is needed. The total cost of the survey is

$$C = C_1 n + C_2 f (n - r),$$

where C_1 = cost per unit for the first phase sample, merely the first
       call costs;

  n = total sample size;

  C_2 = cost per interview of second phase sample of
       nonrespondents over the various number of calls chosen;

   f = fraction of the nonrespondents after call one who are
       chosen for follow-up;

Table 4.8 Survey Results Assuming the Optimal 0.6 Subsampling of Nonrespondents After the First Call

| Call Number | Number of Sample Cases | | | | Total Cost | Cost per Case | Relative Bias | Relative Root Mean Square Error RMSE | RMSE/ Total Cost $\times 10^{-6}$ |
	Still Active	Interviewed This Call	Remaining	Cumulative Interviewed					
1	1000	625	375	625	$3000	4.80	−0.11	0.117	39.0
2	225	76	149	701	4125	5.88	−0.08	0.085	20.6
3	149	36	113	737	4870	6.61	−0.06	0.069	14.2
4	113	21	92	758	5435	7.17	−0.05	0.059	10.9
5	92	13	79	771	5896	7.65	−0.04	0.053	9.0
6	79	9	70	780	6290	8.06	−0.03	0.049	7.8
7	70	7	63	787	6638	8.43	−0.03	0.046	6.9

Source: W.E. Deming, "On a Probability Mechanism to Attain an Economic Balance Between the Resultant Error of Response and the Bias of Nonresponse," *Journal of the American Statistical Association*, Vol. 48, No. 264, 1953, pp. 743–772, Tables 2, 3, and 5.

r = number of respondents in the first phase.

The optimal fraction, f, is found to be proportional to ratios of costs, variance, and bias terms, so that the mean square error is expressed as

$$\text{MSE}(\bar{y}) = A + \frac{B}{n} + \frac{C}{fn},$$

where A = bias term (reflecting the omission of those who will never respond and others who were not reached within the allotted number of calls for the second phase sample);

B,C = functions of element variances for the first and second phase interviews, respectively.

With this expression the optimal f is equal to the square root of

$$\frac{CC_1}{BC_2}.$$

On the first call the mean square error is such that

$$A = E(\bar{y}_1) - \bar{y}_t,$$

where $\bar{y}_1$ = mean on first call interviews;

$\bar{y}_t$ = the sample mean based on all cases except those sample cases with 0.0 probability of responding.

If the group means, variances, and likelihoods of participation are those in Table 4.7, if C_1 = \$3.00 and C_2 = \$5.00, and if a maximum seven-call rule is enforced, then the optimal second phase fraction is found to be about 0.6. That is, after the first call, sample 60 percent of the remaining nonrespondents and pursue for at most six more calls. In Deming's example, if 1000 were sampled initially, 625 interviews would be obtained on the first call. Of the 375 remaining, $0.6 \times 375 = 225$ would be selected in the second phase sample. At the end of seven total calls, 162 of those would be interviewed, leaving 63 of the 225 not interviewed. (We would expect that $0.6 \times 50 = 30$ of the 63 would be members of the hard-core nonrespondent group, with no chance of being interviewed.) Table 4.8 shows a steady decline of the ratio of errors to cost (RMSE/total cost), by roughly a factor of 5 from one-call to the seven-call rule with a 0.6 second phase sampling fraction.

The power of the model is that once the cost and error parameters are known and once the maximum number of calls has been set (seven in this example), the subsampling fraction is known to minimize the combination of sampling variance and nonresponse bias, given fixed costs for the survey. It incorporates both costs and nonresponse errors formally into the design choices.

Despite its elegance the following weaknesses can be observed in Deming's model:

1. It assumes that response probabilities are constant over calls.

2. It does not reflect refusals well either in variance or cost components.

3. It does not offer a way to learn the values of means or proportions in various groups, or variances of measures in those groups.

4. It fixes a design parameter, the maximum total number of calls, which is most often an unknown in many designs.

While Deming's plan permits the subsampling of remaining nonrespondents after the first call, other schemes have investigated the optimal number of calls without a subsampling plan. For example, Birnbaum and Sirken (1950) determine what call rule should be used to obtain a set probability that the bias of nonresponse would not exceed a certain level. They accomplish this for a binomial variable, since such a measure gives them limits on the bias (i.e., the proportion having the given attribute among nonrespondents cannot be greater than 1.0 nor less than 0.0).

Mixed Mode Models with Double Sampling. Another approach, offered by Hansen and Hurwitz (1958), takes advantage of the common fact that different modes of data collection vary both in response rates and in cost. They examine the case of a mailed questionnaire survey, relatively cheap to implement but subject to large nonresponse rates, combined with a personal interview, more successful at obtaining measurements but expensive. Hansen and Hurwitz construct a design that, in its first phase, sends mailed questionnaires to a sample, records returns of completed questionnaires, and then, in its second phase, schedules personal visit interviews with a sample of the remaining nonrespondent cases to the mailed questionnaire. Hansen and Hurwitz address what combination of mailed questionnaires and personal interviews might minimize the sampling error of estimates, given a fixed cost.

The example they use clearly illustrates the principles of their design. Suppose you are faced with a design decision within the context of a mixed mode survey. One option is a survey that would send out a mailed questionnaire to 1000 persons and, for all those who failed to complete the questionnaire, attempt a personal visit interview. The second option is mailing out questionnaires to a larger sample of persons, followed by personal visit attempts on only a sample of the nonrespondents. Could the second design yield lower sampling error for the survey estimates than the first, for the same total cost? The answer to the question is found to depend on the cost of the two data collection methods, the expected response rate of the mailed questionnaire, and the variance of the survey measure for the responders and nonresponders to the mailed survey.

The solution flows from the view of the sample estimate of the population total,

$$x' = \left(\frac{N}{n}\right)(n_c \bar{x}_1' + n_{ni} \bar{x}_2''),$$

N = total number of persons on the frame;

n = total number of questionnaires mailed;

$\bar{x}_1'$ = mean for the responders to the mailed questionnaire;

n_c = number of responders to the mailed questionnaire;

$\bar{x}_2''$ = mean for the persons interviewed in personal visit interviews;

n_{ni} = number of nonresponders to the mailed questionnaire.

The sampling variance of this estimated total is

$$N^2 \frac{N-n}{(N-1)n} \sigma^2 + \frac{N}{n}(k-1)\frac{S^2}{S-1}\sigma_b^2,$$

where σ^2 = element variance in the entire population;

σ_b^2 = element variance among the nonresponders;

S = number of persons in the population who would not have responded if the questionnaire had been sent to all persons in the population;

k = reciprocal of the sampling fraction of nonresponders.

Hansen and Hurwitz show that the optimal k is

$$k = \sqrt{\left(\frac{N^2(S-1)\sigma^2}{S^2(N-1)\sigma_b^2} - 1\right)\frac{C_3 Q}{C_1 + C_2 P}},$$

where C_1 = cost per unit of mailing the self-administered questionnaire;

C_2 = cost per unit of processing completed mailed questionnaires;

C_3 = cost per unit of personal visit interviewing and processing;

P = proportion of the initial sample that returned completed questionnaires; $(Q = 1 - P)$.

Hansen and Hurwitz show that when the cost of mailing is C_1 = \$0.10, the cost of processing completed questionnaires is C_2 = \$0.40, and the cost of personal visit cases is C_3 = \$4.50, the double sample approach is more cost efficient than the following of all nonresponders of the questionnaire. Table 4.9 shows the comparison of prices for the double sampling scheme and a full personal visit follow-up *for designs that yield the same standard error of the mean*. The table assumes that the element variance among responders and nonresponders is the same ($\sigma^2 = \sigma_b^2$). For low response rates the proportion of nonresponder cases that are sampled is high (860/1543 = 56 percent); for high rates (90 percent responding) the sample is a smaller fraction (40/120 = 33 percent). Correspondingly, the cost savings of the double sample procedure is greatest in the case with higher response rates, a cost savings of near 20 percent for a survey with the same standard error.

Hansen and Hurwitz present their results comparing two methods that yield the same standard error. They could have alternatively shown the reduction in standard errors associated with the double sampling

Table 4.9 Cost of Samples with the Same Standard Error of the Mean for Optimal
Double Sampling and Full Personal Visit Follow-up of Nonrespondents

Proportion Completing Mailed Questionnaire	Initial Sample Size n	Nonresponders Sampled r	Cost of Optimal Double Sample	Cost of Full Follow-up
0.10	1714	860	$4110	$4190
0.20	1989	711	3558	3780
0.30	2034	575	3035	3370
0.40	1979	451	2544	2960
0.50	1870	341	2096	2550
0.60	1727	245	1690	2140
0.70	1564	163	1328	1730
0.80	1386	95	1010	1320
0.90	1197	40	731	910

Source: M.H. Hansen and W.N. Hurwitz, "The Problem of Non-Response in Sample
Surveys," *Journal of the American Statistical Association*, December 1958, pp. 517–529,
Table 2.

scheme compared to a full follow-up procedure costing the same amount.
In that comparison an increase in sampling error arising from the
inability to measure a very large sample with expensive methods is
balanced with a cheaper procedure which, although applied only to a
subsample, could purchase a larger total number of cases. The
subsampling procedure does not introduce bias from nonresponse
because, although not all nonrespondents are measured, a probability
subsample of them are. This subsample is guaranteed to have the same
expected values for survey statistics as the entire group of
nonrespondents; thus, the weighted combination of the respondents to the
mailed questionnaire and the personal interview respondents yield proper
estimates of the full population parameter.

The procedure, however, works because of critical assumptions by the
authors. On the error side, only sampling error is considered, not the
nonresponse error attached to those who fail to complete the mailed
questionnaire. Thus, for example, if there are negligible differences in the
survey statistic between responders and nonresponders to the

questionnaire, either the double sampling scheme or the full follow-up would be less cost efficient than less intensive efforts to measure the nonrespondents. Furthermore, the system assumes that the personal interview survey achieves complete cooperation from the sampled nonrespondents to the mailed questionnaire. Indeed, the result of unbiasedness of the resulting survey estimate depends on the full success of the personal interview survey to measure all subsample cases. In the absence of that, the procedure still has some value if the personal interviews obtain *lower* nonresponse error than the mailed questionnaire (a likely result). The optimal subsample size, in such a case, would vary, however, from that cited above. Finally, the procedure ignores any measurement error differences between the mailed questionnaire and the personal visit survey. We investigate such measurement errors in a later section. In summary, these assumptions limit the generality of the Hansen-Hurwitz results, but the approach provides a framework within which an elaboration of the cost and error models could be constructed.

Measurement of Response Probabilities. Another approach used to attack nonresponse with statistical methods has been attempts to measure the probability that a sample person would respond. This follows the notion that sample survey data overrepresent those persons easily accessed by the interviewer. If all sample persons can be reached at some point in the survey (i.e., no one has a zero probability of response), then weighting the interview observations inversely to their ease of acquisition might eliminate nonresponse bias.

Hartley (1946) and Politz and Simmons (1949) were the first to describe a one-call survey design that weighted completed interviews by the chances of finding the sample person at home. The procedure has the following requirements:

1. The time of visiting (or calling) sample households is randomized,

2. Upon contact and interview, several questions are asked of the respondent about his being at home during the past few days. For example, for a survey conducted only in the evenings Monday through Saturday, one would ask someone reached on Saturday the following questions:

(a)   Would you mind telling me whether or not you happened to be at home last night at just this time?

(b)   How about the night before last at this time?

(c)   How about Wednesday night?

(d)   How about Tuesday night?

(e)   How about Monday night?

3. By using answers to the above questions weight groups are constructed which are defined by the proportion of days during the survey period that the sample person was at home.

There are several observations that must be made about the efficacy of this method. First, it addresses nonresponse associated with noncontact, not with refusal (e.g., the questions concern being at home, not being at home *and* being willing to do the interview). Furthermore, to be successful at eliminating nonresponse bias due to noncontact, no person can be absent throughout the scheduled survey period.

Second, it strictly requires questioning about being at home on each of the days of the survey period. Above, it is assumed that the week of the survey period exhibits the same at-home patterns as the week asked about during the interview. If, for example, the respondent is reached on the first call on the first day of the survey period, and then asked about his at home patterns for the prior week, those are used to estimate his at-home patterns for the week of the survey.

Third, it requires persons to be accurate in their reports of their at-home patterns. That is, response errors for these questions can produce erroneous adjustments for nonresponse. It might be likely that such response errors would be higher for those respondents whose at-home patterns are erratic over evenings of the week and whose presence is not linked to the time of the evening. This speculation is guided by notions that similar events may act to interfere with the ability of the respondent to recall any one.

Fourth, the practical problem of having interviewers call on sample households at random moments may cause logistic problems and cost increases. Without randomization, interviewers might tend to call on households *because* someone is at home (or not at home) and thus overestimate (or underestimate) their probabilities of being at home relative to others called on at that time. With randomization, interviewers may be directed to call on sample households at times that are not convenient for them or are perceived by them to place them in some danger.

Ward et al. (1985) attempt a practical investigation of the technique, although they chose not to enforce randomized calling times. Respondents who were not reached on the first call were called two more times. The authors correlated the total number of calls required to obtain contact (1, 2, or 3) and the total number of prior days the respondents reported that they were at home (within the past 5 days). The observed correlation is −0.13, which has the correct sign but is low. On almost all census–based estimates, the three–call rule yielded estimates judged to be closer than the weighted one–call estimates. The authors also showed large increases in the standard error of estimates due to weighting. The conclusions of the Ward work must be limited by their decisions not to randomize cases and the efficacy of the comparison with the three–call rule and to 1980 U.S. Census estimates (the authors do not reveal when the field work was conducted).

The Politz–Simmons procedures are attractive models from a theoretical perspective because they offer a correction for not–at–home bias. Their limitations have been found, however, in application of the randomized calling time requirement and response errors. The difficulty of randomization of time of call is reduced with centralized telephone interviewing using computer–assisted methods. But in most studies such control over calling times might increase survey costs greatly.

The response error threat to the Politz–Simmons procedure appears not to have been fully investigated, although Ward et al. (1985) and Politz and Simmons (1949) themselves discuss the problem. An experimental design with repeated calling on sampling households at the times they would be asked to report on would address this point. Alternative question wording might be investigated. Furthermore, the estimation error involved in the procedure, the use of reports of last week's activities to estimate the full survey period activities, has not been addressed. Although random moments are sampled during the survey period, information is gained only about the past days of the survey period. This information is used to estimate the probability that the sample person would be at home during the entire survey period. Either one uses reports for times before the beginning of the survey period to estimate the probabilities during the survey period or the amount of information obtained for those scheduled earlier in the survey period is less than that for those scheduled late in the survey period.

The Politz–Simmons procedure appears on its surface to be less relevant to the nonresponse pressures facing U.S. surveys in the late 20th century, which arise mainly from refusals. The response rate data cited earlier in the chapter show, however, that more resources are being devoted to reducing the proportion of sample cases not reached, in order to keep overall response rates as high as possible. A combination of double sampling and a Politz–Simmons method might be an attractive approach to nonresponse bias reduction. Such a procedure would use a weighting procedure for not-at-home bias and, with the resources saved from that method would devote proportionately more to refusal bias reduction. The method would determine whether any subsampling of refusal cases was desirable prior to using very expensive methods to obtain their cooperation. The design implies that refusal cases encountered on the first call would be pursued vigorously (more so indeed than would have been possible in a design that also pursued the not-at-home cases).

4.5.3 A Sequential Design Option Incorporating Nonresponse Concerns

The double sampling schemes of Deming and Hansen and Hurwitz require the knowledge of various design parameters *prior* to the survey, in order to achieve optimal gains over traditional nonresponse reduction methods. Rarely does a researcher have this information prior to the survey. This problem has clearly led to the rare use of these techniques.

An alternative approach divides the survey period into two parts. In the initial part expensive callback procedures and refusal conversion efforts are implemented while cost monitoring also occurs. As the first part is completed, cost and error estimates are used to determine the optimal effort for nonresponse reduction. In the second part of the survey, the optimal design features are implemented.

Specifically, the approach implements a replicated sampling scheme in order to provide bias estimates associated with termination at alternative points of efforts to contact sample households or persuade reluctant respondents. Its goal therefore is to give the researcher guidance on the relative gains in specifying high levels of effort to obtain interviews.

A simple error model for nonresponse using a linear statistic, Y, might be presented in four terms:

$$\frac{S_i^2}{I} + \left[\overline{m}_{nc}(Y_i - Y_{nc}) + \overline{m}_{ni}(Y_i - Y_{ni}) + \overline{m}_{rf}(Y_i - Y_{rf}) \right]^2 ,$$

where $S_i^2 =$ unit variance with appropriate design effect adjustment for interviewed cases;

I = number of interviewed cases;

$\overline{m}_{nc}$ = nc/n, proportion of the sample not contacted;

Y_{nc} = statistic for noncontacted portion;

Y_i = statistic for interviewed portion;

$\overline{m}_{ni}$ = ni/n, proportion of the sample with noninterviews for reasons other than noncontact and refusal;

Y_{ni} = statistic for noninterviewed portion;

$\overline{m}_{rf}$ + rf/n = proportion of the sample refusing;

Y_{rf} = statistic for refusing portion.

A cost model attached to nonresponse error might include separate terms for the costs of reducing noncontact, noninterviews, and refusals, as well as costs of interviews. It should reflect the fact (especially in a call scheduling system that is not computer assisted) that sample cases requiring many calls incur increased costs with succeeding calls. For example, in a centralized telephone facility, when initial contact is made on a sample number, interviewers are often instructed to write down notes describing the outcome of the call, citing any problems that were encountered, and offering guidance to the next caller on that sample case. These notes are then reviewed by supervisors and interviewers at later points in order to make decisions regarding assignment to interviewers and by interviewers in determining the desirable approach on the number. This perusal of notes affects every succeeding call on the sample number, whether or not contact is made.

$$C = C_0 + C_i + \left(C_{nc} + b_{nc}D_{nc}\right)(nc) + C_{ni}(ni) + \left(C_{rf} + b_{rf}D_{rf}\right)(rf) ,$$

where C_0 = fixed costs of the survey;

C_i = average cost of interviewing the i respondent cases;

C_{nc} = base cost parameter for calls on noncontact cases;

$C_{nc} + b_{nc}D_{nc} =$ function reflecting the inflation in per unit costs as the number of calls placed to the nc noncontact numbers, D_{nc}, increases;

$C_{ni} =$ average cost of contacting, selecting, and processing the ni cases that cannot be interviewed for reasons of health, mental capabilities, language, or lengthy absence from the household;

$C_{rf} + b_{rf}D_{rf} =$ function reflecting the inflation in per unit costs as the number of calls placed to the rf initial refusal cases, D_{rf}, increases.

This cost and error model combination requires some information about the relationship between the rates of nonresponse and the differences in values of statistics between the nonrespondents and the respondents. One proposed use of this approach is with interpenetration imbedded in a centralized telephone survey:

1. Designate a survey period of at least twice the temporal length required to accommodate the maximum number of calls desired to be investigated (i.e., if 18 days must pass to implement the desired call scheduling algorithm with a maximum of 10 calls, specify a survey period of 36 days tentatively).

2. Construct a replicated sample, with replicates of size sufficient to obtain stable estimates of characteristics of temporarily noncontacted cases and initial refusals (see estimates listed below).

3. Monitor the time that supervisors and interviewers spend on dialing noncontacted numbers, reviewing and dialing initial refusal cases (attempts at refusal conversion), contacting sample households prior to an interview, and conducting interviews. These data should be kept by number of call on the sample case.

4. On the first day of the survey release the first replicate for interviewer processing. Release others as necessary to keep interviewers busy.

5. As the first replicate is approaching completion, begin to assemble estimates of components in the mean square error model, elaborated to reflect the fact that even after maximum callbacks are attempted and full

refusal conversion efforts are used, some sample cases will not provide data.

6. Construct bias reduction tables associated with callbacks and refusal conversion efforts by the call on which the interview was completed. Construct bounding estimates on Y for the remaining refusals and noninterviews, given efforts at persuasion and contact cease at one call, two calls, three calls, four calls, and so on.

Two products result from this data analysis. The first is a table that tracks the point estimates for the interviewed cases by call number, the point estimates for noncontacted cases (those which could be contacted given the maximum call rule), and the point estimates for refusal cases (those which could be persuaded given maximum persuasion efforts). Table 4.10 presents such a table for one variable in the SRC Survey of Consumer Attitudes, whether the respondent reports being better off now financially than 1 year ago; such a table can be created quite simply using any micro-based spreadsheet program. Note that if the survey stopped after one call, the estimated percentage feeling better off now would be 30.1. If the 25-call maximum rule were used, the estimate would be 34.0. Note also that the noncontacted people (but were eventually interviewed within 25 calls) have a higher percentage better off (35.3 percent), but the refusals to the first call are more similar to the respondents (30.2 percent better off).[3] If the survey stopped after the first call, the bias in the estimate would arise from the noncontacts and refusals listed in the table and also from three other groups—the noncontacts who were never interviewed within 26 calls, the permanent refusals who did not cooperate, and the other noninterview cases. Thus, only part of the mean square error can be estimated, given the information in Table 4.10.

The table requires that the point estimate, Y_i, Y_{nc}, Y_{rf}, and the Y_{ni}, be calculated for each call. The Y_i is simply the point estimate for the interviewed cases. The Y_{nc} has two terms, one related to the cases not yet contacted by the particular call and another related to those cases that cannot ever be contacted within the maximum number of calls specified. The first term can be estimated from the first replicate, because all those

---

[3] The "Interviews" column consists of those cases who were interviewed on that specific call number or earlier. The "Noncontacts" are those cases who were not yet interviewed by that call and who eventually yielded an interview without need of refusal persuasion efforts. The "Initial Refusals" are those cases who sometime during the 25-call period provided an initial refusal but then were successfully persuaded to respond within 26 calls. Thus, with each succeeding call, some of the cases in the fourth and sixth columns move to the second column.

Table 4.10 Sequential Design Simulation, Sample Mean and Counts of Cases for
Interviews, Noncontacts and Initial Refusals, by Maximum Number of Calls Made
on Sample Cases

Call	Interviews		Noncontacts		Initial Refusals	
	Mean	n	Mean	n	Mean	n
1	30.1	2,897	35.31	12,117	30.17	1,011
2	31.42	5,828	36.06	9,290	30.42	983
3	31.76	7,909	36.76	7,295	30.6	879
4	32.09	9,496	37.24	5,849	32.28	793
5	32.62	10,825	37.59	4,613	34.51	652
6	32.89	11,771	37.88	3,770	33.09	559
7	33.23	12.613	37.49	3,046	32.89	456
8	33.2	13,207	38.14	2,533	34.62	338
9	33.45	13,737	37.63	2,038	38.52	257
10	33.39	14,127	39.4	1,693	39.64	222
11	33.41	14,482	39.97	1,381	37.29	177
12	33.5	14,780	40.59	1,126	44.78	134
13	33.55	14,991	40.38	946	46.15	91
14	33.6	15,136	40.71	813	63.33	60
15	33.71	15,291	39.79	666	60.83	48
16	33.71	15,403	40.93	562	77.5	40
17	33.81	15,513	40.95	464	84.38	32
18	33.84	15,580	40.95	398	75	20
19	33.92	15,649	38.6	329	78.95	19
20	33.98	15,712	36.8	269	78.95	19
21	34.01	15,752	35.78	232	75	16
22	34.02	15,779	34.93	209	69.23	13
23	34.02	15,815	35.8	176	100	9
24	34.03	15,848	35.66	143	100	6
25	34.04	15,870	37.8	127	100	6

cases were eventually measured within the maximum effort specified. The
second term cannot be estimated using the data. For that term
alternative model-based estimates can be used. The Y_{rf} term can
similarly be divided into two components. Finally, the Y_{ni} term has only
one component, given a decision not to attempt further efforts to measure
the sick, mentally impaired, or absent sample persons.

Thus, at the ith call the mean square error has the form

$$\frac{S^2}{I} + \left[\begin{array}{c} \overline{m}_{nc}(Y_i - Y_{nc}) + \overline{m}'_{nc}(Y_i - Y'_{nc}) + \overline{m}_{ni}(Y_i - Y_{ni}) \\ + \overline{m}_{rf}(Y_i - Y_{rf}) + \overline{m}'_{rf}(Y_i - Y'_{rf}) \end{array} \right]^2 ,$$

where S^2 = unit variance with appropriate design effect adjustment for cases interviewed by the ith call;

I = number of cases interviewed by the ith call;

$\overline{m}_{nc}$ = nc/n, proportion of the sample not contacted by the ith call but interviewed within the maximum calls specified;

Y_{nc} = statistic for portion not contacted by the ith call but interviewed within the maximum calls specified;

Y_i = statistic for cases interviewed by the ith call;

$\overline{m}'_{nc}$ = nc'/n, proportion of the sample not contacted by the maximum call specified;

Y'_{nc} = statistic for portion not contacted by the maximum call specified;

$\overline{m}_{ni}$ = ni/n, proportion of the sample with noninterviews for reasons other than noncontact and refusal;

Y_{ni} = statistic for noninterviewed portion;

$\overline{m}_{rf}$ = rf/n, proportion of the sample with an initial refusal by the ith call but interviewed within the maximum calls specified;

Y_{rf} = statistic for the portion of the sample refusing by the ith call but interviewed within the maximum calls specified;

$\overline{m}'_{rf}$ = rf'/n, proportion of sample with a refusal status at the time of the maximum calls specified;

Y'_{rf} = statistic for the portion of the sample with a refusal status at the time of the maximum calls specified.

The terms $\overline{m}'_{nc}(Y_i - Y'_{nc})$, the bias as component associated with those cases never contacted by the ith call, requires a specification of a bias model. For illustrative purposes we explore four different models:

1. Setting Y'_{nc} at the same value as Y_i, assuming the noncontacted cases have the same characteristics as those interviewed.

2. Setting Y'_{nc} at its minimum value (for a statistic that is a proportion this is 0.0).

3. Setting Y'_{nc} at its maximum value (for a proportion this is 1.0).

4. Assuming Y'_{nc} has the value equal to that of the last nc cases contacted within the maximum rule, where nc is chosen to be large enough to obtain some stability over samples.

Obviously, a whole set of models should be tried. Full design record checks might be appropriate vehicles to estimate these functions for some variables. Four similar models are used for Y'_{rf} ; Y_{ni} is alternatively assumed to be 0.0 or 1.0.

Figures 4.8 and 4.9 present the kind of graphical assistance to the design decision that could be provided by the cost–error model. Figure 4.8 presents the values of different cost components by call number. The top curve (with x delimiters) is a plot of the total interviewer cost of a completed interview. This moves from about $7.50 to $8.50 between the one–call rule and the 25–call effort, a 13 percent increase. The second curve (with squares as delimiters) is a plot of the total costs for cases interviewed by the call. This is an increasing function because the cases interviewed on later calls require several efforts (multiple callbacks and, for some, refusal conversion efforts). The third curve (with plus sign delimiters) is a plot of the costs of noncontact cases eventually interviewed, but at that call still not interviewed. This is a decreasing portion of the total cost per interview because they eventually disappear over the 25 calls. The fourth curve (with circle delimiters) is a plot of the costs of refusal cases which over the 25 calls are eventually converted. These refusals, like the previous category, eventually disappear and form a decreasing portion of the total costs. Finally, the fifth curve (with triangle delimiters) is a plot of the costs of those cases never interviewed (because of permanent refusals, permanent failure contact, or other noninterview conditions). These form an increasing portion of the total per interview cost because efforts to persuade the initial refusals and efforts to contact the uncontacted continue throughout the survey.

Figure 4.9 is a plot of the Mean Square Error (MSE) (sampling variance and the nonresponse bias components) by the maximum number of calls placed to a sample unit. The MSE estimates are based on the four different models described above and in Figure 4.10. The top curve (with diamond delimiters) assumes that remaining noncontact, noninterview, and the permanent refusal cases all have the value 1.0 on the proportion (i.e., $Y'_{nc} = Y'_{ni} = Y'_{rf} = 1.0$). This model yields the highest initial bias estimates, and thus the mean square error is highest for this model for all call rules. The lowest curve (with squares as delimiters) reflects the assumption that there is no bias remaining for the permanent noninterview cases (i.e., $Y'_{nc} = Y'_{ni} = Y'_{rf} = Y_i$). In this case, the bias terms reflect only the failure to interview those cases who can eventually be interviewed with the effort associated with 26 calls. Two curves in the middle of the graph are almost identical decreasing functions of the number of calls. Both of these assume that Y'_{nc} equals the value of other noncontacted cases on the 25th call and Y'_{rf} equals the value of refusal cases on the 17th call. The value of Y'_{ni} is alternatively assumed to be 0.0 or 1.0, with little effect on the overall estimated MSE. Finally, the second to the lowest curve (with plus sign delimiters) reflects the assumption that $Y'_{nc} = Y'_{ni} = Y'_{rf} = 0.0$. It is the only one yielding increasing MSE with higher numbers of calls (implying an optimum of one call).

The simplest form and most desirable outcome for a figure like Figure 4.9 are that all the desirable alternative bias models have curves whose minima are near the same call number. This is unlikely in most cases (impossible for our case, given the model choices). The "no remaining bias" model achieves a minimum at the 21-call rule; the model assuming all noninterviews are worse off ($Y'_{nc} = Y'_{ni} = Y'_{rf} = 0.0$) achieves its minimum at one call. The others are monotonically decreasing.[4]

Failing a satisfactory form of the graph on the first criterion, the second feature worth exploring is the relative increase/decrease in mean square error after various calls. For example, the call intervals for which the mean square error varies less than 5 percent above the minimum appear in Figure 4.10.

For this one survey statistic, a 5 percent tolerance about the minimum mean square error would be obtained with an 18-call rule with three of the four rules. Great care must then be exercised regarding the deviant model, which suggests a one-call rule. In this case since it is a boundary case, it can more easily be rejected. Using this procedure, the researcher is given quantitative information regarding the possible gains of setting nonresponse reduction efforts at the maximum level feasible or of using less intense efforts.

Several attributes of this methodology deserve comment:

---

[4] There is, in all three of these models, an estimated minimum at 24 calls, but this probably reflects instability of the estimated nonrespondent means at that point.

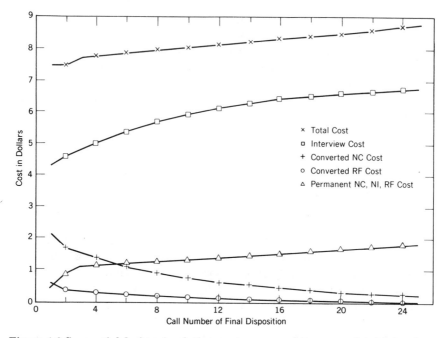

Figure 4.8 Sequential design simulation, cost components per completed interview by maximum number of calls made on sample cases.

1. Graphs like Figure 4.9 should be constructed for all the key statistics in the survey. The multipurpose problem must be faced by the designer.

2. The identity of the models for MSE estimation deserve separate attention. These might be improved by careful scrutiny of the nonrespondents with record check studies, in which records could supply bias estimates.

3. The maximum call rule chosen to be explored determines the burden taken on by the models estimating Y'_{nc} and Y'_{rf} . The more extensive the efforts investigated in the first replicate, the smaller the number of cases in those permanent noninterview conditions.

4. To be more realistic the cost model used could be elaborated to note change in refusal conversion costs for cases initially contacted on early calls and those on later calls.

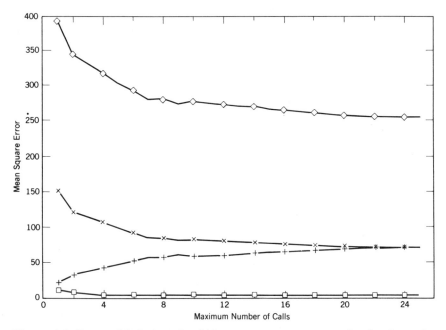

Figure 4.9 Sequential design simulation, mean square error for fixed cost by maximum number of calls made on sample cases.

Model Assumptions	Optimum Effort
"No remaining bias" model	13–25 calls
$Y'_{nc} = Y'_{ni} = Y'_{rf} = 0.0$	1 call
$Y'_{nc} = Y'_{ni} = Y'_{rf} = 1.0$	15–25 calls
$Y'_{ni} = 0$ Y'_{nc} and Y'_{rf} have same values as late interviews	18–25 calls
$Y'_{ni} = 1$ Y'_{nc} and Y'_{rf} have same values as late interviews	18–25 calls

Figure 4.10 Results of sequential design simulation, optimal effort given alternative assumptions.

5. Another cost/administrative model that should be investigated is the dropping of all refusal conversion efforts. The mean square error achieved for fixed cost without refusal conversion should be a competing design.

6. The illustration needs to be elaborated to reflect the stochastic elements of the cost generation process. This is likely to increase the interval of calls within a stated distance of the minima.

7. The administrative requirements of the procedure demand the use of computer assistance in the data collection in order to capture the estimates from the first replicate.

4.6 SUMMARY

The success of household surveys requires a unique combination of societal characteristics: sufficient sharing of language so that probability samples of respondents can be interviewed, freedom for interviewers to attempt contact with sample persons, freedom from fear of harm from strangers, trust on the part of the respondents that their answers will be held confidential, and a labor market of survey interviewers to accomplish the measurement. Nonresponse is a result of losing one or more of these characteristics.

Participation in social surveys by sample households appears to be declining in the United States over time. This is true for government, academic, and commercial surveys. Since response rates can be computed for most survey designs, researchers are alerted to this potential weakness in the method. However, the nonresponse rate is not a good indicator of nonresponse error. For simple statistics the nonresponse error is a multiplicative function of nonresponse rate and the difference on the statistic between nonrespondents and respondents. The latter term is generally unknown.

Do higher nonresponse rates suggest increasing nonresponse error in surveys over time? For the most part, we are forced to speculate on this. The speculation concerns whether the "distinctiveness" of nonrespondents (relative to respondents) increases, decreases, or stays the same as the proportion of nonrespondents grows larger. If the nonrespondents resemble the respondent group more closely as larger nonresponse rates occur, then it is possible that higher nonresponse rates have little ill effect on survey quality. That is a comforting but risky assumption.

This chapter reviewed some design approaches that formally recognize nonresponse error. All of them have the feature that special data collection activities are conducted to learn the costs and error impacts of special efforts to contact and persuade sample persons to participate in the survey. The Deming and Hansen-Hurwitz models describe double sampling or two phase sampling approaches, where the design feature of interest is the sampling fraction at the second phase. The first phase uses cheaper methods (e.g., a one-call rule or a cheaper mode of data collection), subject to larger nonresponse rates. The second phase uses a more expensive collection method which tends to offer higher participation rates. To determine the optimal sampling fraction, estimates of cost differences, element variances, and means are required.

The difficulty with these schemes is the demand for knowledge about the cost and error parameters *prior* to the survey. This is conceivable for ongoing repeated surveys but impractical for one-time surveys. For that reason, the last option in the chapter—using sequential designs in a centralized telephone survey setting—might be more practical. This procedures introduces variation in quality as a deliberate design decision. The highest quality and highest cost options are attempted in early replicates. These are used to provide measures of costs and quality impacts of preferred options. Once that cost and error information is assembled, the design is fixed for later replicates.

All these approaches, however, increase the design burden for the researcher. Bringing nonresponse error into the design considerations is more difficult than ignoring it. Furthermore, the designs offer guidance on the cost efficiency of design features which have already been identified. They do not offer means to create new options for nonresponse error reduction which may be more effective than those now existing. Such construction requires more knowledge about the causes of nonparticipation.

Speculations about the impact of higher nonresponse rates on error also need some theoretical guidance about the causes of nonresponse. The relationships between these causes and likely characteristics on the survey measures then need to be assessed. Here the causes of noncontact nonresponse are likely to be very different from the causes of refusal nonresponse. The difficulty of contact and the reluctance to cooperate once contacted may have very different relationships with the survey measures. Hence, the nature of nonresponse error for noncontacts and refusals is likely to be very different.

Thus, in order to make some progress on these issues, some attention must be given to theories of human behavior that are relevant to the likelihood of failing to measure some sample person. This is the focus of the next chapter.

PROBING THE CAUSES OF NONRESPONSE
AND EFFORTS TO REDUCE NONRESPONSE

I've no wish to take part in things like this any more. It has happened five times and I participated to begin with, but I have got fed up with all those endless, meaningless questions. What do I get out of it, if I take part? I want my private life to be left alone.

Reason given for nonresponse to Swedish Labor Force Survey, in L.R. Bergman, R. Hanve, and J. Rapp, "Why do some people refuse to participate in surveys?", Särtryck ur Statistisk tidskrift, Vol. 5, 1978, p. 347

All measurements of nonresponse error and all postsurvey adjustments for nonresponse can benefit from any knowledge of why nonresponse occurs. Some nonresponse occurs as a result of a design decision made by a researcher (e.g., the schedule of calling on sample cases). Other nonresponse occurs because of decisions out of the direct control of the researcher (e.g., refusals to participate). Knowing the causes of refusal decisions is more complex than understanding noncontacts. However, since nonresponse from refusals is a growing portion of total nonresponse, the need to identify its causes is more important.

We see studies of human behavior which attempt to understand causes of nonparticipation as a necessary component to good statistical models of nonresponse adjustment and designs incorporating nonresponse error in their evaluation. The social science theories related to compliance and persuasion may be applicable to the decisions to participate in a survey. Unfortunately, they have had little impact on survey research. This chapter reviews some parts of those theories which may be applicable to the phenomenon of survey participation.

For the most part the search for causes of nonresponse has not been guided by theories about human behavior, but by the exigencies of practical survey work. The literature offers little guidance on cause and much measurement of correlates of nonresponse, descriptions of what kinds of persons tend to be nonrespondent. Unfortunately, most of these descriptions examine individual attributes of nonrespondents, ignoring multivariate relationships. We learn, for example, that nonrespondents tend to be older people and less educated people, but rarely whether the less educated among the elderly were most prone to nonresponse.

This chapter discusses what kinds of persons tend to be nonrespondent, why they might tend not to be measured in surveys, and what efforts have been made to reduce nonresponse in sample surveys. It separates consideration of nonresponse due to failure to contact the sample person from nonresponse due to refusals.

The causes of nonresponse arising from noncontacts are likely to be somewhat different from those of nonresponse due to refusals and other noninterviews. To describe the influences on noncontact nonresponse, we begin the chapter with a small literature on the patterns of households being at home, and relevant parts of the economic literature on the time use of individuals when they are at home. A parallel treatment of causes of refusals in household surveys draws on the literatures on compliance and persuasion in psychology. The chapter ends with a discussion of methods which survey researchers have employed to reduce nonresponse rates in household surveys.

5.1 EMPIRICAL CORRELATES OF SURVEY PARTICIPATION

The attributes of nonrespondents to surveys have been studied using the following methods:

1. Special studies of nonrespondents (these typically have problems of representativeness but potentially offer rich data).

2. Using information on the sampling frame about nonrespondents (this technique typically does not offer very rich data but when it is possible it is very cheap). This is a popular methodology when a list is used as the sampling frame (e.g., samples of members of an organization, workers in a company). The value in the method is dependent on the accuracy of the data on the records.

3. Asking others about the nonrespondents or having interviewers provide information about them (this avoids the low response rate of seeking out the nonrespondents themselves but threatens other measurement errors in the data). This is now a standard practice of many

large national survey organizations. The DeMaio (1980) work, reviewed below, uses such a nonresponse form as a source of data. Figure 5.1 presents the noninterview form that is sometimes used by SRC. The data obtained by interviewer judgments must be used with great caution. There appear to have been no experimental attempts to validate any of the interviewers' judgments.

4. Comparison of respondent characteristics by call number (this is possible for any survey that records on the survey data the number of contact or interview attempts made on the sample case). The value of this technique is dependent on a model (often only implicit to the analysis) that the nonrespondents still not reached after the nth call resemble those interviewed on the nth call (or some function of those reached late in the survey period). For that reason, the technique does not address the characteristics of refusals very well.

5. Comparison of respondent characteristics to census or other information (this is plagued by the fact that the census data are often out of date, are available only on aggregate units, and are not very rich in variety). This is by far the most frequent method to assess "nonresponse bias." The most recent census data and the survey data are compared on demographic variables (e.g., age, gender, race). If no significant differences are found between the census data and the survey data on those variables, the researcher judges that no nonresponse error exists on any of the survey variables. This procedure often ignores the fact that the survey data and the census data may be subject to measurement errors, that definitions of concepts might differ, and that other survey statistics might be subject to nonresponse error even if the demographic statistics are not.

6. Studying persons who drop out of a panel survey after an initial interview (the problem here is that the nonrespondents at second waves may be very different from the nonrespondents in a one–time survey). Persons not giving an interview in later waves of a panel survey can be described by their first wave data. Thus, the effect of panel attrition can be measured more easily than the effect of failing to measure some sample persons on the first wave.

These different methods can be divided into those that use information external to survey data collection and those that use auxiliary data from the survey or enhance the survey data in some way. All those relying on external data typically suffer from the availability of only a small number of attributes, some that may not be central to the survey's

51. REF (R)	61. NOCAT (DR)	71. HV
52. REF (O)	62. NOCAT (UR)	72. SORE, SV
53. REF (U)	63. NIP (DR)	73. ORE
	64. NIT (DR)	81. SLIP
	67. NIO (UR)	83. VTS
		91. NER

SURVEY RESEARCH CENTER
INSTITUTE FOR SOCIAL RESEARCH
THE UNIVERSITY OF MICHIGAN
ANN ARBOR, MICHIGAN 48106

NR1. Project Number _____ NR2. Total Number of Calls _____ Date of Last Call _____

NR3. Primary Area _____ Segment No. _____ Line No. _____

NR4. Address or Description _____

NR5. Post Office _____ State _____ ZIP _____

NR6. Does this listing describe an HU in the sample segment identified in NR3?

1. YES, SEASONAL	3. YES, OTHER	5. NO ──► TURN TO P. 4, NR22

NR7. This HU is located in a:

☐ 1. MOBILE HOME (TRAILER) IN MH(TR) PARK   ☐ 7. OTHER (DESCRIBE): _____

☐ 2. MOBILE HOME (TRAILER) IN OTHER LOCATION

☐ 3. BUILDING WITH NO OTHER HU'S   GO TO NR7b

☐ 4. BUILDING WITH OTHER HU'S ──────────► NR7a. About how many units? _____

(IF NOT MOBILE HOME OR TRAILER)

NR7b. How many floors are in this building? _____

NR8. Is public access to this HU restricted in any way?

1. YES	5. NO ──► GO TO NR9

NR8a. DESCRIBE: _____

_____

_____

NR9. Was this HU occupied on your first call during the interviewing period?

1. DEFINITELY OCCUPIED	2. PROBABLY OCCUPIED	4. NO INDICATION OF OCCUPANCY, BUT NOT DEFINITELY VACANT	5. DEFINITELY VACANT	7. OTHER
TURN TO P. 2, NR10		TURN TO P. 4, NR22	END OF FORM	TURN TO P. 4, NR22

188

2

NR10. At that time, were <u>all</u> adult (according to listing box instructions) occupants of this HU residing <u>elsewhere</u>?

| 1. YES, ALL ADULTS ELSEWHERE |
TURN TO P. 4, NR22

| 5. NO |

NR11. Were you ever able to talk with someone <u>at this HU</u> (need not be resident)?

| 1. YES | | 5. NO | ⟶ NR11a. On any calls were there people in the HU who did not answer the door (or intercom)?

| 1. YES, DEFINITELY | | 3. SUSPECT SO | | 5. NO REASON TO THINK SO |

NR12. Were you able to identify the primary respondent in this housing unit?

| 1. YES, R IDENTIFIED ON COVER SHEET LISTING |

| 5. NO, HOUSEHOLD LISTING <u>MAY BE</u> OR <u>IS</u> INCOMPLETE |
TURN TO P. 3, NR17

NR13. Age of R _____ (approximate if necessary)

NR14. Sex of R: | 1. MALE | | 2. FEMALE |

NR15. What is the estimated income of R's household?

| 1. LOW UNDER $10,000 | | 3. MEDIUM $10,000-$30,000 | | 5. HIGH $30,000 OR MORE | | 8. IMPOSSIBLE TO ESTIMATE |

NR16. Respondent's Race:

DEFINITELY   PROBABLY

☐   ☐   1. WHITE

☐   ☐   2. BLACK

☐   ☐   3. AMERICAN INDIAN OR ALASKAN NATIVE

☐   ☐   4. ASIAN OR PACIFIC ISLANDER

| 8. DON'T KNOW RACE |

NR16a. Is R of Hispanic origin?

| 5. NO | | YES, 1. MEXICAN-AMERICAN OR CHICANO | | YES, 2. PUERTO RICAN | | YES, 3. OTHER HISPANIC | | 8. DON'T KNOW |

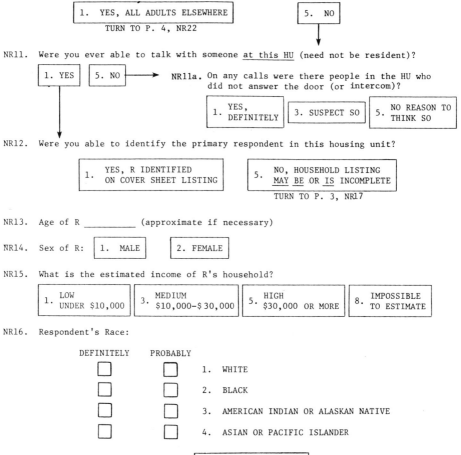
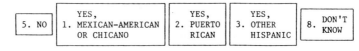

Turn to P.4, NR21

3

NR17. ESTIMATED number of persons 18 years of age or older in the HU.

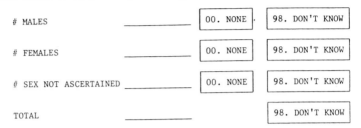

# MALES	_____	00. NONE	98. DON'T KNOW
# FEMALES	_____	00. NONE	98. DON'T KNOW
# SEX NOT ASCERTAINED	_____	00. NONE	98. DON'T KNOW
TOTAL	_____		98. DON'T KNOW

NR17a. ESTIMATED number of these persons eligible according to the study definition. _____

NR18. Is there a married couple in the HU?   1. YES   5. NO   8. DON'T KNOW

NR18a. ESTIMATED age of:

Husband or principal male   1. UNDER 30   2. 30-64   3. 65 OR OLDER   8. DON'T KNOW   0. NO MALE

Wife or principal female   1. UNDER 30   2. 30-64   3. 65 OR OLDER   8. DON'T KNOW   0. NO FEMALE

NR19. What is the estimated income of the occupants of this HU? (NOTE: IF MULTIPLE FAMILIES AND R NOT DETERMINED, ESTIMATE FOR PRIMARY FAMILY.)

1. LOW UNDER $10,000   3. MEDIUM $10,000-$30,000   5. HIGH $30,000 OR MORE   8. IMPOSSIBLE TO ESTIMATE

NR20. What is the race of the occupants of this HU?

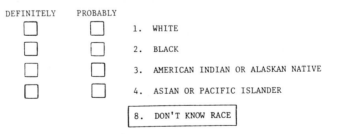

DEFINITELY   PROBABLY

☐   ☐   1. WHITE

☐   ☐   2. BLACK

☐   ☐   3. AMERICAN INDIAN OR ALASKAN NATIVE

☐   ☐   4. ASIAN OR PACIFIC ISLANDER

8. DON'T KNOW RACE

NR20a. Are the residents of Hispanic origin?

5. NO   1. YES, MEXICAN-AMERICAN OR CHICANO   2. YES, PUERTO RICAN   3. YES, OTHER HISPANIC   8. DON'T KNOW

190

NR21. Is the reason for noninterview a permanent condition (e.g. death, mental
incompetence)? NOTE: "Refused and will not change mind" is not a permanent condition.

| 1. YES | | 5. NO |———→ GO TO NR22

NR21a. What is this condition?

☐ 1. DECEASED AFTER LISTING

☐ 2. LANGUAGE (WHAT LANGUAGE?): _____

☐ 3. MENTAL OR PHYSICAL CONDITION (DESCRIBE): _____

_____

☐ 4. MOVED OUT OF RANGE AFTER OCCUPANCY DETERMINED (If new address
or phone number is known, give it in NR22.)

☐ 7. OTHER (DESCRIBE): _____

_____

_____

NR22. Describe in detail the reasons an interview was not taken. Give dates for the
activities and events you describe.

*IF LISTING DESCRIBES SOMETHING WHICH IS NOT AN HU OR WHICH IS LOCATED OUTSIDE
THIS SAMPLE SEGMENT,* indicate what you found.

IF UNABLE TO GAIN ACCESS, what attempts were made? (Try to obtain names,
addresses, and phone numbers of persons to contact re: gaining access.)

IF REFUSAL, indicate who refused and reasons (either given or suspected) and
what efforts (including letters) were made at persuasion.

IF "BUSY", "SICK", ETC., indicate whether you think this is simply an excuse
or a genuine difficulty.

IF R AWAY, check page 98 of the Interviewer's Manual to be sure that (he/she)
should be listed. If so, state when R will return, and if R could be interviewed elsewhere.

*IF YOU HAVE BEEN UNABLE TO DETERMINE WHETHER AN HU WAS OCCUPIED OR VACANT OR
WHETHER AN ELIGIBLE R LIVED THERE OR NOT,* describe the situation -- state what
inquiries and other attempts you have made to determine occupancy status.

IF ALL ADULT OCCUPANTS ARE RESIDING ELSEWHERE, describe situation giving location of other residence, expected length of stay and reason for absence.

_____

_____

_____

_____

_____

191

purposes. The other class of techniques are weakened by nonresponse or measurement error problems that may exceed those of the overall survey.

5.1.1 Patterns of Being At Home

There are two obvious influences on a survey noncontact rate: the number of interviewer calls on a sample house and the "at-home" pattern of the household. Survey researchers have used their own methodology to attempt to predict the likelihood of contact by studying the times at which interviewers contacted respondents. There are several published studies that describe the times at which different kinds of households or persons are at home.

The studies by Weeks et al. (1980), Weber and Burt (1972), Vigderhous (1981), and Groves and Robinson (1983) share the design that data from a sample survey are analyzed to determine when sample households were contacted. For respondent households (on which other interview data are available) the studies describe the at-home patterns of various subgroups. The work by Weeks et al. is based on a 1976 national area probability sample of over 22,000 housing units and reflects the result of the first call on the sample unit. About 4 percent were not successfully contacted after repeated calls. The 1971 Current Population Survey (CPS) data are from a national probability sample of about 6000 households entering the sample for the first time. The 1960 U.S. Census data come from a time study covering all but the most rural areas of the country, about 20 percent of the population. Table 5.1 shows that, collapsing over all days of the week, the proportion of households with someone 14 years or older at home declines over the 16 years represented by the data. In 1976, larger proportions of homes are unoccupied at virtually every hour of the day relative to the 1960 and 1971 data. The authors cite reasons such as increasing proportion of females in the labor force, more multicar families, and smaller numbers of persons per household. Across all three data sets, however, the proportion at home is highest after 5:00 p.m. (generally about 5 to 10 percentage points higher than other hours). Saturday is distinctive from the weekdays in that the evenings have proportions at home similar to morning and afternoon hours (0.55 to 0.65 range generally). Sunday tends to have low proportions at home in the morning (below 0.50), even lower than those during the weekday, but high proportions (over 0.70) after 6:00 p.m.

Data like those presented in Table 5.1 are limited because they are based on specific rules guiding when interviewers called on the housing units. These rules are given to interviewers in training but then are interpreted in many ways over the sample areas. Late evening calls in

Table 5.1 Estimated Proportion of Households in Which One Person 14 Years or Older Was at Home by Time of Day and Day of Week, by Study Source

Time of Day	Proportion by Day of Week							Weekdays Only		
	Sun.	Mon.	Tues.	Wed.	Thurs.	Fri.	Sat.	1960 Census	1971 CPS	1976 RTI Survey
8:00–8:59a.m.	–	–	–	–	–	–	–	0.71	0.57	0.49
9:00–9:59a.m.	–	–	–	0.55	0.28	0.45	–	0.71	0.56	0.39
10:00–10:59a.m.	–	0.47	0.42	0.38	0.45	0.40	0.55	0.69	0.58	0.42
11:00–11:59a.m.	0.35	0.41	0.49	0.46	0.43	0.50	0.62	0.68	0.59	0.46
12:00–12:59p.m.	0.42	0.53	0.49	0.56	0.45	0.55	0.60	0.68	0.59	0.52
1:00–1:59p.m.	0.49	0.44	0.50	0.48	0.43	0.51	0.63	0.69	0.57	0.47
2:00–2:59p.m.	0.49	0.50	0.52	0.47	0.45	0.45	0.59	0.67	0.57	0.48
3:00–3:59p.m.	0.54	0.47	0.49	0.54	0.50	0.50	0.65	0.70	0.67	0.50
4:00–4:59p.m.	0.52	0.58	0.55	0.57	0.57	0.56	0.53	0.72	0.70	0.57
5:00–5:59p.m.	0.61	0.67	0.65	0.67	0.59	0.57	0.56	0.78	0.74	0.63
6:00–6:59p.m.	0.75	0.73	0.72	0.68	0.65	0.64	0.59	0.78	0.75	0.68
7:00–7:59p.m.	0.73	0.74	0.75	0.64	0.61	0.57	0.66	0.80	0.71	0.66
8:00–8:59p.m.	–	0.51	0.51	0.59	0.74	0.52	–	0.76	0.78	0.58
9:00–9:59p.m.	–	–	–	0.64	–	–	–	–	–	0.66

Dash entry refers to a base less than 20 housing units.
Source: M. Weeks et al., "Optimal Times to Contact Sample Households," *Public Opinion Quarterly*, Spring 1980, pp. 101–114, Tables 1 and 4.

personal interview surveys are rare in some areas, more common in others. Weekend calls are avoided by some interviewers but used heavily by others. None of these studies randomizes the time of calling on the sample household, and thus the estimated proportions of at home are biased to the extent that interviewers may visit some areas at times very different from others.

A better source of data might come from time use studies that have respondents report on their activities on a previous day. These data are thus more independent of the time interviewers choose to call on sample households. For example, Hill (1978) uses an "index of wakeful occupancy," which is the proportion of survey respondents who reported being at home and not sleeping. These are based on reports of respondents about their activities in the 24 hour period prior to the survey interview. Figure 5.2 shows this proportion over the different hours of a day, with the characteristic double moded distribution, one mode representing the at-home activities prior to leaving the housing unit in the morning and the second representing those at home upon returning. The index declines in the evening at 11:00 to 11:30 p.m. In addition to knowing that sample persons are home, we learn whether they are awake. Even more information could be obtained, however, with such data. Even though the eligible persons are awake at home, they may be engaged in activities which would prevent them from answering the telephone or likely lead to a noninterview contact. Figure 5.3 shows the proportion of persons who are bathing at various points in the day. This shows that the morning hours, although offering high proportions of persons awake at home, are the most likely times when many people would not answer the interviewer's call.

Such data might be preferable to survey call result data because (1) they may avoid the bias of call data being based on decisions of interviewers on appropriate times to call (thus probably tending to overestimate at-home proportions), and (2) they provide data on what persons are doing when they are at home.

The fact that sample persons may be at home does not necessarily imply that they will consent to the interview. Wilcox (1977) investigates whether the impacts on survey statistics of persons not at home are independent of those for refusals. The work thus offers another example of how survey errors might be correlated: in this case, two sources of nonresponse error having correlations. If the types of nonrespondents do differ from one another, merely concentrating on one (typically noncontact error) could actually increase nonresponse error. Wilcox used a sample of 1125 households, of which 947 were reached after three calls. A short set of demographic questions were asked of the nonresponders; 52 of 178 answered at least one. Note that since there were no refusal

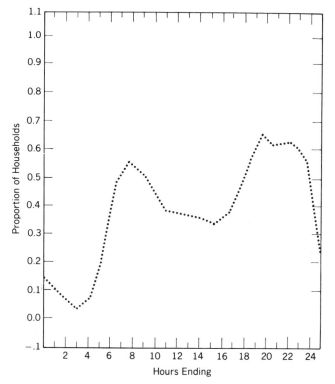

Figure 5.2 Index of wakeful occupancy by hour of the day. From D. Hill, "Home Production and The Residential Electric Load Curve," *Resources and Energy,* **Vol. 1, 1978, Figure 5.**

conversion efforts, the article merely measures the effect of added calls on nonrespondent characteristics. The work found that different kinds of people are home with different frequencies; these same groups have different probabilities of cooperating once contacted. The probabilities of being at home vary differently over calls than do the probabilities of cooperating, given that one is at home. It is even possible that the combination of contact probabilities and cooperation probabilities are such that lower nonresponse bias exists for a survey with fewer calls than one with more calls.

The real interest here is whether subgroup differences in the overall probability of obtaining an interview itself varies over calls. Table 5.2 shows that the probability varies less among education groups on the first call than after two calls. This occurs because those most likely to be home tend to be less likely to cooperate. The opposite finding applies to the

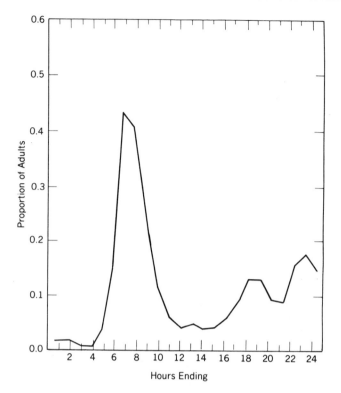

Figure 5.3 Proportion of adults bathing on an average weekday by hour of the day. From D. Hill, "Home Production and The Residential Electric Load Curve," *Resources and Energy*, Vol. 1, 1978, Figure 3.

family size variable; overall probabilities are more variable after one call than after two calls.

The typical approach of survey researchers to decrease the proportion of sample cases that are not contacted is to try repeated calls on the household. Sometimes interviewers are told to make as many calls as possible on the sample household before the survey period ends; sometimes their calls are limited to a fixed number (three or four are popular numbers) and told to vary the time of day (and day of the week, if possible) of the call.

Figure 5.4 presents a call grid used at the Survey Research Center to control the pattern of calls on telephone sample cases that are not easily contacted. The interviewers are told to place a call on the sample number to fill up cells in the grid, avoiding making more than the maximum number in any time period of the grid. Some four-call rules tell the interviewer to make two weekday evening calls, one weekday morning call,

☐ CONTACT MADE

On the grid, please remember to use digital CLOCK TIME. This should be recorded in respondent's time.

	MON	TUE	WED	THU	FRI	SAT	SUN	S/S	
Day	DATE___ TIME___ CALL#___ CODE___	DATE___ TIME___ CALL#___ CODE___	DATE___ TIME___ CALL#___ CODE___	DATE___ TIME___ CALL#___ CODE___	DATE___ TIME___ CALL#___ CODE___	DATE___ TIME___ CALL#___ CODE___	DATE___ TIME___ CALL#___ CODE___	DATE___ TIME___ CALL#___ CODE___	Day
Day	DATE___ TIME___ CALL#___ CODE___	DATE___ TIME___ CALL#___ CODE___	DATE___ TIME___ CALL#___ CODE___	DATE___ TIME___ CALL#___ CODE___	DATE___ TIME___ CALL#___ CODE___	DATE___ TIME___ CALL#___ CODE___	DATE___ TIME___ CALL#___ CODE___	DATE___ TIME___ CALL#___ CODE___	Day
Eve.	DATE___ TIME___ CALL#___ CODE___	DATE___ TIME___ CALL#___ CODE___	DATE___ TIME___ CALL#___ CODE___	DATE___ TIME___ CALL#___ CODE___	DATE___ TIME___ CALL#___ CODE___	DATE___ TIME___ CALL#___ CODE___	DATE___ TIME___ CALL#___ CODE___	DATE___ TIME___ CALL#___ CODE___	Eve.

Day = 9:00 a.m. – 4:00 p.m. (Respondent time)   Eve. = 4:00 p.m. on (Respondent time)

Figure 5.4 Call grid to enforce dialing on various days and hours of day, Survey Research Center.

197

Table 5.2 Estimated Probabilities of Being at Home, Cooperating if at Home, and Providing an Interview, by Education

	Estimated Probabilities		
Education Group	Being at Home	Cooperating, if at Home	Giving an Interview
After one call			
Grade school	0.87	0.85	0.74
High school	0.82	0.94	0.77
College	0.79	0.96	0.76
After two calls			
Grade school	0.96	0.83	0.80
High school	0.97	0.94	0.92
College	0.96	0.96	0.92

Source: J.B. Wilcox, "The Interaction of Refusal and Not-at-Home Sources of Nonresponse Bias," *Journal of Marketing Research*, Vol. XIV, November 1977, pp. 592–597, Tables 1 and 2.

and one weekend call before ceasing attempts on the number. Both of these are based on the logic that persons are at home and away from home at similar times over days and weeks.

The reader will note that all these rules are based on assumptions about the patterns of being at home for people who are difficult to locate. Because of repeated contact attempts, the data presented above on times that people are at home may not be the most useful to a survey researcher. Those data might be used as estimates of the unconditional probability that someone will be at home when the interviewer calls at a particular time. In actual practice, however, conditional probabilities of being at home are needed: that is, given that no one was at home on Tuesday evening at 8:00 p.m., what is the probability that someone will be home on Saturday at 2:00 p.m.? Such estimates would reflect the at-home patterns of households that on initial calls by the interviewer were not home.

Table 5.3 presents a simple form of such estimates. These were derived from the analysis of call records on telephone surveys on the first two calls. For example, for cases where the first and unsuccessful call was made on a weekday evening, it is estimated that 91.5 percent would be at home on a weekday morning. Only 55.8 percent would be at home on a weekday morning among those not reached in a weekday afternoon call. The table demonstrates the lowest conditional probabilities of being at home at times adjacent to those where the respondent was not at home earlier (e.g., those not at home on weekday mornings are unlikely to be at home on weekday afternoons).

Table 5.3 Percentage of Persons at Home During a Current Time Period Among Those Not at Home at a Past Time Period

| | Past Time Period When Respondent Not at Home | | | | | | |
| | Weekday | | | Saturday | | Sunday | |
Current Time Period	Morning	After-noon	Evening	After-noon	Evening	After-noon	Evening
Weekday morning		55.8	91.5	68.9	80.8	62.8	95.8
Weekday afternoon	43.5		89.4	64.8	75.0	57.1	92.5
Weekday evening	62.5	62.5		69.0	89.7	52.4	81.0
Saturday afternoon	83.5	86.1	96.1		66.3	69.4	90.8
Saturday evening	82.7	82.7	97.8	41.0		66.7	89.4
Sunday afternoon	79.5	80.3	87.3	71.4	80.8		90.8
Sunday evening	92.6	88.9	85.2	70.8	80.0	72.4	

Even if such estimates of conditional probabilities were available for all call numbers (e.g., probabilities of fourth call success at different times, given all possible combinations of times for the first three calls), it is unlikely that personal visit interviewers could take advantage of the knowledge. Their times of visitations are limited by concerns about the cost of traveling to a sample area and thus the total number of sample houses that they can usefully visit at one time. Furthermore, the complexity of call procedures based on such empirical guidance would complicate their work.

The most active area of work on noncontact reduction appears to be taking place in the area of computer-assisted telephone interviewing (CATI), where computer algorithms can be used to determine which sample numbers are good candidates for calling during at any particular time. In using these systems, an interviewer hits a particular key on the terminal keyboard to request a new sample number to call. Prior to that request the CATI software would have identified those numbers that should be called during that interviewing period. Call scheduling algorithms now in use vary on several dimensions:

1. *Whether they are deterministic or instead assign priorities to all active sample numbers.* Some call scheduling algorithms use information about the timing of past unsuccessful calls to determine which subset among all active numbers should be called during a given time period. Only those numbers are eligible for calls during the time period. Other algorithms identify priority groups or assign priority scores to all active numbers for a given time period, and numbers are dialed within the time period in order of their priorities. With this approach, if the time period is not productive of interviews, theoretically all sample numbers could be called.

2. *Whether the algorithm attempts to maximize the cumulative probability of a number being answered over the survey period or attempts to maximize the productivity of interviewers at each moment in the survey period.* Some algorithms attempt to identify and schedule for each interviewing shift those numbers that have the highest probability of contact at that time. Others attempt to assure that all numbers are scheduled for calling in a way that they are individually given the best chance of being answered *sometime* during the survey period.

3. *Whether the call scheduling procedures require human intervention for certain types of activity.* Some call scheduling systems have the supervisors assign appointments and refusal conversion cases to interviewers, after a review of the case. Others have complete machine control over all cases, except some that are placed into a group needing supervisory review (e.g., unusual sampling problems or respondent difficulties).

Despite these rather large differences among call scheduling techniques, there is very little written documentation on the rationale for the differences. Furthermore, there has been little experimentation regarding the effects on nonresponse rates of different call scheduling procedures. Once a CATI system implements a procedure, it appears to use it without evaluation. What is needed at this point is an experimentally controlled test of alternative procedures, with comparison of productivity and unit nonresponse rates that result.

To conclude this section on noncontacts as a component to survey nonresponse, we note that noncontacts tend to yield themselves to elimination more easily than refusals. The increases in refusals documented in Chapter 4 have alarmed many researchers concerned about nonresponse bias. Since survey quality is often naively assessed by response rates, researchers have attempted to maintain response rates through intensifying the calls or visits to sample households. For example, the reader will recall that the proportion of total nonresponse associated with noncontact cases decreased over time in most of the data presented in Chapter 4. The decrease appears to have been the result of efforts to minimize the *total* nonresponse rate.[1] As noted earlier, however, a strategy of minimizing overall nonresponse rates, without regard to the different reasons for nonresponse, may do little to reduce nonresponse bias.

5.1.2 Demographic Characteristics of Refusers

Through one of the designs mentioned above, various researchers have obtained information about the characteristics of nonrespondents. Typically, the information is demographic in nature. This section reviews some of the more consistent findings of that literature. It is organized by major demographic characteristics.

There is one analytic problem in describing the characteristics of respondents and/or nonrespondents—the adjustment for probabilities of selection of the sample person. Many surveys use a respondent selection technique that chooses at random one person among those eligible in the sample household. Even if the sample households were selected with equal probability, this procedure yields unequal probabilities of selection for persons by household size. For example, a person living alone would have a higher chance of selection than one living with five other persons. Many demographic characteristics are not distributed uniformly over household size (e.g., women and the elderly tend to live in smaller households). Without compensation for unequal probabilities of selection therefore, many sample surveys overrepresent persons who tend to live in small households. To determine whether the nonrespondents tend to have distinctive characteristics, it is necessary to apply selection weights for statistics describing them as well.

There is also a problem of comparability across studies of nonrespondents. It may not be appropriate to assume that demographic characteristics cause refusals because the topic or the procedures of the survey may have an independent effect on refusal rates. For example, the lower response rates among the elderly may be more pronounced in telephone surveys than in personal interview surveys because of hearing problems on the telephone. When we examine the characteristics of

---

[1] Unfortunately, when survey response rates are reported, it is rare to find separate rates for refusals and noncontacts. Because of this, readers have not generally been alerted to the changing composition of nonresponse over time.

refusers across various surveys, differences across the surveys on these design features may complicate the comparison.

Respondent gender is a frequently studied correlate of nonresponse error. For example, T. Smith (1979) noted a declining proportion of males among respondents to the General Social Survey (GSS) between 1972 and 1978, from 51 percent in 1972 to 43 percent in 1978, using weights to account for the selection of one adult from each sample household. This decline in the representation of males coincided with a change of sample design of the GSS, a move from a "probability sample with quotas" (before 1975) to a "full probability sample" (after 1976). In 1975 and 1976 half of the sample was devoted to one design, half to the other. Smith compares these percentages to 1975–1977 Current Population Survey (CPS) data estimating 47.6 to 47.7 percent of the adult household population was male. He finds that the quota sample method yielded male respondents in the percentages closer to the CPS data. He notes that the quota method controlled the distribution of males among respondents, but that the full probability method only gives males and females in the same household the same chance of entering the sample. On the other hand, Brown and Bishop (1982) find no differences between sexes on whether the interview was completed on the first call, with callbacks or refusals. However, there is some evidence that female telephone interviewers get lower proportions of men than do male interviewers (T. Smith, 1979; Groves and Fultz, 1985), and thus differences across surveys might reflect differences in the proportion of female interviewers used. DeMaio (1980) sees no difference on CPS rates by gender, but Lindstrom (1983) in the Swedish Labor Force Survey finds higher nonresponse among men, especially unmarried men.

The age of the sample person has been found related to nonresponse by many researchers. Cobb et al. (1957) cite refusal rates among older persons as a problem for studies of health conditions. DeMaio found that those over 30 were more likely to refuse than those under 30 in the CPS. Dohrenwend and Dohrenwend (1968) found those over 40 to have higher nonresponse than younger persons (23 percent to 12 percent). Hawkins (1975), in a review of nonresponse over several years in the Detroit Area Studies, finds that the elderly disproportionately refuse to cooperate. Weaver et al. (1975) used the sampling frame for city workers in San Antonio, Texas, to measure the characteristics of refusers to a telephone survey about the 18 year old vote issue. The overall refusal rate among those contacted was 10.9 percent and among those 50 and older, 17.2 percent. The higher refusal rates among the elderly applied across salary groups, and ethnic and race groups.

Mercer and Butler (1967) test the hypothesis that aging involves disengagement from interaction, and that higher refusal rates among the elderly are a product of that tendency. They used a community survey of chronic handicaps to test the hypothesis, with any member of the household 18 years old or older acting as an informant for all household members. Interviews lasted 1.5 hours. The overall response rate was 90.7 percent, but was 84 to 86 percent for those 60 and over. Among the elderly there appeared to be no tendency for males, the wealthy, or those with long tenure in their home to refuse. Among the elderly refusers tended to come from homes without children. There were no differences among the elderly refusers and cooperators on electoral activity, marital status, social activities, and church attendance. To the extent that such indicators are used as a test that the "disengaged" elderly are the most likely to refused, the hypothesis was not strongly supported.

Brown and Bishop (1982) find higher levels of resisting among elderly sample persons in an ongoing telephone survey in the Cincinnati area (Table 5.4). They labeled as "amenable" those respondents who granted the interview on the first contact, as "resisting" those who required several calls before obtaining an interview, and as "refusing" those who initially refused but were persuaded later to cooperate. Their results show that smaller portions of those 60 years of age or older grant an interview on the first contact than is true in the overall set of respondents (75.8 percent among the elderly versus 84.3 percent overall).

Table 5.4 Respondent Distribution of Cooperation Type by Age

	Respondent Age			
Cooperation Type	18–39	40–59	60 and Over	Total
Amenable	88.0%	83.8%	75.8%	84.3%
Resisting	10.2	13.2	16.7	12.4
Refusing	1.8	3.0	7.5	3.3
Total	100.0	100.0	100.0	100.0
n	4219	2157	1726	8102

Source: Greater Cincinnati Surveys 1–4, 6–8, as cited in P.R. Brown and G.F. Bishop, "Who Refuses and Resists in Telephone Surveys? Some New Evidence," University of Cincinnati, 1982, Table 2.

In three SRC surveys with random respondent rules among adults, Herzog and Rodgers (1988) used data about nonrespondents guessed by the personal visit interviewers. Their results do not show monotonic increases in nonresponse rates (combined refusals and other noninterviews) over the higher age groups. The results in Table 5.5 do not separate refusals from other nonresponse, and that fact may dampen the relationship with age. It is interesting to note that in all three studies some of the highest nonresponse rates occur in the older groups.

Kristiansson (1980) in a detailed study of nonresponse to the Swedish Labor Force Survey matched population register and census data to the data from the survey. This match showed that in Sweden there was no tendency for the elderly (those over 70) to have higher nonresponse rates; this was true for males and females, regardless of marital status. The highest nonresponse rates were achieved by nonmarried males (35-49) and nonmarried women aged 55-64. Sweden offers an interesting case study in response rates both because of this difference and the evidence that cooperation with surveys increased during the period 1976 to 1979, after decreasing between 1969 and 1976.

In an interesting match of the Family Expenditure Survey and the National Food Survey data with the 1971 British Census, Kemsley (1975) obtains characteristics of 93 percent of the nonrespondent households. Nonresponse is a common outcome in the survey; about 30 percent of the households fail to cooperate (noncontact rates are about 1 percent). The surveys involve both an interview and the recording by the respondent households of expenditures or food intake during the survey period. Age of the household head is an important correlate, as shown in Table 5.6. Response rates range from the 80s among households with heads and wives less than 30 years of age to the 60s for those over 50. The trend is almost monotonically decreasing across the different age groups. When response rate for the 5 year groups is regressed on age groups, there is a predicted decrease of two percentage points in response rate for each increase in 5 years of age.

There are other studies regarding the relationship between nonrespondent behavior and age of the sample person. The results are not consistent. O'Neil (1979) shows higher resisting rates (using a similar definition as Brown and Bishop above) among respondents aged 55 and older in a telephone survey. Benson et al. (1951) found no differences by age in a study of refusers. DeMaio (1980) has similar results for the U.S. Current Population Survey. Marcus and Telesky (1984) and Benus (1971) find more wealthy give followup interviews after an initial interview. Cannell et al. (1987), in a comparison of telephone surveys with personal interviews, show higher nonresponse among the young and the old (see Table 5.7).

**Table 5.5 Nonresponse Rates by Age of Respondent as Estimated by the
Interviewer, Three Survey Research Center Studies**

	Survey		
		Election Survey	
Age Group	Americans View Their Mental Health, 1976	1978	1980
18–24	26.5	28.4	20.2
25–34	25.3	29.0	22.7
35–44	26.4	32.8	29.5
45–54	37.4	30.6	30.7
55–64	22.6	29.4	30.8
65–74	31.2	31.3	32.0
75–84	37.1	33.5	34.9
85+	48.3	46.1	51.2
n	3173	3332	2242

Source: A.R. Herzog and W.L. Rodgers, "Age and Response Rates to Interview Sample Surveys," *Journal of Gerontology: Social Sciences*, Vol. 43, No. 6, 1988, S200-S205, Table 1.

The relationship between education and response rates is well documented. In survey designs without an interviewer (e.g., mailed questionnaires) it is frequently observed that higher education groups disproportionately tend to return completed questionnaires (see Dillman, 1978). Similar results apply to telephone and personal interview surveys. O'Neil (1979) shows higher resisting rates among lower educated respondents. Robins (1963) in a follow-up study of guidance clinic patients after 30 years found a 12 percentage point higher refusal rate among those without a high school education versus those with a diploma. Dohrenwend and Dohrenwend (1968) find proportionately greater refusing among lower education groups but their small sample size prevented a statistically significant difference. Benson et al. (1951) find lower average education among refusers, in a unique study that converted 32 of 33 refused cases in an urban survey on racial attitudes. Cannell et al. (1987) estimate 58 percent response rates among those with less than a high school education, but 96 percent for those with 13 or more years of schooling. Both Dunkelberg and Day (1973) and Hawkins (1975) investigated the percentage of respondents in different education groups by the call on which they were interviewed. Their results are somewhat contradictory but generally show smaller proportions of higher education groups from interviews obtained on early calls. This reflects apparently the higher employment rates of higher education groups and the need for

Table 5.6 Response Rates by Age of Head of Household and Age of Housewife, British Family
Expenditure Survey, 1971

| Age Group | Response Rates for Expenditure Survey Households | | | | Response Rates for National Food Survey Households | | | |
| | Age of Head of Household | | Age of Housewife | | Age of Head of Household | | Age of Housewife | |
	n	Response Rate	n	Response Rate	n	Response Rate	n	Response Rate
16-20	28	100%	95	88%	26	58%	100	52%
21-25	387	88	620	86	455	65	768	66
26-30	655	85	770	84	844	67	941	68
31-35	754	81	794	77	1017	67	1074	64
36-40	754	74	788	75	1064	59	1165	59
41-45	884	73	896	71	1221	55	1238	55
46-50	971	69	935	67	1355	56	1347	56
51-55	904	67	881	65	1287	53	1166	50
56-60	1026	65	933	63	1386	49	1351	53
61-65	1027	62	962	66	1488	51	1338	49
66-70	916	66	790	65	1231	50	1102	49
71 or more	1520	62	1362	62	2145	42	1929	41
Age unknown	20	45	20	45	32	19	32	19

Source: W.F.F. Kemsley, "Family Expenditure Survey: A Study of Differential Response Based on a
Comparison of the 1971 Sample with the Census," *Statistical News*, No. 31, November 1975, pp. 31.16-
31.21, Table A8, and W.F.F. Kemsley, "National Food Survey—A Study of Differential Response Based
on a Comparison of the 1971 Sample with the Census, *Statistical News*, No. 35, November 1976,
pp. 35.18-35.22, Table A10.

callbacks to reach them when they are free to provide the interview.
Wilcox (1977) separates noncontact sources of nonresponse from refusals.
He finds that the lower response rate of those with less than a high school
education lies more in higher refusal rates than in higher noncontact
rates.

Results on response rates and the race of the sample person are
inconsistent. Weaver et al. (1975) find lower refusal rates for blacks
among all salary groups. O'Neil (1979) finds lower resisting rates among
black respondents. DeMaio (1980) found no difference between the race
groups in the Current Population Survey, using a household informant
respondent rule. Herzog and Rodgers (1981) find whites less likely to
respond in general, but blacks less likely to respond repeatedly in a panel
survey. Weaver et al. (1975), in a telephone survey of city employees in
San Antonio, Texas, found more problems of inaccessibility among blacks
but a smaller tendency toward refusal. Hess and Pillai (1962) combined
data from four SRC national personal interview surveys in 1959 and 1960
that contained a variety of economic and social attitude measures.
Information on nonrespondents was obtained by interviewer observation
and estimation. Over the four studies the response rate among whites was

Table 5.7 Estimated Response Rates for Two Respondent Rules by Various Demographic Categories

	Estimated Response Rate	
Subclass	Knowledgeable Adult Respondent Rule	Random Respondent Rule
Total	81%	75%
Gender		
Male	83	72
Female	81	79
Age		
17–24	82	61
25–44	88	86
45–64	81	81
65–74	66	57
Race		
White	81	77
Nonwhite	82	69
Education		
0–11 years	74	58
12 years	80	72
13+ years	91	96
Marital status		
Married	81	79
Widowed	76	58
Divorced	89	97
Separated	80	86
Single	83	65
Usual activity		
Working	83	80
Keeping house	83	85
Other	77	47

Rates estimated by assuming no nonresponse bias of demographic distributions for Health Interview Survey.
Source: C.F. Cannell et al., *An Experimental Comparison of Telephone and Personal Health Surveys*, 1987, Table 3.3.

estimated at 86 percent (with a standard error of about 0.8 percent); for blacks, it was 90 percent (with a standard error of about 2 percent) Benus (1971), in reviewing national personal interview surveys in the mid-1960s, shows higher nonresponse among blacks for election surveys (by about 5 percentage points) but not on economic surveys. Hawkins (1975), in a nonresponse study of Detroit Area Study samples, finds higher refusal rates among whites. O'Neil (1979), in a metropolitan telephone survey using a respondent rule that required no randomized selection, found that white households more often gave initial refusals than black households.

In short, the empirical results are rather consistent in showing larger nonresponse problems among nonblacks than blacks. These results, bivariate in nature, are somewhat puzzling when added to the discussion on education above. That is, we have learned that lower education groups have lower response rates in general, but blacks, a group disproportionately falling in that category, do not exhibit higher nonresponse.

Another variable, size of the household, is sometimes known both for respondent and nonrespondent sample cases. Brown and Bishop (1982), in eight Cincinnati telephone surveys, find that respondents in single adult households are more likely to resist or break off the interview than respondents in larger households. In another metropolitan sample, Wilcox (1977) notes that single adult households are more likely to be contacted (probably because they tend to be elderly nonemployed persons) but are also more likely to refuse the interview. In two separate surveys, Kemsley (1976) matched survey respondent and nonrespondent households to British census data and found that there were lower response rates among smaller households. Other data show that single adult households tend to disproportionately consist of older persons. Hence, the finding that smaller households have lower response rates reflects the tendency for older persons to refuse survey requests.

Most of the conclusions above come from bivariate analysis of response rate by one demographic variable after another. There are few attempts to build models that identify the causes of the nonresponse behavior. This gap is even more important in view of the variation in results over different studies in the literature. The characteristics of nonrespondents do not appear to be fixed over surveys discussing different topics, entailing different respondent burden, being conducted by different organizations, and using different survey procedures. To make sense of the literature some more succinct theoretical structure must be constructed.[2]

5.2 EFFORTS BY SURVEY METHODOLOGISTS TO REDUCE REFUSALS

Survey methodologists have frequently addressed the problems of incomplete measurement of sample persons. These efforts by and large have had the sole goal of increasing the proportion of the sample

---

[2] This is a clear case of the need for collaboration between the social scientist and the survey statistician and between the "measurers" and "reducers" (see Chapter 1).

measured, without formal concern about the nature of people added to the sample or about response errors they commit. Thus, the implicit assumption in that literature is the higher the response rate, the lower the nonresponse bias. Much of the literature is not guided by theoretical concepts of sociological or psychological influences to respond or not to respond, but rather by ad hoc reasoning. Much of the literature is flawed by lack of full description of all the techniques used to produce the higher response rates and by research designs with confounded variables. Much of it ignores the fact that response rates are survey statistics that have their own sampling and nonsampling error variances.[3]

In the following sections we discuss three procedures investigated experimentally for increasing cooperation among sample persons contacted by interviewers: (1) advanced warning of the interviewer's request, either through telephone calls or letters, (2) manipulation of what interviewers say in their introduction of the survey to potential respondents, and (3) the use of material incentives to sample persons to induce compliance with the survey request.

5.2.1 Contact with Sample Households Before the Survey Request

There are two conflicting hypotheses on whether prior contact with the sample household will help or harm response rates: (1) warning of impending visit allows a household to prepare for the interview and relieves the interviewers from the burden of communicating the legitimacy of the survey by themselves, or in contrast, (2) the prior contact merely permits the household members to prepare a rationale for their refusal, a decision to which they become committed prior to hearing any contrary arguments by the interviewer. The most common hypothesis is the former. Note that prior contact attempts implicitly involve the interplay of survey costs and errors since any prior contact increases the cost of the survey. Thus, for the prior contact design to be the preferable one with a single attempt, the reduction in nonresponse bias must outweigh the increase in sampling error of survey statistics resulting from a smaller sample size.

There are some documented experiments that involve the use of a prior telephone call. Brunner and Carroll (1967), in two personal visit samples of suburban Washington, DC areas, found that the experimental

---

[3] This arises because the likelihood of cooperation varies over persons in the population and hence overall response rates for a survey depend on the sample drawn.

groups receiving a prior telephone call had higher refusal rates than those without a prior call (63 percent versus 33 percent in one city, and 55 percent versus 16 percent in another). There is no description provided about the nature of the prior telephone contact, but it is noteworthy that the interviewers in the survey were students in a marketing research class. Groves and Magilavy (1981) attempted a prior telephone contact before a random digit dialed telephone survey. This contact asked the person answering the telephone two surveylike questions and alerted them to the fact that an interviewer would call later. They found negligible differences between response rates on the experimental group receiving the prior call and those not receiving one (81 percent response rate for the prior contact group and 80 percent for the group with a single contact).[4]

Most of the literature on prior contact involves the use of a letter describing the survey mailed to sample households prior to the visit of an interviewer. There has been little exploration of mechanisms through which advance letters might have their effects. It is assumed by many that letters affect the decision-making process of the sample person regarding the survey request. The letter has been used to legitimize the survey request through the prestige of the organization conducting the survey, to reduce fear of victimization, and to communicate the value of the survey to the respondent or the society at large. In contrast to this focus on the sample person, there is also the possibility of effects acting through the interviewer. Letters might improve response rates by increasing the self-confidence of interviewers, allowing them to refer to an attempt to notify the sample household about their visit (whether or not the letter was actually received or read). Calling attention to the letter might relieve the interviewer from single-handedly legitimizing requests for the respondents' time. The interviewer is more fully an agent sent by an organization that used its resources to forewarn the respondent of the visit. If this hypothesis is true a portion of the effect of the advance letter corresponds to changes in interviewer behavior given knowledge of the advance letter.

Frequency of Reading the Letter. The most striking result in the literature on advance letters is that of Cannell et al. (1965) that 44 percent of the respondents to whom letters were mailed reported that they had not received an advance letter or brochure describing the survey. A total of 33 percent reported reading the letter carefully, and 16 percent quickly. Thus, over 50 percent said they had read neither the letter nor the

---

[4] This experiment is reviewed also in the discussion of the influence of "consistency" in Section 5.4.

brochure. A question raised about these results concerns the actual receipt of the letter; the authors estimate that about 73 percent of the letters were actually received. Among those addresses where there was strong evidence of the letter being delivered, 58 percent of the respondents reported reading the letter. There is some evidence that women read the materials more frequently than men and that lower income groups read more frequently than higher groups. There was no large effect on knowledge about the agency of the interviewer or about the purposes of the research by whether the respondent reported the letter being received.

Affiliation of the Letter–Writer. The affiliation of the letter-writer is communicated on the envelope, the letterhead, and the body of the letter. Brunner and Carroll (1969) find rather dramatic increases in the positive effect of the letter in a first interview when university affiliation versus a market research organization affiliation is made (from a negative effect of about 6 percentage points on the response rate to a positive effect of about 30 percentage points).

Ferber and Sudman (1974) report an experiment in the city of Chicago where the University of Illinois letter increased cooperation from 75 to 89 percent versus a Census Bureau letter from 64 to 81 percent. Interviewers said they were acting as collecting agents for the Census Bureau in the census half sample.

Other Contents of the Letter. Dillman et al. (1976) in a one–time telephone survey used three types of letters: (1) one merely mentioning the study and alerting the respondent to an upcoming call, (2) one providing additional information about the survey, describing its sampling procedures, and inviting people to call with questions, and (3) one containing all the information of the first two types of letter but emphasizing the study's social utility, promising anonymity, briefly describing the kinds of questions to be asked, and offering a copy of results. Names were typed on all; a hand–written signature was used. There was a negative effect of receiving no letter (85 percent response rate versus above 90 percent response with letters, generally). The best form was the middle alternative, although the effects are not large. The authors note that the third option might have created too long of a letter to be read carefully.

In a series of studies, Slocum et al. (1956) included information in letters that (1) described the study, emphasizing its "social utility", (2) explained how the respondent was selected into the sample, and (3) explained why his/her cooperation was important. One personal interview study did this experimentally and found no effect of the letter on response rates but fewer number of calls in the letter group. They then sent a special delivery letter to a sample of nonrespondents emphasizing their role in the success of the study and obtained large effects of the letter.

Instead of emphasizing the importance of each respondent to the success of the survey in the letter, Erdos (1957) argues that the letter should describe how the respondent will benefit personally from the survey.

In a mailed questionnaire of school administrators, Furst and Blitchington (1979) experimentally tested a letter that described how the study was designed, what the researchers hoped to learn, and benefits that could ensue from the results of the research. A 5 percentage point increase in response rates (statistically insignificant) was obtained with the letter.

In a study of older women (aged 66 to 76 years), Koo et al. (1976) required initial screening of households to locate eligible persons, followed somewhat later by a personal interview. An advance letter was experimentally tested on a half-sample. The contents of the letter described the sponsorship of the study (not stated in the article), reviewed purposes of the study, and alerted the respondent of the interviewer's forthcoming visit. Results varied across high and medium socioeconomic status (SES) areas (no effect of letter) and low SES areas (12 percentage point negative effect of letter on response rate). Overall response rates are very low (between 35 and 65 percent).

Cartwright and Tucker (1967) describe a letter experiment on a sample of fathers and mothers of recently born children. The letter identified affiliation (Medical Care Research Unit of Institute of Community Studies), gave little information about the study, sought cooperation of reader, pledged confidentiality, described how sample was selected, and described the Institute. Fathers were approached by male interviewers, mothers by females. One-fifth of the sample were fathers. No effect of letter was found for mothers, but an 11 percentage point effect (23 to 12 percent refusals) was found for fathers. The overall response rates for mothers were higher (by 3 to 15 points) than those for fathers. The letter did have a positive effect on mothers where the head of household was a professional, and a negative effect on those where the head was unskilled.

In short, the dimensions on which letter content seem to vary include the following:

1. How detailed a description of the study is given (from none at all, to a paragraph).

2. Whether the social utility of the respondent's cooperation is emphasized or whether individual rewards on the respondent's part are emphasized.

3. Whether sampling techniques are described.

4. Whether confidentiality of the data is promised.

5.2.2 Descriptions of the Survey at Time of Introduction

Early random digit dialed telephone surveys found that refusals were occurring quite quickly after the telephone was answered. Several research efforts were launched to learn whether the content of the interviewer remarks in introducing the survey requests was related to response rates. For example, O'Neil et al. (1979) in an RDD survey experimentally varied what the interviewer said along three dimensions:

1. The use of surveylike questions immediately after mentioning that the call concerns a research project at The University of Michigan, two questions about the health care and health status of the person answering the telephone. (The hypothesis of increased cooperation with this treatment was based on the notion that such questions would instruct the sample person about the nonthreatening, nonburdensome tasks contained in the survey interview.)

2. Providing a description of the survey organization, the social utility of the project, and the need for a representative sample. (The hypothesis of increased cooperation from this treatment was based on the information about the nature of the need of the survey organization for the sample person's cooperation.)

3. Providing verbal feedback to the person answering the telephone after they responded to initial queries. (The hypothesis of increased cooperation with this treatment involved the communication to the sample person of the successful completion of tasks presented by the survey interaction.)

Experimental variations on these three dimensions produced six treatments (see Table 5.8). The "nothing" group merely involved a short introduction of the caller and an immediate start of the survey request. Both "completion rates," which omit noncontact cases from the denominator, and "response rates," which include them, are presented.

For our purposes, the completion rates might be most informative, and they show only small increases over the "nothing" treatment. The best completion rate is achieved by the group that used the questions, a full explanation of the survey, and feedback to respondent answers. However, this treatment increased the response rate only 3 percentage points over that from the "nothing" group (a nonsignificant difference statistically).

Dillman et al. (1976) ran two experiments imbedded in telephone surveys. The manipulations of the introduction in the first survey included (1) the use of the residents' name (obtained through telephone directory sampling), and (2) a description of the study, sampling technique, and length of interview versus one that did that plus offered study results and noted the sponsorship of the study. No significant differences were found among the treatments (the range of refusal rates is 17 to 22 percent). Another experiment added a note about the social utility of the survey (i.e., "we expect some of these issues to be considered in the future by the State Legislature"). In this experiment the best treatment was one that did not use a personal introduction and did not offer the survey results to the sample person.

Table 5.8 Response Rates and Completion Rates by Form of Introduction

Form	Response Rate	Completion Rate
Nothing	64.7%	76.3%
Questions	69.7	78.1
Feedback	60.1	72.1
Questions, explanation	61.7	74.6
Questions, feedback	54.5	64.7
Questions, explanation, feedback	66.0	78.9

Numerator of response and cooperation rates is the number of completed interviews plus number of partials. Denominator of response rate includes those in the numerator plus refusals, numbers of undetermined status, and numbers never answered. The cooperation rate denominator excludes the numbers of undetermined status and those never answered.
Source: M.J. O'Neil et al., "Telephone Interview Introductions and Refusal Rates: Experiments in Increasing Respondent Cooperation," *Proceedings of the American Statistical Association, Survey Research Methods Section, 1980*, Table 2.

5.2.3 Attempts to Increase the Benefits of Cooperation

Some of the techniques above could be viewed as attempts to communicate to the sample person the benefits of cooperating with the survey. Most surveys cannot offer direct benefits to the respondents, however, and the arguments presented by interviewers generally involve statements about the benefits to the society of knowledge about social problems or attitudes of the populace. Rarely are there logical and direct connections between any single survey and changes in governmental policy. Thus, it is likely that such arguments' effects are mediated by the respondent's valuation of information or of social science research.

There is other research in survey methods that attempts to manipulate the benefits of cooperation directly by offering the respondent incentives. These have included money, gifts, lottery chances, reports of the study results, or newsletters of the survey organization. Cannell and Henson (1974) note that if the survey purposes do not help the sample person achieve his personal goals, then incentives might act as a worthy goal themselves. By cooperating with the survey request, the respondent obtains some small enrichment. With similar reasoning, it is remarked that if the survey task is more like work than a simple and easy conversation, then normal compensation should be extended. Clearly, the value attached to the incentive by the respondent should be related to the effect on the respondent cooperation. There are no studies that have measured perceived value of the gift, but the hypothesis that the poor will be more heavily affected by the gift flows from this reasoning. Compensation has rarely been offered for government surveys. Perhaps this stems from the view that surveys by governments should seek cooperation as part of public service on the part of respondents. In contrast, commercial research agencies earn money because of the cooperation of the respondents and hence should share that reward.

There are negative effects reported in the literature on survey incentives (this might be prevalent among those for whom the only legitimate rationale for the study is social utility). Dohrenwend (1970) obtained lower response rates with compensation for later waves in a panel study—65 percent responded on both waves without compensation; 61 percent with compensation ($5) on both waves; and 53 percent with compensation on wave 2 only. For wave 1 alone the results were the opposite (56 percent responding among the compensated versus 50 percent among noncompensated). None of these results are statistically significantly different. There is experimental evidence that if the payment is too large effects on task performance may be negative; this is consistent with reactance theory described later.

Ferber and Sudman (1974) examined the effect of compensation on cooperation in a consumer expenditure survey that involved the keeping of diaries of purchases for one week at a time. The experimental treatments were (1) "no gift," (2) giving the respondent a plastic folder and ballpoint pen worth about $1.00, (3) giving the respondent a "summary and comparison of purchases," a report of respondent purchases by major categories compared to other panel members, (4) giving the respondent a "large stationery holder," a large padded stationery holder retailing for $5.00, and (5) giving the respondent an "American flag or government publications," or choice of a flag or posters of Illinois history or one of 40 popular government publications. Table 5.9 presents the response rates for the different experimental treatments by different levels of cooperation. The first column cites cooperation among those who received no incentive; without exception, they are lower than corresponding percentages cooperating among the incentive groups. This applies both to initial agreement to keep the diaries and performance on the task. The differences are more dramatic the more diaries are completed. This supports the hypothesis that the larger the burden the larger the effect of incentives. Another experiment offered a $5.00 check for 2 week diaries to a random half-sample. This survey obtained a 79 percent response rate with the $5.00 incentive and 67 percent without. Ferber and Sudman conclude that compensation is desirable only in panel surveys or surveys with great burden.

Table 5.9 Response Rates by Type of Gift

		Type of Gift		
Extent of Cooperation	No Gift	Some Gift	Summary and Comparison of Purchases	Large Stationery Holder	Flag or Book
Agreed to keep diary	85.7%	89.6%	88.3%	92.0%	88.7%
n	113	412	128	125	159
Kept at least one diary	77.1	84.7	85.9	90.4	79.2
n	113	412	128	125	159
Kept at least two diaries	62.1	75.1	71.1	82.6	71.3
n	90	297	90	101	106
Kept at least three diaries	54.1	67.0	66.7	70.2	64.3
n	61	198	63	66	69
Kept four diaries	23.1	48.6	54.8	39.4	51.4
n	30	96	31	31	34

Source: R. Ferber and S. Sudman, "Effects of Compensation in Consumer Expenditure Studies,' *Annals of Economic and Social Measurement*, Vol. 3, No. 2, 1974, pp. 319-331, Table 1.

Table 5.10 Response Rates by Incentive Group

Experimental Group	Response Rate	n
No incentive	70.5%	200
$5 incentive	80.4	214
$10 incentive	85.3	191
Variable	83.3	186

Source: J. Chromy and D. Horvitz, "The Use of Monetary Incentives in National Assessment Household Surveys," *Journal of the American Statistical Association*, No. 363, September 1978, pp. 473–478, Table 3.

Chromy and Horvitz (1978) report the results of an incentive experiment as part of the National Assessment of Education Progress, a sample of persons aged 26 to 35. The experimental groups were defined by (1) no monetary incentive, (2) a $5 incentive, (3) a $10 incentive, and (4) a variable incentive (if one package of the survey were completed, no incentive; if two were completed, a $10 incentive was offered; if three were completed, a $15 incentive was offered; and if four were completed, a $20 incentive was offered). The survey tasks involved the completion by the respondent of self-administered sets of questions (called packages). The experimental treatments are not fully randomized. For example, if an 80 percent response rate were obtained first without incentives (based on four callbacks), cases were classified into the no incentive group. Table 5.10 presents the results of the experiment, showing the highest response rates (85 percent) for the $10 incentive group.

5.2.4 Discussion of Attempts to Reduce Nonresponse from Refusals

There are many other ways that survey researchers have attempted to increase response rates. These include mass media publicity for the survey, use of proxy respondents to report for the sample person (an interesting interplay of response error and nonresponse error, discussed more in Chapter 9), use of multiple modes of data collection for reluctant respondents (this resembles the ideas of Hansen and Hurwitz discussed above), and training all interviewers in persuasion techniques or developing an elite corps of refusal converters.

Several comments on the state of our knowledge about nonresponse reduction are appropriate to summarize this section:

1. Attempts to reduce nonresponse rates might be more harmful than considering $(nr/n)/(y_r - y_{nr})$ simultaneously. Most efforts attempt to reduce the proportion of nonrespondents (nr/n) in an indiscriminate way, without concern for differences between the nonrespondents and respondents on the survey measures. There appear to be no procedures that are targeted on particular subgroups of the population in an attempt to calibrate the $(y_r - y_{nr})$ term to its minimal value.

2. The measures taken to reduce nonresponse often require assumptions of homogeneity of response (e.g., all people will prefer to be given an incentive; no persons would be turned off by an incentive offer). The mixed results that have been obtained might result from heterogeneity in the household population concerning the perceived costs and benefits of survey participation.

3. These measures have just attacked nonresponse error, ignoring links between nonresponse error and measurement errors. There have been no investigations about whether reluctant respondents who cooperate only after extraordinary persuasion provide more or less accurate information to the interviewer.

4. The psychological experimental literature is probably poorly suited to address compliance issues for initial requests because of its laboratory-based methodology. Natural setting experiments are the only way to do this.

5. There is little formal literature on training techniques to minimize nonresponse error. Usually these involve role playing during training, with supervisors playing the role of a reluctant respondent. There are also model questions and answers that are often asked about the survey that interviewers are prepared to deliver.

6. There is little commentary on the design of refusal conversion efforts, but these often entail the use of interviewers who are different in sex and age from the initial interviewer. Similar lack of research exists on the content of persuasion letters from the study directors, fuller descriptions of the purposes of the project, and answers of respondent questions.

5.3 SOCIOLOGICAL CONCEPTS RELEVANT TO SURVEY NONRESPONSE

The attempts to reduce nonresponse described above often are silent on the question of why refusals occur. The perspective taken in a small sociological literature on nonresponse is that occupational and social roles, the strength of social networks, concerns about privacy, the saliency of a survey topic to an individual, and the degree of symmetry in the exchange of benefits between respondents and researchers all combine to influence the decision to cooperate with a survey request. This work takes a very different viewpoint than that of the survey methodological literature. It focuses on the individual sample person and examines what influences might come to bear on a decision to participate in the survey. The literature ignores to a large degree the influences arising from decisions of the survey designer (e.g., more callbacks, incentives).

Surveys are most often conducted through sponsorship of large institutions within the society, by the government, profit-making companies, or academic institutions. For that reason, they can be viewed as the monitoring of the society by those currently in positions of power over the course of large-scale social events. Or stated more forcefully by Marsh (1982, p. 128), "those who can privately own information about society potentially gain many advantages of increased control over individuals and situations." While early characterizations of surveys emphasized their democratizing features through giving a voice to every person (Gallup, 1948), an alternative view is that they are used to strengthen the positions of those in power.[5] Thus, when special studies find that groups with lower socioeconomic statuses tend to refuse surveys (Goyder, 1987), some infer that this is consistent with other acts of disengagement of those groups from societal norms (e.g., voting in elections, participating in community affairs).[6] When the elderly who have reduced their social contacts over time are also found to refuse surveys, some take this as another example of the same principle. Goyder (1987) thus hypothesizes that those more "peripheral" to the society are less likely to participate relative to those more "central" to it. The purposes of most surveys can be seen as irrelevant to those with self-perceptions as out of the mainstream.

---

[5] Since governmental survey data are freely shared and publicized to everyone, the value of the information is not monopolized by the research organization. This may help explain higher response rates to surveys conducted by U.S. government agencies.

[6] This reasoning, however, does not explain the finding that in some surveys those with higher socioeconomic status tend to refuse as well as those with lower status.

A theoretical structure frequently taken to interpret survey participation is that of social exchange (see e.g., Dillman, 1978; Goyder, 1987). In this view, sample persons are seen to decide to cooperate with a survey request based on judged costs and benefits to them of that behavior. The costs of the survey participation include the time lost to other activities, the loss of privacy or control over information about oneself, the potential of being asked to reveal embarrassing attributes of oneself, and the engaging in an interaction whose agenda is controlled by another. The benefits of the interview include the supplying of information that might improve society, the provision of assistance to the interviewer herself/himself, the opportunity to discuss a topic of personal interest, or the pleasure of interaction with another person. Tests of exchange theoretic hypotheses regarding surveys are difficult because they are based on values of the respondents themselves, which are seen to vary over individuals. Nonrespondents should have different valuations of the interaction with the interviewers, but their reactions cannot be measured, by definition. Dillman (1978) and others interpret several design decisions as likely to be universally valued by the sample person—for example, a personalized advance letter or commemorative stamps on mailed questionnaire envelopes. Such features, the argument goes, activate the exchange influence and yield more cooperation from sample persons.

Goyder (1987) notes several features of the survey interview atypical of other exchange relationships. The exchange in a survey is brief (extended somewhat by an advance letter or repeated callbacks) and thus probably relies on an "existing, natural, pre-conditioning and normative structure" (p.176). There is typically no opportunity to incorporate the survey request into a larger set of interactions between the researcher and sample person. It does suggest, however, that attempts to do so might be interpreted by some as enriching the relationship. For example, Dillman argues that follow-up of nonrespondents to seek their cooperation can itself more powerfully evoke exchange obligations by rewarding the nonrespondent with more attention from the researcher (Dillman, 1988). Those follow-ups may be effective for some reluctant respondents because they successfully communicate to the sample person the importance that their participation is given by the researcher.

Another unusual feature of the exchange relationship in a survey context is the imbalance of power between interviewer and respondent. The interviewer initiates the interaction always; the interviewer often clearly communicates that repeated efforts at contact will result from an initial polite refusal; the interviewer clearly intends to set the agenda for the interaction. This asymmetry may make attempts by the researcher to increase the benefits of cooperation patently manipulative (e.g., monetary incentives are seen as a token gift to obtain information that is much more valuable). If this is true, sample persons may not interpret the "favors," "gifts," and "acts of kindness" on the part of the interviewer as genuine. The perceived intent of the actions may be to increase cooperation. The exchange principle might then not be invoked.

A constituent concept in the exchange theory sketched above is the value placed on privacy by the sample person. Privacy is defined as "the right of an individual to keep information about herself or himself from others" (National Academy of Sciences, 1979, p.1). In order to make a decision about survey participation, each sample person must balance their right to privacy against the benefits of providing the desired information to the interviewer. From a content analysis of letters to the editors of a sample of newspapers in Canada and the United States, Goyder (1987) shows that issues of privacy are often mentioned in complaints against surveys and censuses. Without obtaining measures of nonrespondents' valuation of privacy rights, however, we cannot know how important they are as influences on cooperation.

Finally, many people believe that low response rates are caused by the society being "oversurveyed." They usually note that the populace has been saturated by survey attempts, and they cite market research surveys and political polls as chief offenders. Some acknowledge that their definition of surveys include telephone sales calls which begin with questions about the household.

There is little evidence assembled to address this at a societal level. It is no doubt true that an increase in the number of surveys over the past 20 years in the United States has accompanied the declining response rates. Whether there is a causal link between the two trends is more difficult to discern. Goyder (1987) in a study of residents of a Canadian town does cite evidence that the proportion of survey requests refused increases with the number of requests received. Underlying this finding is probably the belief that surveys fail to achieve benefits either to their respondents or to some larger population. We might expect that the influence of "oversurveying" might depend on whether sample persons discriminate among different kinds of surveys (e.g., those seeking attitudes on products versus those measuring societal needs).

In summary, sociological discussion of survey participation treats possible effects of the frequency of social measurement on cooperation, of relative positions of interviewer and respondents in the social order, and of the value of privacy to the respondent. These in turn help shape reactions to the exchange relationship possible with the interviewer. Although this is the focus of the literature, there are auxiliary observations about the importance of the saliency of the topic to the sample person. This is a design feature chosen by the researcher, which also probably acts to influence the attraction of the exchange relationship.

The first and most obvious value of these concepts to applied survey methods is that they acknowledge that nonrespondents can have well-founded rationales for not cooperating with a survey request. In contrast to the view that their actions are in some sense based on ignorance, the work attempts to identify costs and benefits of responding from their perspective. What can survey researchers do with this knowledge? The first possibility to pursue would be an attempt to tailor survey design to different groups that differ on relevant values. For example, those judged more "peripheral" to the goals of the survey and to societal norms relevant to survey participation might be given special benefits for their participation or some cost-reducing feature (e.g., shorter interviews). Implicit in this idea is the balancing of nonresponse error with costs of data collection.

The second possibility is to take measurements on the concepts reviewed above. These might be observational or question-based measures on respondent and nonrespondent cases in order to build predictive models of survey participation, useful for postsurvey adjustment (see Chapter 3). The challenge of this approach is determining whether such measures can be made in practical survey conditions.

5.4 PSYCHOLOGICAL ATTRIBUTES OF NONRESPONDENTS

The discussion above reviews various attributes of refusers to surveys, but it is unlikely that any of those attributes are the immediate *causes* of the refusal. Many of the sociological interpretations of survey refusals suggest that the person would be inclined toward certain psychological states which make cooperation less likely. Two literatures in social psychology study behaviors that resemble those of sample persons contacted by a survey interviewer. The first are studies of altruism and compliance. "Altruism" is operationalized as the provision of aid or assistance to another person without a formal request to do so. "Compliance" is the consent to a request for assistance by another. The second are studies of process of persuasion, how people respond to arguments for or against some belief or action on their part.

5.4.1 The Literature on Altruism and Compliance

The literature on compliance has explored a wide variety of behaviors from charitable contributions to consenting to donate bone marrow. Table 5.11 provides an idea of the scope of activities that have been conceptualized as helping behaviors. Many of the experiments measure only the verbal consent to help the requestor, not the actual provision of the assistance. Many of the experiments use college students enrolled in psychology courses as subjects, with no concern about possible variation across subgroups in helping behavior.

The diverse types of helping behavior studied make it difficult to determine the best way to apply the findings to the situation of the sample person who is asked to complete an interview. Using ratings from subjects who were students in the behavioral sciences in Australia, Amato and his colleagues (Pearce and Amato, 1980; Smithson and Amato, 1982; Amato and Saunders, 1985) have explored the dimensions of similarity and dissimilarity of various helping situations. Using a factor analytic approach, they distinguish "planned, formal" versus "spontaneous, informal" help; "serious" versus "nonserious" help; "giving, indirect" versus "doing, direct" help; cognitive familiarity with the situation (how uncertain the subject is about what to do) and personal versus anonymous help. On these dimensions, granting an interview to a researcher is viewed as a relatively spontaneous, nonserious, anonymous act. Among the acts studied by social psychologists, the interview is seen as closer to "giving what change you have to a student who approaches you in a library asking for 50 cents to make some xerox copies" or "breaking up a fight between two college students" than to "donating bone marrow to a seriously ill person (a stranger)." Thus, the utility of the experimental research to the survey researcher is probably highest for those investigating requests for research interviews, granting more involvement in an experiment, and signing a petition. Other experiments, for example, those concerned with blood donations, might be less applicable.

This problem of generalizing from experimental to natural conditions exists for both the literatures on altruism and compliance and that on persuasion. The context of persuasion experiments is often laboratory environments with the subject reading a text arguing a specific position on some issue. Thus, the survey researcher must make judgments about whether the concepts found influential to subjects' decisions in those settings have relevance to the survey setting.

One way to assist in identifying the most useful concepts is to organize them into higher-order concepts and then logically test their applicability in a wide variety of settings. The work of Kahneman and Tversky (1971, 1974), describes a set of cognitive procedures which allow

Table 5.11 Some Activities Investigated in Psychology of Helping Behaviors

Behavior	Authors
Contributions to charitable organizations	DeJong and Funder (1977), Jackson and Latane (1981), Benson and Catt (1978)
Signing a petition	DeJong (1981)
Volunteering for a research project	Balson et al. (1979), Thomas and Balson (1981), Hill et al. (1979), Jones (1970), Horowitz (1968), Thompson et al. (1980)
Signing one's name	Harris and Meyer (1973)
Giving a ride to someone	Shotland and Stebbins (1983)
Helping someone who dialed a wrong number	Holahan (1977), Korte and Kerr (1975), Shotland and Stebbins (1983), Boice and Goldman (1981)
Correction of overpayments by store clerks	Korte and Kerr (1975)
Mailing of "lost" letters	Korte and Kerr (1975), Forbes and Gromoll (1971), Benson et al. (1976)
Giving money to a stranger	Merrens (1973), Kleinke (1977)
Granting an interview in a public place	Korte et al. (1975)
Granting an interview at home	Freedman and Fraser (1966), DeJong, (1981)
Helping a person in physical distress	Tice and Baumeister (1985), Shotland and Heinold (1985)
Providing directions to a stranger	Merrens (1973), Korte et al. (1975)
Giving a stranger change of money	Merrens (1973)
Returning a dropped object	Korte et al. (1975), Strenta and DeJong (1981), Cunningham et al. (1980)
Guarding packages	Uranowitz (1975)
Consenting to donate blood	Zuckerman et al. (1977)
Consenting to donate bone marrow	Schwartz (1970)
Helping another experimental subject	Schopler and Matthews (1965), Goodstadt (1971), Brehm and Cole (1966)
Helping a problem customer in a store	Schaps (1972)
Reading to the blind	Schwartz and Howard (1980), Clark (1976)

humans to make quick decisions based on insufficient information. These procedures, relying on what they call "heuristics," many times serve their users well. As Cialdini (1984) notes, many of the notions of heuristics apply to decisions to cooperate with a request from salespeople, advertisers, waiters, and others seeking some action from us. Cialdini organizes the influences on compliance into six different concepts:

1. **Reciprocation**, the tendency to favor requests from those who have previously given something to you.

2. **Commitment and consistency**, the tendency to behave in a similar way over situations that resemble one another.

3. **Social proof** or behavioral norms, the tendency to behave in ways similar to those like us.

4. **Liking**, the tendency to comply with requests from attractive requestors.

5. **Authority**, the tendency to comply with requests endorsed or given by those in positions of legitimate power.

6. **Scarcity**, the tendency for rare opportunities to be more highly valued.

These influences have a tendency to be overused by the requestee and abused by the requestor to achieve his/her ends. For example, with regard to social proof, consumers may incorrectly judge that a popular automobile is well-built merely because others have purchased it. The mistake is that popularity may not be based on quality but on price. Conversely, advertisers can attempt to evoke the use of social proof by staging testimonials by "average" people about the quality of the product. They hope to influence a judgment that many people have carefully evaluated the product and found it superior. Hence, the viewer can be spared the burden of a detailed evaluation.

Reciprocation. Reciprocation has its clearest applicability in incentives that are offered sample persons in surveys. The research involving reciprocation suggests circumstances under which survey incentives are likely to be more effective. An underlying premise of reciprocation effects is that of a social norm requiring that acts of assistance or kindness be answered in kind. This appears conceptually distinct from, for example, the relationship between employer and employee, whereby the grant of a wage is cause for the worker to offer his/ her labors. Those arrangements are more immediate in their initiation (once the job contract is made, the relationship is defined), specific in their extents, and limited in time. Hence, the giving of money to sample persons prior to the interview may evoke a different set of behaviors than, for example, a gift for the home, of a nature similar to that brought by visiting guests. This may explain some of the variation in effects of different kinds of incentives.

Reciprocation has been studied with a broad set of operationalizations. For example, one experiment (Jackson, 1972) placed a student subject in a situation in which he was asked to collaborate with another student subject to complete a particular task. The other subject was in fact a stooge of the experimenter. For a random half of the subjects, at one point during a break in the experiment, the stooge subject left and returned with two soft drinks, one for the himself, one for the subject. The stooge noted that the experimenter allowed him to buy one, and that he decided to bring another for the subject. For another random half of the subjects, the stooge returned empty handed. In both half-samples, after the experiment was over the stooge subject mentioned that he was selling raffle tickets and would receive a prize if he sold the most. Each ticket cost $0.25. More tickets were purchased by subjects receiving the favor than by those in the other half-sample.

Reciprocation effects are likely to fail if the subject perceives that the intent in providing the gift is to encourage the reciprocation. For example, one could speculate that if the subjects in the Jackson experiment above were given full information about the experimental design, ticket sales might not have been higher among the group offered the soft drink. The effect of the reciprocation is enhanced when the initiating act of kindness is not directly linked to the later request.

This has implications for the performance of incentives in surveys and the performance of all sorts of interviewer behaviors that might invoke the reciprocation effect. Most survey incentives are directly linked to the survey request. Sometimes the offer of the incentive is made directly following the request for the interview. The experimental literature suggests that research might be conducted to check on the heightened effects of incentives when they are separated (in time and nature) from the interview request. The separation in time of the incentive and the request has cost implications for the research. It demands some prior contact or prior delivery of a gift before the visit of the interviewer. The separation in kind or nature implies that the incentive should not remind the sample person of the later request for the survey. For example, for some populations (e.g., the elderly) gifts which commonly accompany a visit of a friend might be used on a first visit in order to seek to activate the reciprocation norm.

Current thought in ethnographical interviewing is relevant to concepts of reciprocation (see Werner and Schoepfle, 1987). Ethnographers often use another member of the culture to introduce them to a person they wish to interview, utilizing the established relations of reciprocation to benefit the research. The request for the ethnographic interview is thus made by a friend of the subject, on behalf of the ethnographer. Such techniques are clearly more difficult when the sample is randomly selected and when there are confidentiality concerns. It is also obvious that the technique requires longer field periods. There seems to be little use of the strategy therefore in household survey settings.

Commitment and Consistency. The concepts of commitment and consistency are used to describe how persons value their past behaviors in making decisions about future behavior. A large set of experiments in the so-called "foot-in-the-door" procedure claim that if a person consents to a small initial request, the likelihood increases that they will consent with a larger request given later. The seminal work, for example, of Freedman and Fraser (1966) initially asked some persons to sign a petition promoting safe driving and followed the request with one for display of a large sign in their front yard which said "Drive Carefully." They found that those receiving the first request had higher compliance rates with the second larger request than a control group which received only the second request.

There is a relatively large set of experiments of similar type which have varied the relative sizes of the two tasks requested, the length of the time interval between the tasks, and the similarity of the two tasks. These factors have been found to alter the experimental results. Furthermore, most have been seen to do so by affecting whether the subject does indeed recall his/her past behavior as relevant to the second request. This gets directly at the notion of whether the first request successfully induces self-perceptions of being a person who would comply with the second request.

Groves and Magilavy (1981) note that some methodological aspects of the foot-in-the-door literature reduce its applicability to survey nonresponse: (1) often only one attempt was made to reach subjects, thus not exposing the technique to persons difficult to reach; (2) most experiments dropped from the second request those who did not comply with the first request; and (3) no persuasion techniques were combined with the experimental conditions, as would be common in a survey setting.

An attempt by Groves and Magilavy to test the foot-in-the-door technique in a telephone survey setting failed to show gains of a two-contact procedure over a one-contact procedure. However, the design was flawed by a failure to contact the same household member on some of the two contacts. It also used a short two-question task at the initial contact. An unresolved issue is whether the two-question interview was sufficient to evoke the self-perception of one who cooperates with survey requests. Finally, there was a lack of control over the length of time between the initial request and the second request.

Social Proof. The influence of "social proof" on compliance with requests might be thought of as a proxy form of consistency. If we learn that others who resemble us have chosen to comply with the request, we are more likely to comply ourselves. The converse is that if others like us fail to act, we tend toward inaction.

The influence toward inaction is related to a large set of studies examining whether observers come to the aid of victims of crimes, heart attack victims, or others in need of assistance. The helping experiments that appear most relevant to survey research involve nonemergency situations (e.g., requests for aid to a researcher). There *is* one result from the experimental literature on helping in emergency situations that might have some relevance to survey response rates. For example, Latane and Darley (1968) and Shotland and Heinold (1985) found that helping among bystanders to an emergency decreased in proportion to the number of bystanders present. One rationale for the result was that persons present felt no *unique* ability or responsibility to provide help. These traits were shared with several other persons, each of which could provide the help. In contrast, Schwartz and Clausen (1970) note that when such unique capabilities are induced in the group, for example, when some members have medical training to help the victim, they are not affected by the size of the group as others are.

We might generalize the concepts above to apply them to the survey context. If the sample person feels no unique abilities to provide the survey information, if they perceive that there are simple substitutes who could provide the assistance the interviewer seeks, then they might be less inclined to cooperate with the survey request. One anecdote and one empirical result support this interpretation. Years ago at the Survey Research Center, an interviewer consistently obtained higher response rates than others in the same urban territory. In an effort to discover the secret of his success, the field director visited the interviewer and quizzed him about his techniques. The interviewer noted that his secret was simple. After describing why he was calling on the house but before asking for the interview itself, he asked every household, "Do you have children living with you?" If the person said, "yes," the interviewer replied quickly, "Then we want to interview you for our research." If the person said "no," the interviewer said exactly the same thing. Apparently, the identification of a specific characteristic of the sample person as the reason for diffuse responsibility to other possible substitutes is limited by such a rationale for the request.

A similar underlying variable may explain a result in Groves and Kahn (1979) of higher response rates for a telephone sample design that required the screening question, "What county do you live in?" prior to the interview request. The sample design that used this procedure was based on a stratified multistage area probability sample that yielded a self-weighting national sample of household telephone numbers. When compared with an RDD national design that did not require the screening question, it received higher response rates. Perhaps the exclusivity implied by the screener question addressed the sample person's concern about why his/her telephone number was selected.

Interviewers are often instructed to tell respondents that the survey organization knows nothing about them, that the household was chosen at random. This message sometimes yields the response, "Why don't you call someone else then?" The diffusion of perceived responsibility implied in this response might follow the reasoning that the whole community shares the responsibility of providing information for the research and that the chosen respondent has no unique abilities or information to provide. Others are easy substitutes. The interviewer should seek them out. The rebuttal by the interviewer that the tenets of probability sampling do not permit such substitution are complex messages that are rarely understood fully by the interviewer and less rarely believed by the reluctant respondent. In reaction to this problem, rationales for response might identify specific characteristics of the sample persons in an attempt to decrease attempts to diffuse the responsibility to others.

The influence of the concept of "social proof" could work in other ways in surveys as well. The influence is another shortcut to careful consideration of a request and is based on the belief that the others either know something we do not or have themselves considered the costs and benefits of compliance and have chosen to comply. Those outside survey research are often surprised when they learn that a majority of sample persons do agree to be interviewed and that for some surveys over 90 percent of the sample agrees.

There appear to be few practical uses of this information to benefit response rates. Some interviewers are trained to note that "most people enjoy doing the interview," as a way to encourage cooperation. There does not seem to be frequent instructions to interviewers to tell people that the majority of the sample does indeed give the interview. Experiments regarding the effect of this information would be interesting in the light of the experimental evidence.

Liking. The influence of "liking" on compliance demands some prior exposure of the subject to the person making the request. Subjects having positive feelings toward the requestor tend to comply. Related to liking, however, are effects of physical attractiveness and similarity to the subject. These can be based on more immediate reactions by the subject upon first seeing the requestor.

In experiments using the "lost letter" technique, envelopes containing an application to graduate school, together with a picture and a stamped envelope, have been left in public spaces (e.g., telephone booths, public benches) (see Benson et al., 1976). Relatively more applications with pictures of persons judged to be attractive were returned. The inference from these experiments is that people tend to help more attractive persons.

Sometimes effects of liking occur by merely associating the request with an attractive person. This is the ploy used by advertisers using movie stars or athletes to promote their products. They hope that the liking of the person will lead to more purchases of the product.

Are any of these findings relevant to cooperation with surveys? There do not appear to be any formal studies of the effect of interviewer attractiveness on cooperation rates. There is a literature on the effects of "rapport" on the quality of data provided in interviews (Weiss, 1968; Henson et al., 1976), but apparently not a corresponding literature on effect on cooperation. Indeed, one possible fear of manipulating survey characteristics related to liking or attractiveness is that the same variables may affect the quality of data. Improved cooperation may come at the cost of heightened effects of interviewer on survey responses.

Authority. "Authority" in this context concerns the perceived knowledge of the requestor, the legitimacy of him/her to make the request, and the perceived power the requestor has through money or prestige. Requestors who are such authority figures tend to obtain higher compliance. The authority of a requestor is conveyed through job titles, educational titles, dress, and mannerisms. The concept of authority demands certain limits. An authority figure to one person may not be an authority figure to another. Roles and social context affect the extent of authority of any person.

Survey researchers use this principle frequently. Advance letters are sent to sample households signed by the director of the institute, the head of the funding agency, or a well-known government official. In pretests of a survey on the prevalence of the AIDS infection, interviewers carried videotape players to show a video of the Surgeon General urging their cooperation.

Scarcity. "Scarcity of opportunity" refers to the fact that some requests cannot be repeated at a future time. If the subject does not act positively at the time the request is made, there may be no opportunity to do so in the future. Cialdini (1984) points out how this influence is used in retail sales—"This is the last red full size Chevrolet that I have left"—as a way to heighten the attraction of the object being marketed.

These concepts help explain some naturally occurring behavior among survey interviewers. Interviewers sometimes change their persuasion techniques in the last days of a survey period, when they are attempting to close out the final cases: "There are only two more days left in the survey period; if I don't talk to you now, you may not be able to participate at all." The interviewer is applying the notions of scarcity to enhance the attractiveness of cooperation at the time of the call. The tactic is probably most effective is the absence of describing to the respondent exactly what the interview will entail.

The influence toward cooperation from "scarcity of opportunity" is countered by another influence against cooperation in such settings: reactance (Brehm and Cole, 1966). Reactance theory notes persons value the freedom to choose their own course of action in any situation. One seeks to maintain the option of refusing to engage in an activity. External pressure to provide help (e.g., that applied by an interviewer) thus might induce both tendencies to comply with the request *and* simultaneously a resistance to the reactance across different experiments. Jackson (1972) had the experimenter provide an unsolicited favor (e.g., a soft drink) to the subjects prior to asking for their assistance. In this experiment those not receiving the favor showed more sensitivity to the needs of the researcher (i.e., the importance of the request as seen by the researcher) than those receiving the favor. That is, there was an interaction effect between the researcher's dependency and the providing of a favor prior to the request. If the subject perceived some pressure to help the researcher, cooperation was not affected by the level of need of the researcher (it was uniformly higher/lower than the group that received no favor). In another experiment, Batson et al. (1979) sought more help for a second experiment from subjects. Half of them were encouraged by a confederate to do so because "the guy really needs help"; the other half because it "sounds like we're being volunteered...I guess we really don't have a choice." All subjects in both groups performed the second experiment. Those in the latter group, however, subsequently volunteered at a lower rate than the first group to still another request for volunteer work at a local service agency. This is an example of complete cooperation with a request, preceded by attempts to force a rationale for action that has subsequent effects on helping. The interpretation of the findings are consistent with reactance theory—that persons seek to avoid the loss of freedom to choose, the feeling of being a compliant person. Those induced to feel this way were more reluctant to provide further help to the researcher.

One problem facing the survey researcher and examined by the psychological experiments is whether the rationale for cooperating with the survey request should be eliciting a positive state or avoiding a negative state. For example, Cunningham et al. (1980) asked for contributions for the "World Children's Fund" either "to keep the children smiling" or because "you owe it to the children." The positive request generally obtained more support than the negative approach. Only in a condition when the subject was induced to feel guilty and there was no other manipulation to induce a positive mood did the negative request appear to be superior. Benson and Catt (1978) in a simpler design (albeit with other experimental variables) found that a United Way appeal among a mixed-race lower to middle class neighborhood was more successful when the subject was told "it would make you feel good to help those less fortunate than yourself" than "it's really your responsibility to help those less fortunate than yourself." This effect was magnified when it was noted that the funds would benefit "innocent victims" versus those who "have themselves to blame." More loosely interpreted, the rationale that "it would make you feel good" is effective when recipient of aid really deserves it. A different interaction was obtained by Horowitz (1968). Some student subjects were told that they had to participate in another experiment to get credit; others were told it was optional. They were later asked how much time they were willing to give for the second experiment, from merely writing a few comments on the experiment to 8 hours in a sensory deprivation room. Those given a free option volunteered more time than those without a choice. But in contrast to the Cunningham et al. work, the freedom to choose was more important when the "victim" they were helping was himself responsible for the difficulty he was facing.[7]

The application of reactance theory to survey requests could be explored by an experiment which offers explicitly the rationale that the sample persons "would feel good" if they "helped" the interviewer. Such a rationale would be contrasted with one that notes the "importance of the

---

[7] Here it should be noted that, in one sense, all subjects volunteered to participate in the experiment to some extent. The dependent variable measures the extent of their volunteered participation. For application of the results to the survey environment, one must decide which experimental task resembles most the survey interview.

research" or how "every sample person must cooperate for the research to be a success."

Reactance effects appear to arise when the cause of the limitation on choosing to cooperate is the requestor himself. When the opportunity to cooperate will disappear shortly because of some external force (e.g., someone else buys the last remaining car), reactance effects appear less likely to arise.

Other Psychological Explanations for Compliance. The consistent finding that those living in large metropolitan areas produce lower response rates for surveys leads researchers to ponder about what psychological influences might be related to that behavior. One attribute that has been offered as a common cause of nonresponse behavior is fear of crime. The argument is presented that the survey interview involves the attempt to gain access (either through a telephone call or through a personal visit) to one's house. The attempt is made by a stranger, a person whose contact was not elicited by any household member, a person who is not fulfilling a request for a service of some kind for the house, someone who seeks access for his/her own purposes, for a "research project" sometimes only vaguely described. Among sample people whose personal contacts with strangers are infrequent, such requests might be met with reluctance and even fear that the interaction might produce physical harm, property loss, or at the very least some unpleasantness or embarrassment.

House and Wolf (1978) examined response rates for the National Election Studies between 1960 and 1972. The unit of analysis of the work was the primary sampling unit (PSU), about 100 counties or groups of counties from which sample households are drawn. As the reader would expect from examining the Steeh (1981) data, House and Wolf find that the large urban PSUs consistently achieve lower response rates than do the smaller PSUs. They attempt, however, to examine what properties of the large urban areas might be connected with the lower response rates. Table 5.12 presents the results of multiple regressions with response rate as the dependent variable. The analysis starts with the relationship between two variables, type of place of residence and year of survey, as explanations of response rates. "Type of place of residence" is an eight category variable:

1. Old, large, high–density city area.
2. Old, large, lower density city area.
3. Old, intermediate city area.
4. New, large city area.
5. New, intermediate city area.

Table 5.12 Increase in R^2 Associated with Different Predictors in Bivariate and Multivariate Regression Models Predicting Refusal Rates, 1960 to 1972

Predictor	R^2	Population	Density	Increase in R^2 Due to Predictor After Controlling for					
				Nonwhite (%)	Foreign (%)	Violent Crime	Property Crime	Total Crime	Total Crime and Density
Place of residence	0.195	0.094	0.088	0.198	0.182	0.090	0.067	0.064	0.043
Year	0.072	0.066	0.070	0.072	0.086	0.023	0.011	0.012	0.019
Place, year	0.258	0.154	0.150	0.264	0.257	0.124	0.088	0.084	0.069
Place, year, and place × year interaction	0.324	0.220	0.215	0.332	0.317	0.181	0.147	0.141	0.132
Interaction, net of additive effects	0.066	0.066	0.065	0.068	0.060	0.057	0.059	0.057	0.063

Source: J.S. House and S. Wolf, "Effects of Urban Residence on Interpersonal Trust and Helping Behavior," *Journal of Personality and Social Psychology,* Vol. 36, No. 9, 1978, pp. 1029–1043, Table 3.

6. Small city area.
7. Small town and rural area.
8. Rural area.

The second column of Table 5.12 shows the square of the multiple correlation coefficients for various combinations of year and place of residence. Year, by itself, explains about 7 percent of the variance in response rates at the PSU level; place of residence about 20 percent. The two variables jointly (main and interaction effects) explain about 32 percent of the variance in response rates. Of interest to House and Wolf was how much of that explanatory power could be associated with crime–related variables. Table 5.12 shows results from matching crime rates for police jurisdictions to the corresponding sample areas (the match was not perfect because of boundary differences). The last column shows the increase in squared multiple correlation coefficient associated with year and place of residence, when total crime rates and population density are already in the regression model. It is uniformly the case that the explanatory power of the place of residence variables is greatly reduced in the presence of the crime rate predictors. That is, the effects of place of residence are "explained" by variation in crime rates and population density. This finding is viewed by the authors as support for the identification of fear of crime as the cause of nonresponse behavior.[8]

5.4.2 The Psychology of Persuasion

Another view of the decision to participate in a survey notes that prior attitudes about surveys, survey organizations, the topic of the specific survey, and survey interviewers may affect the likelihood of cooperation. These attitudes represent "predispositions" to act in one way or another. Improving cooperation may involve either the refutation of negative attitudes or the invoking of positive attitudes.

Recent research (Petty and Cacioppo, 1981, 1986) has shown that full cognitive processing of the arguments of the advocate (in our case, the survey interviewer) occurs in some situations but is absent in others. In the latter case, acceptance or rejection of the arguments flows from reaction to cues which lie outside the direct arguments being presented. The first mode has been called the "central route" to persuasion because it

---

[8] The reader will note that fear of crime was not measured in the analysis, only crime rates. Thus, the use of the data to argue that fear of crime is the cause of nonresponse behavior goes beyond the analysis itself although it is completely compatible with the findings.

results from the careful consideration and integration of the various arguments. The second is labeled the "peripheral route" since it avoids direct address of the arguments.

Under some circumstances, perhaps surveys of obvious importance but entailing large burden on the part of the respondents, central route processing may be prevalent. In those cases, some identification of the likely counterarguments may be useful. It has been repeatedly shown experimentally that refutation of counterarguments prior to the persuasive message heightens agreement with the message (McGuire, 1969). If central route processing is being used by sample persons, then refutation of those arguments counter to survey participation may lead to greater participation when the survey request is actually given.

In some circumstances, for example, rapid decisions to refuse telephone surveys (see Chapter 11), it is likely that peripheral route processing is being used. Reactions to sounds of the interviewer's voice (see Oksenberg et al., 1986), to individual words used, and in the face to face situations to the dress and appearance of the interviewer, might be cues to specific attitudes of the respondent. On the basis of those attitudes, the decision to participate is being made (Chaiken, 1987). In these cases, the authority and attractiveness of the interviewer may affect compliance. Alternatively, attitudes based on past experiences with surveys may be salient (Bem, 1967) or the shear number of arguments for participation may be gauged (Petty and Cacioppo, 1984).

In addition to failure of the recipient to want to consider strong persuasive arguments (i.e., those changing attitudes when they are fully processed), there are other circumstances in which they fail. Distractions during the persuasion attempt may occupy respondents so that full attention to the arguments cannot be given. If the arguments being presented are strong ones, distractions reduce attitudinal change. If they are weak ones, easily refuted by the respondent, then distractions may actually help attitudinal change.

There is also evidence that if the arguments are presented by several sources independently (Harkins and Petty, 1981), there is greater motivation to process the messages. One interpretation of this is that the respondents are more highly motivated to determine the reasons for support by multiple others. Key to the effect appears to be the fact that the multiple sources are independent of one another. Almost by definition, conveying a sense of independence among sources of the persuasive message is difficult in a survey setting. One example concerns the argument that an ongoing survey is important to the society. Interviewers in such surveys often carry with them copies of newspaper articles giving the results of the survey, an independent demonstration that the data are of wide interest or use.

5.5 SUMMARY

Is nonresponse error a major problem in U.S. household surveys? We have now seen that nonresponse *rates* are increasing and that certain subgroups tend to refuse requests for survey interviews. There appears to be little documentation, however, for different kinds of surveys of the levels of nonresponse error (i.e., the combined effects of nonresponse rates and the differences between nonrespondents and respondents). The absence of external information about nonrespondents is the insurmountable hurdle to routine measurement of nonresponse error.

Nonresponse rates vary across groups defined by urban/rural status, gender, and age. There is evidence that socioeconomic status and race are related to response rate on some survey topics but not others. Nonresponse appears to vary by various survey characteristics: length of the interview, topic of the survey, and number of callbacks on sample households.

Survey researchers have used a variety of tools in attempts to improve response rates. These include advance notification of the survey request, financial and other incentives to cooperate, and various persuasion attempts. There is some evidence that different techniques are successful for different parts of the household population. Many organizations use multiple methods at the same time to increase response rates. All these methods have impacts on the cost of the data collection effort.

Every attempt to reduce nonresponse error through increasing nonresponse rates or postsurvey adjustment requires some model of the process by which decisions are made to participate in surveys. This is true of advance letters sent to sample units (e.g., hypotheses about the importance of establishing legitimacy through means other than the interviewer). It is also true of postsurvey weighting adjustments (e.g., hypothesizing constancy of values of survey variables within weighting classes). To be realistic these models should reflect the causes of positive and negative decisions to participate in the survey.

In other words, realistic models for statistical adjustment and survey administration require theories of survey participation. As we have seen in this chapter, theories of human behavior related to survey participation could come from sociology, from social psychology, and from cognitive psychology. The sociological explanations describe how groups outside the mainstream culture which is sponsoring the research may judge that the costs of their participation are not balanced by benefits to them either directly or indirectly. They frequently utilize notions of exchange but note the relative powerlessness of the sample person to affect the behavior of the interviewer.

The psychological literature is mainly experimentally based. Its conclusions flow from changes of cooperation obtained when attributes of the requestor, the task, benefits given to the subject, and aspects of the environment are varied. They focus on what influences one person to aid another, to comply with a request of another, or to agree with the persuasive efforts of another. This literature has not frequently been used to motivate nonresponse rate reduction efforts or the form of postsurvey adjustment. It studies, however, processes that are similar to those of the survey situation.

There has been an unfortunate lack of application of these concepts to the study of survey participation. The application requires the development of indicators of these concepts in practical survey settings. A test of these theoretical concepts requires measurement of both respondents and nonrespondents. Development of practical nonresponse adjustment models using these theories demands the same. The greatest challenge is the development of measurements that can be taken on nonrespondent cases. These must be based on observations made by interviewers, data collected from another source, or answers to questions on special nonrespondent subsamples, utilizing the double sampling designs described in Chapter 4.

Survey research with few problems of nonobservation requires a special set of societal conditions: a population that can be contacted in a reasonable time period, a population measurable using a small set of common languages, a population willing to discuss a wide range of subjects with strangers, and a population trusting pledges of confidentiality of the answers they supply. In the United States and several Western European countries, some may question whether these conditions still apply. It is more imperative than ever that survey researchers begin to understand the influences on a person's decision to participate in a survey, both to improve cooperation and to use the knowledge to qualify conclusions from surveys subject to nonresponse.

CHAPTER 6

COSTS AND ERRORS ARISING FROM SAMPLING

Just as an experienced cook can tell how a large pot of soup will taste by trying only a spoonful, it is possible to learn the opinions of a very large group of people by talking to a small number of them.

Why We Ask You, *Survey Research Center booklet for sampled persons*

6.1 INTRODUCTION

This chapter is unlike others in this book because it can appeal to a set of well-founded principles of error measurement and reduction. Many fine sampling texts cover this material in greater depth than will be possible in this chapter (e.g., Hansen, et al., 1953; Kish, 1965; Cochran, 1977; Wolter, 1985). There is no attempt to duplicate that material, and readers desiring a thorough description of sample design and estimation should review those texts. Instead, this chapter offers a treatment of sampling error which parallels the treatments of other errors discussed in this book. We begin by describing the nature of sampling error in surveys, then identify its causes, review the procedures used to reduce sampling error in surveys, and finally discuss methods of cost modeling and sampling error modeling that permit the minimization of sampling error for a survey of specified cost. Whenever possible we reinforce the perspective of other chapters. Above all, we repeatedly note the ways in which sampling error estimates can be gross underestimates of the total error of survey statistics.

239

6.2 THE NATURE OF SAMPLING ERROR

Sampling error is an error of nonobservation. Survey estimates are subject to sampling error because not all members of the frame population are measured. If they were, sampling error would be eliminated. The notion of sampling error generally includes the possibility that the entire selection process that produced the sample at hand could be repeated, yield a different set of sample persons from the same population, and hence produce different estimates of the frame population's characteristics.[1]

Most work in sampling error concerns sampling *variance*. An estimator is subject to sampling variance to the extent that its achieved values would vary over different samples drawn with the same design. Sampling *bias* exists to the extent that the estimates from all the samples of a given design are subject to the same departure from the frame population parameter. This occurs when a distinctive subset of the population is given no chance of selection.

In the notation of Chapter 1, sampling variance of a survey statistic, y, can be described as average squared deviations of individual sample values of the statistic and its own average value:

$$E[y_r - E(y_r)]^2 \, ,$$

where y_r is a sample statistic on the rth distinct sample of the sample design (r, for realization of the design) and $E(y_r)$ is the expected value of y_r over all samples of the given design. The sampling variance is thus a feature of a distribution over all possible samples that could be drawn with a particular design. Each observation in that sampling distribution is the result of one sample of the given design. Figure 6.1 presents a sampling distribution of a correlation coefficient, $\hat{\rho}_{xy}$, between two survey variables. This particular statistic is *biased*; that is, the average value over all samples is not equal to the population value of the correlation. In notation, this is

$$E(\hat{\rho}_{xy}) \neq \rho_{xy} \text{ or } [E(\hat{\rho}_{xy}) - \rho_{xy}] \neq 0.$$

The variance measuring the dispersion of this distribution is the sampling variance of the sample estimate of the correlation. The set of $\hat{\rho}_{xy}$ values

---

[1] The frame population is the only population relevant to sampling error, given our definition of coverage error as arising from discrepancies between the frame population and the target population (see Chapter 3) and the specification of the sample as a subset of the frame population.

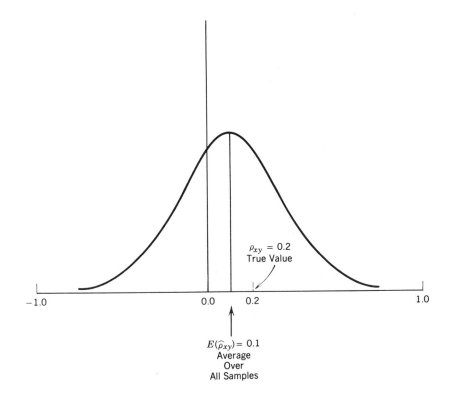

$\rho_{xy} = 0.2$
True Value

-1.0 0.0   0.2   1.0

$E(\hat{\rho}_{xy}) = 0.1$
Average
Over
All Samples

Figure 6.1 Sampling distribution of correlation coefficient estimated on a sample.

that appear in the sampling distribution are all possible samples (say, $r = 1, 2, ..., R$) that could be drawn given the design.

Why does sampling error exist? The heterogeneity of the frame population itself produces the variation over samples in estimated values. If the population were homogeneous, if all the frame population members shared the same traits, then all samples (of any size) would yield the same value for the survey statistic. It would not be subject to sampling error. Sampling error therefore arises because sample units (ultimately people in household surveys) are different from one another and because only a subset of the population is measured in a sample survey. The statistic of interest for the researcher (e.g., a sample mean, a regression coefficient) will vary in value over different possible samples of persons, given the sample design.

There are thus three types of distribution that should be kept conceptually distinct when considering sampling error. The first is the distribution of variable (the characteristic to be measured in the survey) in the frame population, the **population distribution**. This distribution is unknown to the researcher but does exist and contains observations for all members of the frame population. Figure 6.2 presents the population distributions for X and Y (two variables to be measured in the survey). It also shows the joint or bivariate population distribution of X and Y. Population distributions of elements form the first type of distribution; they have N points, for each of the N elements in the frame population. Sample surveys draw a set of points from these distributions; these are the samples of size n, based on a particular sample design.

The second type of distribution is the **sample distribution**. Imagine one of samples which could be drawn given a particular sample design. Call it the rth sample. Figure 6.3 presents three sample distributions, each based only on the n sample elements from the rth sample. They mimic the corresponding population distributions of Figure 6.2, but they are based on smaller numbers of elements. Indeed, the more each of the possible samples from this design (recall that Figure 6.3 presents only one of a possible large number) resembles its corresponding population distribution, the lower the sampling error of a related survey statistic.

The third type of distribution is the **sampling distribution of a sample statistic**. Sampling error concerns the variability of values of statistics over the different samples that could be drawn. Figure 6.4 shows the sampling distribution of the sample mean of x and y and the sample-based correlation coefficient. These are just three statistics which might be computed using x and y. Sampling distributions correspond to specific survey statistics. Sampling error is a measure of the dispersion of this distribution. Sampling variance measures average squared deviations about the expected value of $\hat{\rho}_{xy}$. Note that sampling variance *does not* measure differences, for example, between the average sample value $E(\hat{\rho}_{yx})$ and the true value ρ_{xy} in the frame population. This is bias in the sample correlation coefficients, an error common to all of them (and thus not measured by comparing them).

6.3 MEASURING SAMPLING ERROR

Sampling variance is the variability of a statistic over all possible samples using the same design, but the vast majority of surveys are conducted once, using only one sample. The challenge to the sample designer is thus to choose one sample in such a way as to be able to measure the sampling variance due to the design. The keys to measuring sampling variance are *randomization* and *replication*.

A *probability sample* is one for which all members of the frame population have a known, nonzero chance of selection. The word

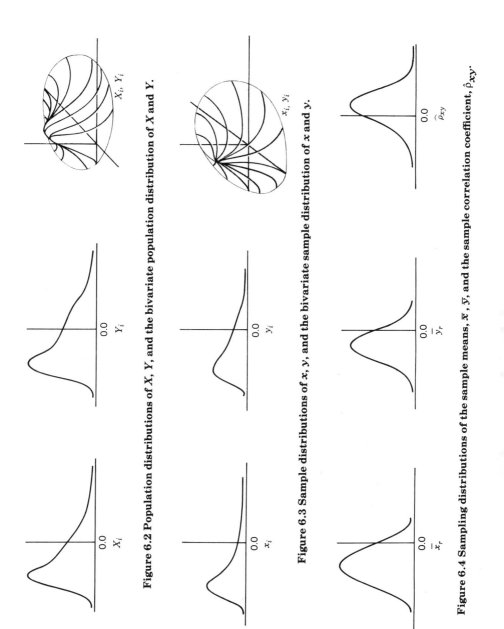

Figure 6.2 Population distributions of X, Y, and the bivariate population distribution of X and Y.

Figure 6.3 Sample distributions of x, y, and the bivariate sample distribution of x and y.

Figure 6.4 Sampling distributions of the sample means, $\bar{x}$, $\bar{y}$, and the sample correlation coefficient, $\hat{\rho}_{xy}$.

243

"chance" in this definition incorporates the notion of randomization. When probability samples are drawn with two or more independent selections, the sampling variance of many statistics can be estimated from only implementation of the design. The "two or more" phrase in the last sentence incorporates the notion of replication. The replication permits comparisons among sample units as a method of estimating the variation over different samples.

In many cases, randomization also leads to a particular sample statistic being an unbiased estimator of the frame population statistic [i.e., that $E(\bar{y}_r) = \bar{Y}$]. This unbiasedness property exists only in the absence of all the other errors that this book discusses. For example, if there is nonresponse among the sample cases, the unbiasedness property does not apply. In addition, if the randomization is not applied uniformly over all frame population elements, sometimes the unbiasedness property is lost (even with 100 percent response rates). This can occur when some elements in the frame are given no chance of selection. This induces sampling bias. For example, some nonprobability samples of households have interviewers in a single visit contact and interview a predefined number of persons having specified demographic characteristics among the residents of sample blocks. Those not at home at the time the interviewer calls have no chance of selection. Sampling biases arising from such practices cannot generally be measured from the data in the sample itself. External information about the population elements that were differentially treated by the design is necessary to assess the impact of their treatment.

To illustrate the estimation of sampling variance of a survey statistic from a single probability sample, we can examine the case of a simple random sample of size n elements. A simple random sample is the result of selecting a fixed number of individual elements from the frame population, one by one, independently, without giving the same element a chance of being selected twice (called "sampling without replacement"). Simple random samples are often selected by numbering the elements in the frame population from 1 to N, and drawing random numbers from the range 1 through N, until n distinct numbers have been chosen. The elements in the frame population assigned those numbers are selected into the sample.

One simple statistic of common interest for a variable y in a survey is the mean:

$$\bar{y} = \frac{\sum_{i=1}^{n} y_i}{n}.$$

The sampling variance of that estimator, under a simple random sample design of size n, is

$$\text{Var}(\bar{y}) = \frac{(1-f)}{n} S^2 ,$$

where $(1 - f) = 1 - n/N$, the finite population correction factor;

n = number of sample elements chosen in the sample;

N = total number of elements in the population;

S^2 = population element variance,[2] $\sum_{1}^{N}(Y_i - \overline{Y})^2/(N-1)$.

Survey analysts more frequently work with the square root of the sampling variance, called the standard error. Note that the population element variance, S^2, is a property of the frame population that was sampled. It is not a property of the sample design. It cannot be altered by the sample design. The formula above means that sampling variance of the sample mean is a function of variability among the selection units and the sample size. If there is great heterogeneity in the frame population (large S^2), a larger sample size will be required to achieve a desired level of sampling error. Regardless of the value of S^2, the larger the sample size the smaller the sampling error. In this case sampling error is a simple function of sample size and heterogeneity in the population. For more complex sample designs (ones involving stratification, clustering, and perhaps unequal probabilities of selection), the functions become more complicated. Complexity in sample design is introduced either to reduce sampling error or to reduce survey cost.

Note that the formula above for the sampling variance of the mean contains only one effect of the population size, the finite population correction factor, $(1 - f)$, where $f = n/N$. This factor is bounded by 0 and

---

[2] Most sampling variance formulas are expressed in terms of S^2, the squared variation divided by $N - 1$, instead of σ^2, the squared variation divided by N, because the sampling variance expressions are slightly simplified using this form. All the expressions could easily be rewritten in terms of σ^2.

correction factor, $(1 - f)$, where $f = n/N$. This factor is bounded by 0 and 1. It always acts to reduce the sampling variance. The factor $(1 - f)$ is close to 1.0 when the sample size is small relative to the population size. This is almost always true in large surveys of the household population. Indeed, most social surveys measure very tiny parts of the population, so the term is negligibly different from 1.0. In practice it matters very little whether the population is restricted to a geographical region or not. For example, a sample of 1500 persons from among the 250,000,000 in the United States would have a sampling error very similar to that of a sample of 1500 from the 1,500,000 people in a large city, other design aspects being similar. Sampling error is relatively unaffected by population size. It is directly affected by sample size.

Finally, note that the standard error or sampling variance above reflects variability that arises because of differences among samples. It does not reflect the two other variable errors of nonobservations we have discussed earlier, coverage error and nonresponse error. For example, standard errors are often used to construct confidence intervals about sample estimates (e.g., the monthly unemployment rate is 3.4 percent plus or minus 0.1 percent). These should be interpreted as measuring variability about the expected value over all possible realizations of the sample design. That expected value might be very different from the true population parameter because of failure to include some persons in the frame (noncoverage) or failure to measure certain kinds of people when they fall in the samples (nonresponse). Furthermore, to the extent that the different surveys on the same sample may vary in the amount of noncoverage (because of difficulty of listing a housing unit in the frame) or nonresponse (because of variation in the reluctance of sample persons to comply with the survey request) or measurement error (because interviewer differences affect respondent behavior), the estimated confidence interval may be too narrow. That is, the standard error measures variability due to sampling different elements, but not variability due to differing nonresponse or noncoverage over samples. This means, in short, that the sample estimate can be biased due to the nonsampling error, and the standard error is usually an underestimate of total variability about the expected value over all samples.

6.3.1 Nonprobability Samples

Not all sample surveys use a probability selection process where each element of the frame has a known, nonzero chance of selection. Some designs have selection rules that exclude certain kinds of persons purposely but control the mix of types of persons who can enter the sample. These were formerly called "quota samples," although the phrase "model-based samples" is becoming more common.[3] These methods have in common one of two concerns on the part of the designer: (1) attempts to limit the cost of the data collection or (2) attempts to limit nonsampling errors at the point of the selection process. One difficulty with probability selection processes is that once the selection takes place, the interviewer must commit all efforts to the selected persons. Some of these can be difficult to contact, reluctant to cooperate, or physically or mentally limited in their abilities to respond. Large costs are spent attempting to get measurements on such respondents. Furthermore, consistent with point (2) above, some sample persons are never interviewed, and thus probability samples may disproportionately miss certain kinds of sample persons.

How can model-based samples reduce costs and nonsampling errors? Under restrictive assumptions some persons can be viewed as exactly equivalent to others. For example, Sudman (1967) describes a design called "probability sampling with quotas," which is a multistage technique used at the National Opinion Research Center (NORC) as late as the mid-1970s. This was a design that started with a sample of counties or county groups with known probabilities of selection and continued in successive stages using probability techniques to sample cities within counties, census tracts within cities, and finally city blocks (or equivalents in rural areas). At the final stage, however, the probability selection mechanism was halted, and interviewers were given orders to obtain certain counts of interviews (quotas) for different demographic groups. For example, an interviewer might be instructed to obtain

> two interviews with males aged less than 30,
> two interviews with males aged 30 or older,
> one interview with females who are employed,
> two interviews with females who are not employed.

These quotas are based on age, gender, and employment status. Others could have been used. According to Sudman (1967), the quotas are determined such that "the probability of being available for interviewing is known and is the same for all individuals within the stratum, although varying between strata" (p. 8). Although the term "availability" is somewhat ambiguous, it appears clear that the notion does not incorporate willingness to respond, as Sudman acknowledges that the

---

[3] We exclude from this discussion the completely uncontrolled selection procedures that merely obtain measurements on any person, regardless of his/her characteristics.

nonresponse bias of the method is similar to that of full probability sampling (see Section 3.2 for a discussion of this). However, the model could easily be extended to include cooperativeness and patterns of being at home as criteria for forming the strata. The assumption of homogeneity with strata would obviously be more attractive if the strata were defined using a larger number of attributes, but the larger the number of attributes the more expensive the interviewer search to fill the quotas.

With probability sampling with quotas the interviewer is instructed to begin calling on housing units of a sample block in a prescribed manner, perhaps trying to fill the quotas that are most difficult first (the young males). If a household is absent from the unit at the time of the visit, the interviewer is instructed to go to the next unit to seek an interview. Under this procedure each sample block obtains the specified set of interviews.

What are the error properties of such nonprobability designs? Survey statistics (e.g., a sample mean) can be demonstrated to be unbiased estimates of the frame population statistics only under restrictive assumptions. If within each final sampling unit (the block in the example above) persons within the same quota group are completely homogeneous on the survey attributes, then the estimates can be shown to be unbiased. This requirement of an assumption is the basis of the label "model-based sampling." If the model of the frame population is true, it is likely that the sampling procedure achieves lower errors for fixed costs than ones based on probability methods. If the assumption is false, properties of unbiasedness of statistics and of their sampling variance estimators cannot be demonstrated.

The great attraction of nonprobability methods is the potential cost savings. Sudman (1967) presents a cost comparison of several surveys conducted by NORC in the 1960s, which differ in their sampling methodology. Table 6.1 shows costs per completed interview for six probability sample surveys and four quota sample surveys. As we note elsewhere, it is difficult to compare average costs across surveys because there are potentially scores of design differences across the surveys that can affect the comparison (e.g., number of callbacks, length of interviewing period, length of interview, number of interviewers used). With such qualifications in mind, we can see that the average cost of the probability sample interviews is about three times that of the average cost of the quota sample methods. It is difficult to imagine that uncontrolled design differences among the surveys would account for such large differences.

Table 6.1 Cost Comparisons per Completed Interview for Full Probability and Probability Sampling with Quotas, NORC, 1960s

Study	Direct Field Costs	Supervision	Other Costs	Total	n
		Costs per Interview			
		Probability Sample Designs			
Study 1	$13.36	$3.40	$72.72	$89.48	2380
Study 2	9.63	13.53	48.71	58.35	2810
Study 3	8.86	2.23	42.45	53.54	2200
Study 4	6.58	3.29	41.31	51.18	760
Study 5	8.80	3.80	12.56	25.16	2500
Study 6	11.27	4.00	17.66	32.93	1500
		Quota Sample Designs			
Study 1	7.42	1.58	13.33	22.33	1200
Study 2	6.60	1.13	12.40	20.13	1500
Study 3	6.54	0.92	10.85	18.31	1300
Study 4	6.00	1.27	9.87	17.13	1500

Figures differ from Table 2.8 because of rounding.
Source: S. Sudman, *Reducing the Cost of Surveys*, Aldine, Chicago, 1967, p. 25, Table 2.8.

6.3.2 Comparing Probability and Nonprobability Samples

The comments above on nonprobability samples place into words statistical properties of nonprobability samples. The question remains whether in practice such methods give very different results than do full probability methods. There are few studies comparing probability and nonprobability samples. Indeed, since nonprobability samples can have an infinite variety of features, the question in general cannot be addressed. We can, however, discuss how a particular nonprobability method compares to a full probability method. Even then, however, the criteria for the comparison need careful definition. Are the two procedures to be compared on sampling variance alone? If so, how will the variance be estimated for the nonprobability method? Are they to be compared on sampling bias properties? If so, how will this be assessed?

Will they be compared on nonsampling error differences? Can estimates of these be separated from estimates of sampling bias? There are two comparisons involving the "probability sampling with quotas" method described briefly above.

T. Smith (1979) in an analysis of the sex ratio among respondents to the General Social Survey offers data relevant to the nonresponse differences between the full probability method and the "probability sampling with quotas" method. Table 6.2 shows an apparent lower percentage of males among respondents using the full probability method than the quota method. In using the quota method, there are an estimated 48 percent of the household population who are male; with the full probability method, the estimate is 43 or 44 percent. Are these differences larger than those to be expected from sampling variance alone? The standard error of the full probability method is about 2.0 percentage points. If the standard error for the quota method were similar (we cannot assert that without the assumption of homogeneity with quota groups within sample clusters), then the differences between the two methods are not large enough to reach standard levels of statistical significance (.05) based on sampling error alone. The estimated proportion of males in the household population, based on Census Bureau demographic and survey estimates, was 47.6 percent during this period. Smith cites other studies to bolster the argument that lower proportions of males in the full probability sample are typical of the method. He attributes the lower proportions to differential nonresponse among males, a problem that is avoided with the quota method.

Stephenson (1979), in a broader analysis of differences between the two techniques used the same data as did Smith. He found that the quota method tends to underrepresent persons who are typically difficult to interview. These include males who are working fulltime. The method, however, does tend to represent neighborhoods that are typically difficult to measure, since it requires some interviews from the areas.

The question of choice between methods therefore is not fully answered by the comparisons available in the literature. Indeed, a researcher fully subscribing to the probability method would not find the available evidence persuasive, since it compares one quota sample (in each of two years) to one probability sample. The sampling variance of the probability sample can be estimated. From classical sampling theory, the sampling error of the quota sample cannot. Therefore, the real test would require replication of the quota sample survey to build up an empirical sampling distribution in order to address sampling bias and variance properties.

On the other hand, a researcher from the other school can attack probability methods for the weakness of their execution. Although each

Table 6.2 Estimated Percentage of Males, Following Weighting by Number of Adults, Quota Sample Versus Full Probability Sample, by Year of General Social Survey

	Percentage Male	
Year of Survey	Quota Sample	Full Probability Sample
1972	51.2 (1613)	
1973	47.4 (1504)	
1974	49.0 (1484)	
1975	48.3 (755)	44.2 (735)
1976	47.9 (755)	43.3 (742)
1977		46.6 (1530)
1978		43.4 (1525)

Source: T. Smith, "Sex and the GSS," General Social Survey Technical Report No.17, September 1979, National Opinion Research Center, Chicago.

person in the frame population may have a known nonzero chance of selection, each person rarely has a known nonzero chance of measurement, because of differential nonresponse. The quota method does not avoid this problem entirely but does "substitute" neighbors with similar gender and age or employment characteristics for those nonrespondents. The criterion for choice between the two methods should thus be based on a balancing of cost differences and applicability of the model of homogeneity within quota groups for the last stage selection units. This decision will obviously differ for each variable measured in the survey.

The literature on quota methods that appears in social science journals has not presented new material since the initial work of Sudman, but the issue has been revived, albeit at a higher level of statistical sophistication, in the survey statistics literature. In this work, the models on which the selection procedure is based have been stated more explicitly and have been made more complex. Royall (1970), Royall and Herson (1973a, b), Royall and Cumberland (1978), and others have explored model-based samples that can be demonstrated to offer unbiased estimates and lower sampling variance, if the models are true. In a readable and important summary of the debate between classical probability sampling and model-based sampling, Hansen et al. (1983) and the following commenters note that the key to the choice of method remains the appropriateness of the model that guides the selection process. Kalton (1983b) notes that probability sampling techniques sometimes are based on hypotheses about the nature of differences within the frame population (note the discussion of stratification in the next section) but rarely posit these hypotheses so strongly that they give some members of the population no chance of entering the sample. Such a procedure (inherent to all model-based methods) prevents the testing of those hypotheses, a method that can alert the researcher that the hypotheses may be wrong. For that reason, the probability sampler chooses to avoid such procedures.

6.4 FOUR EFFECTS OF THE DESIGN ON SAMPLING ERROR

Although sampling error is partly a function of variability in the population studied, the sampling error in a statistic is under the control of the sample designer. In contrast to other errors described in this book, sample designs can often be mounted to obtain a specific level of sampling error. The sample design features which are most important in this regard are (1) stratification, the sorting of the population into separate groups prior to selection, (2) assignment of probabilities of selection to different kinds of elements in the population, and (3) clustering, the selection of groups of elements together instead of independent selection of separate elements. The fourth feature of interest is the sample size itself.

   Figure 6.5 presents a $2 \times 2 \times 2$ table categorizing different sample designs using the three major dimensions we discuss. The base sample design is a simple random sample, which is one example of an unstratified, equal probability (or *epsem*), element sample. Few large household surveys use this design, but almost all software packages used to analyze the survey data make the assumption that such a design was used (exceptions to this are SUPERCARP, the SESUDAAN routines compatible with SAS, and the sampling error functions in OSIRIS.IV). The most common designs are also the most complex—stratified, clustered, unequal probability samples (like the National Health Interview

	Element Samples		Cluster Samples	
	Unstratified	Stratified	Unstratified	Stratified
Equal probabilities of selection	Simple random sample	Proportionate stratified element sample		Many area probability samples of households
Unequal probabilities of selection		Disproportionate stratified element sample		Many area probability samples of persons

Figure 6.5 Eight major types of samples.

Survey and the Current Population Survey). They occupy the lower rightmost cell in Figure 6.5.

6.4.1 Stratification

Stratification is a step of sample design that is carried out prior to selection of the sample.[4] Stratification is the sorting of the frame population into separate subgroups (called "strata"). Each element of the frame population belongs to one and only one stratum. After the groups have been defined, separate samples are drawn from each group. In a sense, stratification transforms the sampling problem from selecting elements of *one* population to selecting elements from *many* populations (many strata).

Stratification by and large tends to reduce sampling error. This can be seen both intuitively and through the formulas for sampling variance of statistics from stratified samples. From an intuitive perspective it is clear that each possible sample drawn with a stratified design will have representatives from each of the strata. This property is not guaranteed

---

[4] We concentrate here on the use of explicit stratification. Some effects of stratification can be introduced through the manipulation of the sort order of the sampling frame and use of systematic selection (a "1 in N" selection). Stratification effects can also be obtained after the survey is completed through poststratification. Interested readers are referred to standard sampling texts for discussions of these techniques.

without stratification. In a design without stratification (e.g., a simple random sample like that above) by chance some samples may fail to include cases from all possible groups. To the extent that the characteristics of elements in the different strata vary greatly, the values of a statistic will depend on what strata contribute sample cases and what proportion of the sample comes from each stratum. When different samples of the same design have different proportions of cases from the various strata, they will also tend to have different values on the statistic in question. These differences lead to high sampling variance for the statistic.

To illustrate this with a simple case we can change the simple random sample case to a stratified random sample design. This is done by sorting the frame into strata prior to the numbering of elements from 1 to N. If H different strata are constructed, let N_h be the number of elements in the hth stratum, where $h = 1, 2, ..., H$. In this case the population mean is viewed as

$$\overline{Y} = \frac{\sum\limits_{i=1}^{N} Y_i}{N} = \sum\limits_{h=1}^{H} \left(\frac{N_h}{N}\right) \overline{Y}_h .$$

This is sometimes expressed as

$$\overline{Y} = \sum\limits_{h=1}^{H} W_h \overline{Y}_h ,$$

where $W_h = N_h/N$, is the proportion of the population in the hth stratum, and $\overline{Y}_h$ is the mean of the hth stratum.

If simple random samples were drawn from each of the H strata separately, then the sample–based estimator of the population mean is

$$\overline{y} = \sum\limits_{h=1}^{H} W_h \overline{y}_h.$$

Note that this estimator weights each stratum mean by a population parameter. That is, the W_h's are the proportions of the full frame population lying in the hth stratum, not the proportion of the sample in the hth stratum.[5] If simple random samples were drawn in each stratum, the sampling variance of the sample mean is

$$\sum_1^H \frac{(1 - f_h)\,W_h^2 S_h^2}{n_h},$$

where f_h = sampling fraction used in the hth stratum, n_h/N_h;

W_h = proportion of the frame population in the hth stratum, N_h/N;

S_h^2 = population element variance in the hth stratum;

H = total number of strata;

n_h = sample size in the hth stratum.

The reader will note that this expression is different from that for the simple random sample case, in which the element variance for the whole population was a factor in the sampling variance. Here, only the within–stratum variation in values (the S_h^2 terms) contributes to the sampling variance of the survey statistic.

The full variability of the measure, y_i, in the frame population might be separated into components due to within–stratum variability and between–strata variability. The expression above shows how stratification acts to reduce sampling error. The between–strata component of variability is eliminated as a factor in the sampling error. This is sometimes illustrated by rewriting the expression above, in the case of equal sampling fractions in all strata $(f_h = f$ for all $h)$ as approximately equal to

$$\frac{(1 - f)S^2}{n} - \frac{(1 - f)}{n} \sum_{h=1}^{H} W_h(\overline{Y}_h - \overline{Y})^2 .$$

The first term in this expression is merely the sampling variance of the mean for a simple random sample. The second term shows the gain in precision (reduction of sampling variance) resulting from the stratification. Note that the gain is a function of how different the strata means are from the overall population mean. In short, stratification

---

[5] In the case of equal probability selections, $W_h = n_h/n$ so that $\Sigma\Sigma\, y_{hi}/\Sigma n_h = \Sigma\, W_h \bar{y}_h$.

generally reduces the sampling error of survey statistics to the extent that the strata defined have different values on the survey statistics in question.

It is important to note that the expression above is limited to cases in which the same sampling fraction is used in every stratum, a proportionate stratified design. In stratified designs with unequal probabilities across strata, either higher and lower sampling variances are possible than with a simple random sample. The direction of the change depends on the relationship between the within-stratum element variance of particular strata and the sampling fraction in the strata. If higher sampling fractions are used in the strata with higher element variances, then reductions in sample variance are likely relative to a proportionate design.

Stratification can affect all statistics calculated in the sample. The effect of stratification, however, is not the same for all statistics. Subclass statistics (i.e., statistics calculated on a subpopulation), for example, can sometimes receive no benefit at all from stratification. For example, if a mean is estimated on a single stratum in a stratified element design, the sample is merely a simple random sample. There are no gains of stratification for that statistic although there might be large gains on the estimate of the overall mean. Similarly, it is often found that the reduction in sampling error enjoyed for statistics on the total sample is diminished somewhat for subclass statistics. For example, in a sample of undergraduate students, stratified on year in school (freshman, sophomore, junior, and senior), it is to be expected that the standard error of the estimated proportion with graduate school plans might enjoy more benefits of stratification than, say, the proportion with graduate school plans among those interested in computer science careers. This occurs because there was no control over the number of freshman computer science students, the sophomore computer science students, and so on. This lack of control over the allocation means more variability over different samples.

The discussion of stratification has focused on only one variable measured in the survey, but most surveys measure hundreds of variables. The effects of stratification vary across different measures. Those highly correlated with the stratifying variables enjoy the benefits of stratification. Those not highly correlated generally do not. The problem of choosing stratifying variables in multipurpose surveys has no simple solution.

6.4.2 Probabilities of Selection

A simple random sample assigns equal probabilities of selection to each element in the frame population. There are many practical reasons, however, to depart from this design. First, the costs of measuring some members of the population may be very high. For example, a sample of Hispanic persons based on the frame of households generally requires a sample of housing units which are screened to identify those containing Hispanics. In areas with few Hispanics (e.g., Alaska) such screening costs are very high. Lower probabilities of selection are often used in those areas to reduce the costs of the survey. Second, there may be a desire to study a subgroup of the population intensively, with smaller sampling errors of statistics desired than those expected from a sample based on probabilities equal to those of all other persons. For example, the National Health Interview Survey in the 1980s contains a sample of black persons drawn at a rate higher than that of nonblacks. This gave to the analysts interested in analysis of black health behavior a larger sample of blacks. Third, in some infrequent cases, there is prior information about the within-stratum variability on the survey measures (i.e., the S_h terms, the element variances within strata). In such cases oversampling of the strata with higher element variances can reduce the sampling error in the overall mean relative to a design using equal probabilities of selection to achieve the same total sample size.

One estimator of the population mean from a sample with unequal probabilities of selection is

$$\frac{\sum_1^n (y_i/p_i)}{\sum_1^n (1/p_i)},$$

where p_i = probability of selection of ith person.

Note that $\sum (1/p_i)$ is an estimate of the population size, N. Furthermore, note that for estimating the population mean the p_i do not have to be exactly the probabilities of selection but could be within a multiplicative factor of these probabilities and the adjusted mean would have the same value. This is not true for estimates of other population quantities (e.g., the population total).

The weighted sample mean generally has a higher variance than the unweighted mean except when different sampling fractions are chosen so that there is a correlation between weights and element variance of individual observations. If higher selection fractions are used in groups with high element variance, then better precision can be obtained relative to an equal probability design. If not, then inflation of the sampling error, relative to an equal probability sample of the same size, would occur.

The beneficial use of disproportionate sampling is typified by the optimal allocation solution in a stratified element problem. This entails an explicit use of cost and error models. Suppose the total cost of a survey is expressed as

$$\text{Total Cost} = \text{Fixed Cost} + \text{Variable Strata Costs}$$

$$C = C_o + \sum_{1}^{H} C_h n_h \, ,$$

where C_o = fixed costs of the survey (e.g., the cost of developing training materials for interviewers, of developing the questionnaire, of designing the sample); these are costs that must be paid regardless of what sample size is chosen;

C_h = cost of selecting, contacting, and interviewing a single sample element from the hth stratum; this would be an average cost blending the costs of respondent and nonrespondent cases;

n_h = number of sample elements chosen from the hth stratum.

If this cost model accurately describes the relationship between sample size in the strata and total resources needed for the survey, then the sampling error of the estimated population mean is minimized when

$$f_h = \frac{n_h}{N_h} = \frac{kS_h}{\sqrt{C_h}} \, ,$$

where k is a constant of proportionality (see Chapter 2 for the derivation of this f_h).

This means that larger sampling fractions should be used in strata with high internal variability (large S_h) and smaller in those with low internal variability, other things being equal. Similarly, if the element variances are the same, larger sampling fractions should be used in strata with low costs for data collection. The first result makes intuitive sense since the sampling error of the overall mean is a function of the element variances in each stratum. The second statement strikes some as a threat to precision in order to save money. The error in that reaction is that in each stratum the sampling variance is a function of the sample size, as well as the element variance. If there are no differences in element variances across strata, then the precision of the estimated mean can be maximized by increasing the sample size by the largest amount possible within the budget. This can be accomplished by buying more cases for the stratum with the cheapest measurement costs.

The optimal allocation formula above applies to an estimate of the overall mean, not to other statistics in the survey. For example, the sampling error of the difference between strata means is minimized when the sample sizes (instead of the sampling fractions) are proportional to the ratio above. Second, just as the gains of stratification for sampling required more information about the frame population, the optimal allocation requires even more knowledge. Here the element variances must be known for each stratum. This is an unlikely event for a survey topic that has never been investigated. It might be possible in cases of ongoing surveys on relatively stable phenomena. Third, the optimal allocation applies to only one variable's mean. It is an allocation that is optimal for only one statistic in the survey. Most surveys calculate hundreds or thousands of statistics and each would have its own optimal allocation. Fourth, the optimal allocation is conditional on the given strata. It is possible that the use of another stratifying variable can reduce the sampling error of the overall mean more dramatically. In situations in which bad guesses are made about element variances within strata (S_h's) or cost differences, the achieved sampling error of the mean might be much higher than that from an equal probability design. Thus, departing from equal probability designs entails a risk for the designer.

6.4.3 Clustering

Sometimes instead of selecting persons directly from the frame population, groups (clusters) of persons are first selected and then individuals within the clusters are interviewed (e.g., national samples based on counties, samples of classrooms of students). In practice, cluster samples tend to produce higher sampling errors for statistics than element samples of the same size (i.e., the same number of sample elements). However, there is nothing inherent in the design that statistically requires this to happen. The loss of precision arises from a sociological fact that most natural groupings of persons (e.g., residential neighborhoods, classrooms, office buildings, work groups, carpools, voluntary organizations) contain persons who are similar to one another on many variables that are measured in surveys. Another way of stating that is that there are large between-cluster differences in the survey variables. Sampling error is reduced to the extent that every sample represents all the diversity that exists in the population. When that occurs, estimates based on different samples do not vary as much as when different types of persons disproportionately fall in some samples and not in others.

Assume that we plan a survey about the costs of housing in a city. One statistic of interest is the proportion of persons who live in rental housing. A sample of blocks in the city is chosen. As is typical in many cities, rental housing tends to be separated from owner-occupied housing, so that sample blocks (clusters of persons) tend to be either all rental housing or all owner-occupied housing. Now consider the problem of adding to this existing sample some new sample persons, say 100 new persons. Would it be better to spread the sample to 100 different blocks or use the same blocks as are in the sample already? To reduce sampling error of the estimated population mean, we want to represent the diversity in the population in each sample. This is not well done by adding more sample persons from the existing sample blocks, because persons in the sample blocks resemble one another on the survey attribute—whether they rent or own their home. Instead, the diversity of the population is better represented by spreading the sample over many blocks. This reduces the clustering of the sample. Clustering induces higher sampling error.

Despite the loss of precision for survey statistics from cluster samples, every major U.S. federal government household survey and almost all academic and commercial household surveys use cluster designs. A reader attuned to our emphasis on balancing costs and errors will guess that the reason for this frequent use of cluster samples is the desirable cost features they typically have. Cluster samples tend to be cheaper than element samples. When household surveys are done with personal interviews, the interviewer must travel (usually by car) from sample household to sample household in order to contact the chosen persons. If the households are spread widely geographically, these travel costs can form a large component of the total survey costs. Cluster samples reduce that component because the interviewer can attempt to interview many sample persons with a trip to a single location (the sample cluster).

The sampling error of the estimated population mean for a cluster sample is inflated by two factors: the correlation of values among persons in the same clusters and number of sample elements chosen from a cluster. This is sometimes expressed as

$$\text{Var}(\bar{y}) = \frac{S^2[1 + \rho(b-1)]}{n},$$

where $S^2 =$ the population element variance;

$\rho =$ intracluster correlation;

$b =$ number of sample persons chosen from each cluster;

$n =$ total number of persons in the entire sample.

This expression applies to a simple random sample of equal–sized clusters (each with b persons). Here the intracluster correlation can be presented as

$$\frac{N-1}{NS^2}\left(\frac{1}{A}\sum_{\alpha}^{A}\right)\frac{2}{b(b-1)}\sum_{\beta<\gamma}^{b}(Y_{\alpha\beta}-\bar{Y})(Y_{\alpha\gamma}-\bar{Y}),$$

where $N =$ total frame population size;

$S^2 =$ population element variance;

$A =$ total number of clusters in the population, each with b persons;

$Y_{\alpha\beta} =$ value of the survey variable for the βth person in the αth cluster;

$\bar{Y} =$ overall population mean.

The intracluster correlation measures covariation of pairs of persons in the same cluster, calculated by deviations from the overall mean. If elements in the same cluster have similar deviations from the population mean, then ρ will be positive, and the sampling error of the estimated mean from a cluster sample will be inflated over that from an element sample of the same size.

Cost and error modeling are combined to address one feature of cluster sampling, the optimal number of elements to select from each cluster. This is done in the context of two–stage cluster samples, where in the first stage of selection, sample clusters are selected and in the second stage, elements within those clusters are chosen. Kish (1965) shows that in the case of unrestricted random sampling of clusters and simple random sampling of elements within clusters the optimal number of sample elements per cluster has a simple solution. Consider the cost model that separates costs of a cluster from costs of each sample element in the cluster:

Total Cost = Fixed Cost + Cluster Costs + Element Costs

$$C = C_o + C_a a + C_b ab \, ,$$

where C_0 = fixed costs of doing the survey, independent of the number of sample clusters or sample elements per cluster;

C_a = cost of selecting, listing, and locating to each cluster those costs independent of the number of elements selected per cluster;

a = number of sample clusters;

C_b = cost of selecting, contacting, and interviewing a single sample element from a cluster; this would be an average cost blending the costs of respondent and nonrespondent costs;

b = number of sample elements per cluster.

If the sampling error of the estimated population mean can be expressed as above, then the optimal number of elements per cluster is

$$\sqrt{\frac{C_a(1-\rho)}{C_b\rho}}\;.$$

That is, large numbers of sample elements should be taken from clusters that exhibit internal high homogeneity (high positive ρ's) on the survey variable. Small cluster sizes should be taken with low homogeneities. This optimal cluster size has some of the same character as the optimal allocation to strata listed above. That is, given the relative costs of processing a cluster and collecting data from each person, and the levels of homogeneity within clusters, then the cluster size above will minimize the sampling error of the estimated population mean relative to all other cluster sizes. If the estimated costs or ρ are not accurate, relative losses in precision can occur.

The limitations of this optimal solution are also similar to the limitations of the optimal allocation solution (described in the Section 6.4.2). It applies to one statistic (the overall mean) calculated on one survey variable. It requires more knowledge of the survey problem, the costs of data collection, and a covariance property of elements in the population. These are not typically known in one-time surveys but might be known in ongoing repeated surveys. However, if different variables in the survey have very different levels of the between–cluster component, then there will be no single optimal cluster size, but a whole host of them.

Cost and error models can also be used not to locate an optimum design but to study effects of alternative design options in an objective manner. For example, as is shown in Waksberg (1970), a periodic redesign of the Current Population Survey involved assessing a choice of a final cluster size of eight households in a multistage design. These clusters are called "segments." Assume a survey of 60,000 housing units is needed, some with telephone interviews, some with personal visit interviews. A time allocation study is performed by the field staff to collect data on cost model parameters:

1. The number of segments visited by segment size (some need no visit because of telephone interviewing).

2. The number of trips from home to a segment by segment size.

3. The number of trips from another segment to a segment, by segment size visited.

4. The distance from home to segment.

5. The distances from segment to segment.

6. The time spent traveling from home to segment and segment to segment trips.

7. The time spent at segment interviewing per household.

The overall cost model in a simplified form is

Total Cost = Fixed Cost + Segment Costs + Household Costs

$$C = C_o + C_a a + C_b ab,$$

where C_o = fixed costs;

C_a = costs of each segment (i.e., the increase in cost expected by adding another segment to the sample, keeping the overall number of interviews the same);

C_b = cost of processing and interviewing each household;

a = number of segments selected in the sample;

b = number of households selected per sample segment.

Although the equation looks simple, the C_a term is the sum of several components:

Segment Costs = Mileage + Salary

$$C_a = (T_h D_h + T_s D_s)C_m + \left(\frac{T_h D_h + T_s D_s}{R}\right)S,$$

where T_h = number of trips from home to segment;

D_h = distance for trips from home to segment (miles);

T_s = number of trips from segment to segment;

D_s = distance for trips from segment to segment (miles);

C_m = mileage reimbursement cost for interviewers per mile;

R = average speed on trips to segments (miles per hour);

S = interviewer's salary per hour.

The error model is set in terms of a design effect, a function of segment size, the only term that is linked to the cost model:

$$\text{Deff} = 1 + \rho(b - 1),$$

where ρ = intrasegment correlation;

b = average segment size.

The deff or design effect of a statistic is defined as the ratio of sampling variance reflecting all the complexities of the design to the sampling variance expected from a simple random sample with the same number of elements. Hence, deff values greater than 1.0 are generally expected for clustered samples, because clustering generally inflates standard errors.

Since ρ values vary across the measures in the study, four alternative ρ values are examined, 0.00, 0.03, 0.05, and 0.10. Table 6.3 illustrates how information assembled from field experience might be combined with data on within- and between-segment variances to simulate the effects of alternative designs. The first row of Table 6.3 presents estimates of the proportion of segments that would have to be visited at least once. With very small numbers of sample households per segment it is possible that all sample units might be interviewed by telephone. With large samples from each segment it is more likely that at least one unit will not have a telephone or will refuse to cooperate with a telephone interview request. Hence, personal visits will be required for 70 percent of the segments if two houses per segment are selected but 85 percent with eight houses per segment. Similarly, on the next row the number of trips from home to segment increases with larger numbers of units per segment (from 0.43 per completed segment with segments of 2, to 1.42 with segments of 8). In contrast, with small segments there are *more* trips from one segment to another. This makes sense because there would be more segments in the sample. The trips would each have to be shorter, however, than with a few dispersed larger segments. From the trip counts and distances shown at the top of the table, estimated costs can be computed (in the middle rows of the table).

These costs are expressed as costs per household, but they have been arbitrarily adjusted for illustrative purposes to produce a $20.00 cost for the design with segments of six households. Note that the sampling costs decrease as larger segments are chosen. This reflects the fact that more segments have to be sampled for a fixed sample size with small segments. Travel costs are highest with the small segments, because there is more segment to segment travel. Interviewing costs are constant, regardless of segment size. The result of this is that the total cost per interview varies from $23.33 with segments of 2 to $19.77 with segments of 8, a reduction of about 15 percent in total costs for a survey of 60,000 interviews.

However, cost comparisons alone do not provide the solution to the cluster size problem. A sample of 60,000 in 7500 segments of size 8 will have higher sampling error than one with 30,000 segments of size 2. We need the design effect model to judge the effect. Table 6.3 presents design factors adjusted so that the design effect of the design with segments of 6 equals 1.00. Other design effects are relative to that of segment size 6. They can be interpreted therefore as proportionate change in sampling variance associated with change in segment size. They reflect both change in the $[1 + \rho(b-1)]$ design effect factor and the change in sample size because of cost differences associated with the given segment size. For example, if the statistic in question is a mean on the total sample and has a ρ value of 0.10, then the segment size of 2 has the lowest sampling variance among all alternatives (relative design factor of 0.85), given fixed resources devoted to the survey. If instead, however, the ρ value is much smaller (in the extreme if $\rho = 0.0$), larger clusters sizes are more cost efficient. The best among those examined is a segment size of 7 (achieving a relative design factor of 0.98). The best cluster size depends on the variable being studied and how strongly homogeneous it is within clusters.

Despite the fact that the best cluster size depends on the type of variables being measured, only one segment size can be chosen for the sample. If a segment size of 6 were chosen there would be a 17 percent increase in sampling variance (about 8 percent increase in standard errors) over that expected from a segment size of 2. This occurs in the worst case, with a ρ of 0.10. With lower ρ values the sampling error with a segment size of 6 departs only in minor ways from that of the optimal segment size. Hence, it might be an attractive choice for segment size if expected ρ's are lower than 0.10 in general.

Table 6.3 Cost and Error Parameters and Relative Variance (to that of Segments of 6) for Various Segment Sizes, Current Population Survey, 1970s

Activity	Segment Size in Housing Units						
	2	3	4	5	6	7	8
Proportion of segments visited	0.7	0.76	0.78	0.81	0.83	0.84	0.85
Trips, home to segment (T_h)	0.43	0.59	0.77	0.93	1.09	1.29	1.42
Trips, segment to segment (T_s)	1.15	1.04	0.94	0.86	0.81	0.71	0.68
Distance between segments (D_s)	4.8	5.7	6.5	7.1	7.7	8.3	8.8
Distance, home to segment (D_h)	8.8	9.6	10.3	11.1	11.9	12.7	13.5
Sampling cost	$2.84	$1.99	$1.66	$1.43	$1.20	$1.07	$.97
Travel cost	$9.40	$8.46	$7.89	$7.71	$7.71	$7.51	$7.71
Interviewing cost	$11.09	$11.09	$11.09	$11.09	$11.09	$11.09	$11.09
Total cost	$23.33	$21.54	$20.64	$20.23	$20.00	$19.67	$19.77
Relative Variances Given Fixed Costs							
$\rho = 0.10$	**0.85**	0.86	0.90	0.94	1.00	1.05	1.12
$\rho = 0.05$	0.97	**0.94**	0.95	0.97	1.00	1.02	1.07
$\rho = 0.03$	1.04	**0.98**	0.98	0.98	1.00	1.01	1.04
$\rho = 0.0$	1.16	1.07	1.03	1.01	1.00	**0.98**	0.99

For illustrative purposes the total cost of the design with segments of 6 was set at $20.00.
Source: J. Waksberg, "Optimal Size of Segment—CPS Redesign," U.S. Bureau of the Census memorandum to Joseph Daly, November 18, 1970.

6.4.4 Sample Size

Sample size is the design factor related to sampling error which has the most intuitive rationale. Most consumers of survey results realize that the larger the number of people in repeated samples, the smaller the variability in the sample statistics. This fact remains true in all sample designs, although the "sample size" might be defined in more sophisticated ways. For example, in a cluster design larger sample sizes obtained by increasing the number of sample clusters generally improve precision more dramatically than increasing the number of sample elements per cluster. Thus, a survey data set with 10,000 interviews drawn from 100 clusters should be viewed very differently from one of 10,000 interviews from 1000 clusters. "Sample size" thus has impact on sampling variance as a function of the number of independent selections at each stage of the sample and the relative within- and between-unit variability at each stage. Counts of records in data sets are not sufficient indication of these design properties to judge likely sampling error.

6.4.5 Combinations of Sample Design Features

Most large-scale household samples are complex combinations of stratification, clustering, and alternative probabilities of selection. For example, national household samples for personal interviews often are multistage cluster designs. At the first stage of selection counties or county groups are stratified on census and administrative data (e.g., population density, major industry, racial composition) and then selections are made from each stratum. Within the selected first-stage selections (called primary sampling units or PSUs) cities, villages, minor civil divisions, or census enumeration districts are stratified by size of population, location, and other variables. Then these second-stage units are sampled, separately for each PSU. This process continues for one to seven or more stages, depending on the design chosen. It sequentially selects smaller and smaller area units. The second to the last stage of selection is a unit of size small enough that individual housing units can be listed by a field worker visiting the site. Then in the last stage of selection, individual housing units are sampled from the list made for each penultimate unit. The expressions for sampling error for these designs are much more complex than those given above for simpler designs. The basic principles of those designs, however, still apply.

6.5 THE EFFECT OF NONSAMPLING ERRORS ON SAMPLING ERROR ESTIMATES

Many casual readers of survey statistics interpret estimates of sampling error as estimates of total variable error for a statistic. This book is devoted to describing all the other errors that affect survey data, and other chapters illustrate how they can affect the total error of a survey statistic. The practical question remains, however: if classical sampling error computation formulas are used, but the data are also subject to nonsampling errors, how are the sampling error estimates affected? Do the estimated sampling errors reflect some nonsampling errors as well?

Under very restrictive circumstances, sampling error estimates will reflect some nonsampling errors. Most simply stated, this will occur (1) only for variable nonsampling errors, (2) only when the mechanisms producing the nonsampling error coincide with the sampling unit definitions, and (3) only for nonsampling errors that are independent of the true values of the observation unit. We illustrate one case below.

Assume that there exists in a one-time survey estimating the variable y a measurement error process (similar to that of the classical true score model) such that the answer obtained is x, such that

$$\text{Response} = \text{True Value} + \text{Error}$$

$$x_i = y_i + \varepsilon_i,$$

where x_i = answer supplied to the question about y by the ith respondent;

y_i = true value for y for the ith respondent;

ε_i = response error committed by the ith respondent.

The data set analyzed for the survey consists only of the x variables; we have no separate indicators of y or ε. If the error ε_i is completely unrelated (no covariance) of the ε committed by other respondents, then all sampling error calculations of means and proportions on the x variables will include the measurement error variance associated with the ε terms. The conditional clause in the last sentence forms the assumption that

$$E(\varepsilon_i \varepsilon_{i'}) = 0 \text{ for all pairs of } i \neq i'.$$

If, on the other hand, there is some correlation between errors committed by one respondent and those by others, there is no guarantee of such results. For example, in Chapter 8 we discuss the ways by which interviewers might influence all respondents assigned to them to make similar response errors. This process induces a correlation among the ε terms of these different respondents. Most sample and data collection designs are such that, in those circumstances, the sampling errors do not reflect all the measurement error variance. First, assume a simple random sample were drawn of size n and k interviewers were each given m respondents in a random assignment. Assume the measurement error model is now

$$\text{Response} = \text{True Value} + \text{Interviewer Effect} + \text{Error}$$

$$x_i = y_i + \alpha_j + \varepsilon_i,$$

where x_i = answer supplied to the question about y by the ith
     respondent;

   y_i = true value for y for the ith respondent;

   α_j = error induced by the jth interviewer on all her
        respondents;

   ε_i = remaining response error committed by the ith
        respondent.

Here $E(\alpha_j \alpha_{j'}) \neq 0$; that is, there is a covariance between the interviewer
effects across different respondents. Those in the same interviewer's
group are subject to the same effects. Fellegi (1964) and others show that
the sampling variance of the mean would underestimate the total variance
of the mean by approximately the factor

$$\frac{(m-1)\rho_{int} + \sigma_r^2}{\sigma_y^2 + \sigma_r^2 \left[1 + \rho_{int}(m-1)\right]},$$

where σ_y^2 = variance of y_i in the full population;

   σ_r^2 = variance of $(\alpha_j + \varepsilon_i)$ over all possible combinations of
        interviewers and respondents, ignoring the clustering of
        respondents into interviewer workloads;

   ρ_{int} = an intrainterviewer correlation, the correlation of the
        $(\alpha_j + \varepsilon_i)$ terms within an interviewer's workload.

Thus, in this circumstance, sampling errors are underestimates of total
variance, as Bailey et al. (1978) and others have demonstrated empirically.
   There is one design under which traditional sampling error
calculations fully incorporate measurement error and other nonsampling
errors associated with interviewers. Assume a cluster sample is drawn,
one interviewer assigned at random to each sample cluster, and that
interviewer completes all interviews in the cluster. Then traditional
cluster sample standard error computations will reflect both the sampling
and nonsampling error variance associated with different interviewers.
This is an example of the data collection feature associated with the

variable error in question (the interviewer) being completely coextensive with the sampling unit (the sample cluster).[6]

Most data collection and sample designs do not induce perfect correspondence between the units creating nonsampling errors and the sampling units. For that reason, what portion of the nonsampling error variance is reflected in the sampling error calculations is unknown. It is probably safe to say that sampling error calculations often reflect some portion of nonsampling error variance, but not all.

6.6 MEASURING SAMPLING ERRORS ON SAMPLE MEANS AND PROPORTIONS FROM COMPLEX SAMPLES

In keeping with the practical emphasis of this book, this section presents estimates of sampling variance in a format that permits the reader to explore the causes of increases and decreases in sampling error. This format permits us to make the observation again that the error properties of statistics require both statistical and social science knowledge for understanding. We concentrate on the three dimensions of sample design discussed above: stratification, clustering, and assignment of probabilities of selection.

We present real data from complex sample surveys in search of empirical tendencies for design effects across different statistics from the same survey: (1) the inflation of standard errors from clustering tends to be larger for a mean for the total sample than the same mean for some subclasses that form part of all strata; (2) design effects are similar for the overall mean and individual strata means; (3) design effects for differences of subclass means tend to be smaller than design effects for the individual means in the difference; and (4) inflation of sampling error due to weighting does not tend to decline on subclass statistics.

To make these points it is useful to introduce some new terms. Kish et al. (1976) make the distinction between subclasses that form part of all strata, labeled "crossclasses," and those that tend to be isolated in only a subset of the strata, labeled "segregated classes." In this section we use the word "subclass" to mean any subgroup of the population, either crossclasses or segregated classes. For crossclasses it is argued that the model

---

[6] Note that although the sampling error calculations would automatically reflect the nonsampling error variance associated with interviewers, it could not be estimated separately from sampling variance.

$$\text{Deff}(\bar{y}) = 1 + \rho(b - 1)$$

is more appealing than for segregated classes. Empirically, the design effects for crossclass means tend to be those expected if the b is set to the number of crossclasses and the ρ is that obtained on the full sample. For example, Kish et al. (1976) present results from a U.S. national area probability sample, based on nine geographical strata. A systematic sample of 152 PSUs (formed by 10,000 contiguous housing units based on census enumeration areas) was taken. Within each PSU, 13 blocks or enumeration districts were selected and an expected 18 housing units sampled from each. Sampling errors were calculated on 36 means and proportions for the total sample, concerning fertility experiences, contraceptive practices, attitudes toward fertility, and demographic characteristics of the respondent women. The authors describe results for both black women and nonblack women in the sample. We concentrate our remarks on the nonblack women. The authors present square roots of design effects (called *deft's*), which can be interpreted as the ratio of standard errors. The average over the 36 statistics was 1.37, which corresponds to a ρ value of 0.0375. The values of the design effects vary greatly over the different statistics. For example, the deft for the estimated proportion who would approve of an abortion for a woman who cannot afford another child is 2.039 (a ρ of 0.09), but the deft for the proportion of women currently pregnant is 1.028 (a ρ of 0.002).

Early advice to survey analysts about design effects on means (e.g., Stuart, 1962) recommended inflating the standard error of means, computed under simple random sample assumptions, by a factor 1.4 or so. This would reflect a design effect (ratio of sampling variances) of 2.0. Many survey analysts continue to use this method, despite the clear evidence, like that above, that design effects vary greatly within the same survey over different statistics.

Recall that the design effect will reflect all the departures from simple random sampling. This study is both stratified (by geography and city size) and clustered (into geographical units). Such stratification will have differential power over statistics to reduce sampling errors. If the survey characteristic varies greatly over the strata, then the stratification is likely to reduce the variation of the mean over different samples. Each possible sample will have representatives of each of the different groups (strata). If, on the other hand, the survey statistic is relatively constant over strata, then we would expect small marginal benefits of stratification.

In spite of the potential beneficial effects of stratification on the design effects, they are generally overcome by the loss of efficiency due to clustering. This is evidenced by the design effects in the U.S. Fertility Survey all being greater than 1.0. With both stratification and clustering affecting the design effects (and hence the estimates of ρ), the variation across the statistics in deft cannot be attributed solely to homogeneity in neighborhoods.

Our first observation about empirical tendencies of design effects is that they are smaller for means on crossclasses than for the same mean on the total sample. Kish et al. (1976) calculated design effects on the same statistics for subclasses based on education, husband's occupation, family income, the number of live births for the woman, duration of the marriage, and age. The education, occupation, and income subclasses are likely to form different proportions of the metropolitan strata than of rural strata. These socioeconomic groups are closer to "segregated classes" than the demographic subclasses, based on number of live births, duration of marriage, and age.[7]

Let us examine age subclasses first. For example, of the 5597 women in the sample, 1063 (19 percent) were aged 30 to 34 years. The average deft for the 36 means and proportions examined for this subclass is 1.12. This is to be compared to the average 1.37 on the total sample. In terms of the design effect calculations, on the total sample, deff $= 1 + 0.0515(5597/152 - 1) = 2.845$, or deft $= 1.687$. On the 30 to 34 age subclass, the average deff $= 1 + 0.041(1063/152 - 1) = 1.246$ or deft $= 1.116$. This shows that both the ρ and b terms can differ from the total sample results when subclass means are considered. The ρ for the subclass can vary because it is easily possible that the between–cluster variation for the subclass is a larger component of total variation than is true for the total sample. Kish et al. (1976) found that ρ estimates tend to be higher for subclass statistics than for the same statistic calculated on the entire sample. Rust (1984) shows that this is a function of the relationship of the subclass to the survey statistic itself. Typically, the reduction in b for subclasses overwhelms any possible increase in the ρ terms, so the design effect for the crossclass means is lower than the design effect for the full mean. This implies that the impact of the design, the departure from simple random sampling, is smaller for subclasses than for the total sample.

Table 6.4 presents a summary of sampling errors on subclass means for the 1970 U.S. Fertility Survey. Each row in the table represents the average of 36 sampling error calculations, one for each of 36 means or proportions on the subclass listed. The rows in the table are sorted by the sizes of the subclasses, from the smallest subclass to the largest. If the

---

[7] One measure of the segregation of the subclasses is the ρ value for the proportion of women in the subclass. The average ρ value for the socioeconomic subclasses is ρ = 0.12, but for the demographic ones, only 0.02.

average ρ values were the same for all subclasses, one would expect that the average deft's in the table (in the last column) would be monotonically increasing down the rows. Such is not completely the case. For example, although those women in families earning less than $7000 form only 17 percent of the sample, the subclass has a larger than expected average ρ value, 0.086. Hence, the average design effect is larger, 1.21, than most other subclasses of that size. In contrast, the subclass containing the women with two live births, although forming 26 percent of the sample, has an average deft of 1.14. There are three observations of importance here. First, between-cluster variability in the characteristics of some subclass members may be relatively larger than the between-component for others. Clustering effects vary by subclass. Second, as Rust (1984) and Skinner (1986) note, the design effect of subclasses is a function of the relationship between the subclass membership and probabilities of falling in the class that the proportion represents. Third, to the extent that a subclass is not uniformly spread over clusters the variance of the subclass means tends to increase. This may be part of the cause of the higher design effects for the income classes, where the proportion of low income women is highly variable over PSUs.

Another lesson about sample design and standard errors on descriptive statistics can be drawn from comparisons of sample designs measuring the same attributes. The 1982 National Election Survey pre-election survey was conducted on a stratified multistage cluster sample with an equivalent of 102 primary areas. The 1984 edition of the survey was conducted on a new design, reflecting the 1980 Census of Population measures. Two factors are of interest in comparing the sampling error of the two designs. First, the new primary selections in the 1984 survey reflected the changed distribution of the U.S. household population between 1970 and 1980 (even with that, however, the estimates of population were 4 years out of date). The effect of this difference is expected to be a reduced variation in actual number of selections per cluster. This should improve the precision of the 1984 estimates relative to those in 1982. Second, the number of clusters used in the 1984 survey are vastly reduced over the number used in 1982, from 102 to 52. If the same sample size were used in both sample designs, one would expect higher design effects, other things being equal, because of the larger cluster sizes inflating the $1 + \rho(b - 1)$ term.

Table 6.5 presents a comparison of the design effect statistics on the estimates of eight different total population means and proportions. The reader will note that design effects for the 1984 survey, with larger cluster sizes, are higher (1982 average, 1.18, versus the 1984 average, 1.41). The ρ values, however, are quite similar, 0.032 average for the 1982 sample and 0.026 for the 1984 sample. Note that the cluster sizes for the 1984 sample

Table 6.4 Average Design Effects and Intraclass Correlations for 36 Means on 24 Subclasses, Ordered by Size of Subclass (1970 U.S. Fertility Survey)

Subclass	Proportion of Total Sample in Subclass	Average Number of Nonblack Women per PSU	Average ρ Value	Average deft[a]
Husband's occupation: farmer	0.05	1.8	0.145	1.06
Education: 0–8 years	0.07	2.6	0.108	1.08
Marriage duration: 20 or more	0.14	5.2	0.054	1.11
Family income: <$7000	0.17	6.3	0.086	1.21
Education: 12 years	0.18	6.6	0.035	1.09
Number of live births: 3	0.18	6.6	0.033	1.09
Age: 30–34 years	0.19	7.0	0.041	1.12
Family income: $15,000 or more	0.2	7.4	0.046	1.14
Number of live births: 4 or more	0.21	7.7	0.049	1.15
Marriage duration: 5–9 years	0.22	8.1	0.049	1.16
Age: 25–29 years	0.22	8.1	0.058	1.19
Husband's occupation: skilled	0.23	8.5	0.029	1.10
Family income: $7000–9999	0.25	9.2	0.031	1.12
Age: <25 years	0.25	9.2	0.053	1.20
Education: >12 years	0.26	9.6	0.076	1.29
Number of live births: 2	0.26	9.6	0.035	1.14
Husband's occupation: operatives and laborers	0.27	9.9	0.051	1.21
Marriage duration: 0–4 years	0.28	10.3	0.065	1.27
Number of live births: 0–1	0.34	12.5	0.048	1.25
Age: 35–44 years	0.34	12.5	0.041	1.21
Marriage duration: 10–19 years	0.36	13.3	0.048	1.26
Family income: $10,000–14,999	0.38	14.0	0.051	1.29
Husband's occupation: white collar	0.45	16.6	0.055	1.36
Education: 9–11 years	0.48	17.7	0.039	1.28

[a]Deft is the ratio of the standard error of the mean, reflecting the sample design, to the standard error of the mean assuming simple random sampling.
Source: L. Kish, R. Groves, and K. Krotki, *Sampling Errors for Fertility Surveys*, World Fertility Survey, Occasional Paper 17, 1976.

are over three times those in 1982 (41.4 to 13.2). Despite the higher design effects, the sampling variance of those statistics from the 1984 survey tend to be lower, because of the larger number of interviews collected. The 1984 sample is relatively less efficient than the 1982, but this is relative to a simple random sample of the same size. For example, for a proportion of about 0.5 we would estimate an average standard error from the 1982 survey at $[1 + 0.0321(13.4 - 1)](0.5)(1 - 0.5)/1371 = 0.0002549$ or a standard error of 0.016. A standard error for a similar statistic from 1984 would be $[1 + 0.0255(41.4 - 1)](0.5)(1 - 0.5)/2150 = 0.000236$ or a standard error of 0.0154. This illustrates the influence of clustering on sampling variance, even with apparently small ρ values. The total sample size of the 1984 survey is 57 percent larger, but the standard errors are only about 3 percent smaller. Sampling variance is not a linear function of sample size in complex samples.

Given this loss of efficiency when the design reduces the number of primary units, why would a researcher choose such a design? There are two answers to this question. First, the costs of the design with reduced number of primary units can be lower. The 138 interviewers in 1982 completed an average of 10.3 interviews. The 140 interviewers in 1984 completed an average of 16.1 interviews. The average cost per interview declined by 11 percent from 1982 to 1984. Ratios of sampling variance to cost are lower in 1984 than in 1982. Second, most analysts of the data are not interested in the estimates of overall population means and proportions. If the loss of relative precision with the larger clusters diminishes for these statistics, then the new design might be preferable. Table 6.6 shows that in both designs the design effects of subclass means and proportions do indeed decline, as expected by the model $[1 + \rho(b - 1)]$. For example, examine the subclass of males, which form slightly less than 0.5 of the respondents. For the 1982 survey the average design effect over the eight statistics is 1.10 (versus 1.18 on the total sample). For the 1984 survey the average design effect for males is 1.22 (versus 1.41 for the full sample). For a 10 percent subclass in the 1982 survey we would expect a standard error for a proportion of value about 0.5 of $[1 + 0.0485(1.34 - 1)](0.5)(1 - 0.5)/137 = 0.00186$ or a standard error of 0.0431. A similar subclass in the 1984 sample would have an expected sampling variance of $[1 + 0.0341(4.14 - 1)](0.5)(1 - 0.5)/215 = 0.00129$ or a standard error 0.0359. Recall that for the overall means, the larger 1984 sample enjoyed only a 3 percent advantage in standard errors. For these subclass means, however, the proportional gain is a $1 - 0.0359/0.0431$ or about a 17 percent gain.

In short, the loss of precision due to clustering is greater for means and proportions on the total sample and relatively less important for subclass analysis. If the goals of the survey include much analysis on

Table 6.5 Deft's (Square Roots of Design Effects) for the National Election Surveys, 1982 and 1984

Statistic Proportion	1982 Election Survey		1984 Election Survey	
	deft	ρ	deft	ρ
Interested in campaign	1.18	0.0311	1.26	0.0141
Strongly approved of Reagan	1.05	0.00845	1.44	0.0278
Approve of Reagan	1.26	0.0496	1.65	0.0441
Care very much about election outcome	1.18	0.0319	1.19	0.0103
Mean thermometer rating on Reagan	1.25	0.0439	1.48	0.0284
Better off than 1 year ago	1.03	0.00451	1.39	0.0227
Voted in last election	1.26	0.0461	1.65	0.0461
Mean over 8 statistics	1.18	0.0321	1.41	0.0255
Number of first-stage units	102		52	
Average sample size	1371		2150	
Average cluster size	13.4		41.4	

Source: S. Heeringa, personal communication, 1986.

subclasses of the population, clustered samples are often cost efficient alternatives.

Many analysis plans which include estimation of subclass means also involve the comparisons of subclasses. This is most simply accomplished by estimating the difference of subclass means. The effect of sample design on this statistic is different from that of the subclass mean itself, since

$$\text{Var}(\bar{y}_1 - \bar{y}_2) = \text{Var}(\bar{y}_1) + \text{Var}(\bar{y}_2) - 2\text{Cov}(\bar{y}_1, \bar{y}_2) ,$$

where $\bar{y}_1$ and $\bar{y}_2$ are two subclass means and $\text{Cov}(\bar{y}_1, \bar{y}_2)$ is the sampling covariance of the two sample means over all samples of the given sample design. The covariance term arises from the joint distribution of the two subclass means in each of the sample clusters. The clusters are the units in which the dependencies of selection probabilities exist. If the two subclass means tend to be positively correlated within clusters, then the covariance term is positive. In many real cases, the term is positive. Since this positive covariance term is subtracted from the sum of the variances of the two subclass means, it acts to reduce the variance of the difference of the two means. The design effect of the difference of two subclass means is defined as

Table 6.6 Design Effects and ρ Values for Subclass Statistics, 1982 and 1984 National Election Surveys

Subclass	1982 Election Survey			1984 Election Survey		
	Average Deft	Average Cluster Size	Average ρ	Average Deft	Average Cluster Size	Average ρ
Males	6.0	1.10	0.0422	18.3	1.22	0.0281
Females	7.4	1.11	0.0372	23.1	1.30	0.0311
Democrats	5.9	1.18	0.0819	15.2	1.20	0.0307
Independents	4.0	1.02	0.0120	14.0	1.20	0.0337
Republicans	3.3	1.04	0.0392	11.3	1.23	0.0505
Without college	7.5	1.16	0.0543	23.7	1.21	0.0200
Some college	5.9	1.07	0.0306	17.4	1.23	0.0312
Income < $20,000	5.9	1.08	0.0322	17.1	1.11	0.0147
Income $20,000–$34,999	3.7	0.997	−0.00230	10.8	1.30	0.0701
Income $35,000 or more	2.3	1.07	0.109	8.8	1.11	0.0310
Age 18-24	7.3	1.09	0.0311	23.4	1.30	0.0305
Age 45 or more	6.1	1.12	0.0488	17.5	1.12	0.0159
Catholic	3.0	1.11	0.115	10.6	1.24	0.0570
Non-Catholic	9.2	1.18	0.0471	27.1	1.36	0.0328
Average ρ			0.0485			0.0341

Source: S. Heeringa, personal communication, 1986.

$$\frac{\mathrm{Var}(\bar{y}_1) + \mathrm{Var}(\bar{y}_2) - 2\mathrm{Cov}(\bar{y}_1, \bar{y}_2)}{\mathrm{Var}_{srs}(\bar{y}_1) + \mathrm{Var}_{srs}(\bar{y}_2)},$$

where $\mathrm{Var}_{srs}(\ )$ is the sampling variance of the statistic ( ) under assumptions of simple random sampling. The denominator of the design effect does not enjoy the effects of a positive covariance term. Hence, it is often the case that the design effects of differences of subclass means are lower than the average design effect of the two subclass means themselves. The more the subclasses resemble one another within clusters, the larger the reduction in the design effect.

From empirical studies of design effects on overall means, subclass means, and differences of subclass means, Kish and Frankel (1974) and Kish et al. (1976) offer the rule of thumb that

$$\text{Deff}(\bar{y}) > \text{Deff}(\bar{y}_1) > \text{Deff}(\bar{y}_1 - \bar{y}_2) ;$$

that is, the inflation of standard errors due to clustering is relatively higher for total sample means than for subclass means on the same variable. Subclass means, in turn, tend to have their standard errors inflated by clustering effects more than differences of subclass means.

Sampling statisticians often view the statistic of the difference of subclass means as the simplest example of an analytic statistic, a measure of the relationship between two variables. The tendency for design effects of these statistics to approach 1.0 relative to other descriptive statistics is used by some to justify the assumption of simple random sampling in calculating standard errors of subclass differences.

This reduction of design effects for analytic statistics also has great importance at the design stage of survey projects. If the main purposes of the project involve the estimation of differences of subclass means, then relatively larger clusters can be taken without smaller increases in standard errors of those estimates than would be suffered by overall means. This is the first suggestion that surveys with such analytic purposes seem to be less affected by clustering in the sample design.

6.7 THE DEBATE ON REFLECTING THE SAMPLE DESIGN IN ESTIMATING ANALYTIC STATISTICS

In Chapter 1 the distinction was made between "describers" (those using surveys to measure characteristics of finite populations) and "modelers" (those using surveys to test hypotheses about causal relationships). In no other area do these two groups have more distinctive perspectives than on complex statistics from complex sample designs (e.g., regression models on complex survey data). The debate between them centers on the nature of response models estimated on survey data. For a simple example, consider a model predicting negative reaction to airport noise among residents in a 10 mile radius of the airport. Imagine that the psychological theory being tested could be expressed as

$$y_i = a + bx_i + e_i,$$

where y_i = score for the ith respondent from 0 to 100 measuring
the amount of annoyance with the noise of airplane
takeoffs and landings;

a = expected annoyance among those on the airport
property, given that the model is true;

b = change in annoyance scores expected under
the model with a 1 mile added distance
from the airport;

x_i = distance in miles between the residence of
the ith respondent and the airport;

e_i = deviation of the ith person's annoyance score
from the expected annoyance under the model.

The simplest issue on which the two sides disagree is the reflection of
unequal probabilities of selection. Imagine that a sample of residents was
chosen from the population living within 10 miles of the airport. Assume
the design gave twice the probability of selection to those within 1 mile of
the airport as to those farther away.

The modeler might approach the estimation of the model parameters
in the following manner. The psychological theory guiding the analysis
describes a process that creates the annoyance. It is stochastic or random
in nature, varying in some ways that are outside the theory. The stimulus
for the annoyance is the sound of the airport noise. This sound level is
measured by the distance from the airport. Those close to the airport are
expected to suffer from high annoyance from the noise; those far away
suffer less annoyance.

This process produces the annoyance values in the population around
the airport. That statement was true at the time of the survey, it is true
today, and for any time for which the process will still apply. Indeed,
although each person is expected to have annoyance levels that may differ
somewhat from those expected under the model, they are all determined
by the same process. It is the process (as formally stated by the model)
which is of interest to the researcher.

Suppose the researcher chooses to use ordinary least square
regression to estimate the parameters in the model. That is,

$$\hat{b} = \frac{\sum_{i=1}^{n}(x_i - \bar{x})(y_i - \bar{y})}{\left(\sum_{i=1}^{n}(x_i - \bar{x})^2\right)^{1/2}} \quad \text{and} \quad \hat{a} = \bar{y} - \hat{b}\bar{x}.$$

These estimates have properties of unbiasedness and minimum sampling variance under the following assumptions:

$E(e_i) = 0$ that is, the model is true;

$E(e_i e_j) = 0$ for $i \neq j$, that is, the random disturbances about the expected values are not correlated;

$E(e_i^2) = \sigma^2$ that is, the element variance of the disturbances throughout the population is constant.[8]

Given these assumptions, the estimators of a and b have the desired properties. The modeler is fortunate because those are precisely the assumptions she is willing to make. She views the particular set of annoyance values as the result of a random process which has those properties. Indeed, the population relevant to this process is infinite in size. The process did produce the set of annoyances observed at the time of the survey measurement, but it goes on producing annoyance levels according to this model forever. The population size is actually infinite because of the temporal extents of the process.

Note that the modeler ignores the sample design in estimating the a and b (and their sampling variances). She does this because the *selection* method for the observations is irrelevant to the fact that the values of the annoyance variables were produced by the random process. The annoyance-generating process produces the random variability about the model, not the randomization of the selection process. Indeed, any collection of observations would be well suited to the estimation task (i.e., yield unbiased estimates of a and b), given that the model is true. The collections differ only in the stability of the estimated coefficients (more on that later).

The survey statistician (describer) approaches the problem very differently. His job is to describe the relationship between distance and

---

[8] Clearly, if the researcher doubted the latter two assumptions then weighted least squares or generalized least square procedures might be used.

annoyance in the population sampled. This consists of the residents in the frame population at the time of the survey. Although hoping that the experiences of the residents at the time of the survey are informative about larger groups at later times, the describer is aware that unanticipated differences between that population and later or different populations may require a change of theoretical perspective. Certainly, he claims no statistical support for inference to a future population.

The describer is concerned that his sample observations do not mirror the frame population distribution. The residents close to the airport were oversampled by a factor of 2. If this fact were ignored, their observations would disproportionately contribute to the impression formed about the relationship between distance and annoyance. To counteract that, the describer uses selection weights, as in the case of estimating the population mean. In doing this the estimator of b is

$$\hat{b} = \frac{\sum_{i=1}^{n} w_i (x_i - \bar{x}_w)(y_i - \bar{y}_w)}{\left(\sum_{i=1}^{n} w_i (x_i - \bar{x}_w)^2 \right)^{1/2}} \quad \text{and } \hat{a} = \bar{y} - \hat{b}\bar{x},$$

where $w_i = 1/p_i$, the reciprocal of the probability of selection of the respondent, p_i, into the sample;

$$\bar{x}_w = \frac{\sum (x_i/p_i)}{\sum (1/p_i)} \, , \text{ the weighted mean for the } x \text{ variable;}$$

$\bar{y}_w =$ the weighted mean for the y variable.

This is an attempt to estimate the population regression coefficient, b, which is

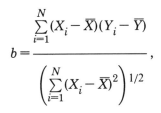

$$b = \frac{\sum_{i=1}^{N}(X_i - \overline{X})(Y_i - \overline{Y})}{\left(\sum_{i=1}^{N}(X_i - \overline{X})^2\right)^{1/2}},$$

where N = total number of persons in the frame population;

$\overline{X}$ = population mean for distance from the airport;

$\overline{Y}$ = population mean for annoyance.

This b, strictly speaking, is not identical to the modeler's b. It is a fixed property of a finite population.

The differences in approaches illustrate the gulf between analytic and descriptive uses of survey data. To the modeler the survey is a convenient data collection tool to obtain observations on the stochastic process of interest. The process is one that produces observations identical to one another, given their distance from the airport, except for random variability. Hence, to estimate the coefficients in the model, she need only know the two values, x_i and y_i. How the person happened to be chosen for observation is irrelevant. The describer is more interested in the frame population than the process. He knows his sample, in an unadjusted form, is a poor reflection of the population. Given a probability sample, however, he is able to know the nature of the discrepancies and to adjust for them with selection weights. He does so.

Can both researchers be correct? Let us eavesdrop on a likely conversation.[9]

Modeler: "What, in ordinary least squares estimation, prompted you to use these weights?"

Describer: "I've oversampled those persons close to the airport and I don't want that fact to affect my estimates of the regression coefficient, b."

Modeler: "Why would you think it would?"

---

[9] The literary form of the dialogue mirrors the presentation of Brewer and Mellor (1973).

Describer: "Well, I don't know for sure. I'm really protecting myself."

Modeler: "What do you mean?"

Describer: "It could be that the people close to the airport are different from those far away. Let me illustrate. Look at the scatterplot of the sample cases (Figure 6.6). See how the sample is heavily loaded close to the airport?"

Modeler: "Yeah, so? How could that affect the regression coefficient?"

Describer: "If these people close to the airport were different somehow... say, the scatterplot was more like this (draws Figure 6.7), then ...."

Modeler (sensing the kill): "Oh, I see. Because your model might be wrong, that's why you're concerned with the oversampling?"

Describer: "I don't think of it that way."

Modeler: "If that were the case, it looks like you need a new theory, say one that posited rapidly declining annoyance as a function of distance for those close to the airport and slowly declining for those further away. That's a problem for your theoretical development, but I still don't see what it has to do with the sample design."

Describer: "But I may not know that beforehand."

Modeler: "Look, the ordinary least square estimators make sense when the model is correct. The random process underlying the model is what we're about studying."

Describer: "OK, so is sampling completely irrelevant to you?"

Modeler (moving to the blackboard): "No, not completely. I know that the sampling variance of the estimate of the regression coefficient can be affected by the sample. Its value is

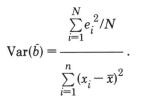

$$\text{Var}(\hat{b}) = \frac{\sum\limits_{i=1}^{N} e_i^2 / N}{\sum\limits_{i=1}^{n} (x_i - \bar{x})^2}.$$

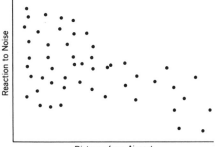

Figure 6.6 Measured reaction to noise by distance from the respondent's residence to the airport showing disproportionate allocation of the sample to those close to the airport.

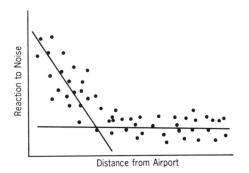

Figure 6.7 Alternative fitted lines, assuming declining annoyance among those near the airport and uniform annoyance with increasing distance among those beyond a certain distance from the airport.

The denominator is affected by the size of the sample and the average squared deviation from the mean distance, $\bar{x}$. This tells me that I want a large sample first of all. Second, if I really want to reduce the standard error of the estimates, I'd want to find x_i's that were far apart from one another. The extreme would be like this (draws Figure 6.8). This would then maximize the deviations of...."

Describer (shocked): "Whoa. You mean you'd omit all the people who lived, say, between 2 and 9 miles away from the airport?"

Modeler (in a soft voice): "Yes. You see from the formula for the sampling variance of the regression coefficient, $\hat{b}$, that such would be the case."

Describer (more aggressively): "But what if the full population scatterplot looked like this (draws Figure 6.9)? If you omitted all the people in the middle you would certainly get a wrong impression of the relationship of the two variables in the population. This seems ludicrous to me."

Modeler (annoyed): "I don't understand you people. You always try to change the topic! I thought we were talking about the linear model, $y_i = a + bx_i + e_i$. Why do you suddenly start talking about a different process, one that produces a

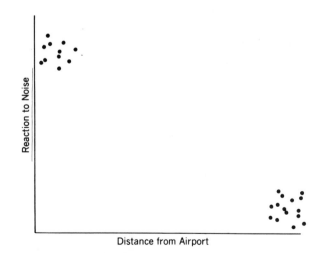

Figure 6.8 Optimal allocation of the sample to minimize standard error of regression coefficient under linear model.

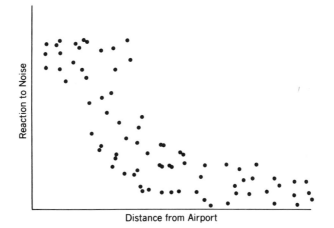

Figure 6.9 Scatterplot of reaction to noise by distance from the airport illustrating curvilinear relationship between reaction and distance.

curvilinear relationship? I *said* that given the model, the linear model, *the one we've been talking about for some time,* that taking extreme points on the distance variable makes sense. Of course, if you have a different process you're studying you would need a different sampling design to minimize the standard error."

Describer: "But you have more problems than just the standard error. Your regression coefficient is misleading — it doesn't describe the population relationship between distance and annoyance."

Modeler: "My friend, you are confusing the specification of a model and the selection process for observations. My job as an analyst is to form my theory correctly so that the model reflects the process I'm studying. If your process produces a curvilinear relationship, then your model should reflect it. Your problem is a misspecified model. I don't know the properties of the ordinary least squares estimators (or any others, for that matter) if the model is not properly specified. Your sample can't save you if you have the wrong model."

Describer: "Yeah, but have you ever seen a perfectly specified model? We're always in the situation of imperfectly specified models. I want to make sure that the observations I collect will alert me to poorly specified models. Without those points in the middle I would never be alerted to the curvilinearity."

Modeler: "Well, ok. If you're worried about that possibility, include them. But that still doesn't change my mind on weighting because you oversampled the nearest neighbors to the airport."

Describer: "But don't you see. It's the same point. If the sample doesn't reflect the population (I would add in known ways) then I'm not assured that I *will* be alerted to poor models."

Modeler: "It's not the same point at all. Look, let's go back to your original point. You've oversampled people near the airport. You want to reduce their contribution to the value of the regression coefficient. You say you want to do that because you suspect that the relationship between distance and annoyance is different among that group than among others. This sounds to me that you suspect your theory is wrong—a linear relationship is a misspecification— but you want to estimate one anyway. The weights don't let you ignore the fact that you have a misspecified model. I don't follow your reasoning."

Describer: "I may want the linear component of the relationship estimated only."

Modeler: "Why?"

Describer: "Well, as a first approximation, to check whether there's any relationship between the two variables in the population."

Modeler: "But that sounds like you're not really interested in understanding the psychological dynamics that create the relationship between distance and annoyance. It sounds alot to me like you're letting your data build your model. I guess I just don't understand you. Our job is to build theory, test causal models, to understand these stochastic processes. You're sounding more and more like a technician not a scientist. Don't you see that you can't avoid your responsibility of specifying the model correctly?"

Only the publisher's constraint of making this book of finite length forces the ending to this debate. It has no end. Nor, I expect, will it ever end completely. However, there are some key issues in this dialogue that need elaboration. First, with sufficient discussion the modeler and the describer would agree that *if the model were correctly specified* the sampling variance of the slope estimator is minimized by omitting the persons living mid-range distances from the airport. Such an allocation would be optimal in the sense that the standard errors of the coefficients would be minimized for any given total cost of the research. Alternatively stated, one could achieve the same standard error as that of the current design at much lower cost, if the cases in the mid-ranges were omitted from the sample. Costs can be reduced radically without increase in error if the model is correct.

There is a second consequence of omitting cases in the middle. Those cases might be needed to obtain unbiased estimates for some other population attribute. For example, assume that in addition to estimating the coefficients of the linear model of annoyance on distance, the survey was used to estimate the relationships between self-perceived psychological stress and salary levels of workers in the area. In this case, people who live in the deleted areas may have distinctive income levels not found in the other parts of the target population. Problems of "selection bias" described in Chapter 3 might affect the coefficients. These would be acknowledged as problems by both the describer and the modeler. The conclusion is that optimizing the sample under the assumption that one model is true may improve the estimation of parameters in that model but might severely damage other uses of the data.

Furthermore, the discussion above is not the only issue involved in the use of weights in regression models on complex surveys. Estimates of coefficients utilizing the selection weights often have higher sampling variance attached to them than those not utilizing weights. Modelers often point out to describers that even if they are correct in saying that the weights correct some biases of misspecified models, the estimated coefficients are less stable (higher standard errors). Following that, a mean square error perspective on error is desirable. That is, both bias and variance of the estimators must be discussed. The discussion above merely reflects bias features of the estimators.

The dialogue above concerns unequal probabilities of selection. The disagreement centers on whether the sample design should be reflected in regression estimators or ignored. A similar debate exists for clustering of the sample selections. Does this violate the assumption of independence of error terms? Are the estimators of sampling variance from ordinary least squares still appropriate? Another issue concerns stratification and unequal variances of residuals across strata. The debate always contrasts formal reflection of the sample design in estimates of the model coefficients or of their variances versus ignoring the sample design entirely. In short, each of the three sample design dimensions discussed above—assignment of probabilities of selection, clustering, and stratification—are part of the controversy in estimation of analytic statistics and of their sampling error.

The two groups differ on their willingness to rely on the assumption of a correctly specified model for selection of cases into the sample, for estimation of model parameters. In addition, however, they sometimes have fundamentally different reasons for estimation of the coefficients in the equation. The describers sometimes seek description of a finite population relationship, or at least its linear components through a model like $y_i = a + bx_i$. In some sense they view the slope coefficient as another descriptive statistic on the finite population (see Pfefferman and Nathan, 1981).

The modelers sometimes seek to identify causes and model their effects through similar analysis. In these efforts the two groups have very different populations of inference. The modelers either implicitly or explicitly view the one finite population from which the sample might have been drawn as only one possible population from an infinite set, a so-called "superpopulation." Each of the populations is equivalent because the same stochastic process being studied produced the y_i values in all of them.

Is there a right or wrong answer to the reflection of the sample design in estimating model parameters? The two camps both claim many followers. Almost all analytic statistics books promote the modelers' perspective. Many survey statisticians believe the describers' perspective is correct. After the smoke clears from the initial salvos, there is a narrow field of agreement. The describers will acknowledge that for weights to have an effect on model coefficients, the different probabilities of selection must be related to the dependent variable. As Dumouchel and Duncan (1983) note, a comparison of weighted and unweighted estimates may alert the analyst to possible misspecifications. Faced with large differences, analysts are urged to reexamine the theory leading to the model specification. The procedure would lead to respecification until convergence, then testing on an independent set of data. This approach is integral to sequential model fitting on survey data but is rather foreign to the single-model, single-test operation.

6.8 REFLECTING THE SAMPLE DESIGN WHEN ESTIMATING COMPLEX STATISTICS

At this writing, the case of those who argue for reflecting the design in model parameter estimates is hurt by (1) the absence of closed form expressions for unbiased estimators of sampling error (i.e., direct calculation procedures as opposed to iterative procedures) and (2) the retarded development of statistical software to do the job. Through a variety of replication techniques, sophisticated users can reflect stratified cluster unequal probability designs to estimate regression coefficients (through the REPERR program in OSIRIS.IV, SESUDAAN compatible with SAS, SUPERCARP, an independent program), to estimate parameters in weighted least squares analysis of contingency tables (PSTABLE in OSIRIS.IV), and for maximum likelihood estimation of log linear models (CPLX). All the programs require more work by the analyst to specify the design attributes properly and to move among several statistical software systems.

One practical question about reflecting the sample design in estimating complex statistics is "What difference does it make?" This essentially ignores the debate of the previous section and focuses on empirical effects of the different positions. The question can be interpreted as asking what happens to parameter estimates and their estimated standard errors. There are no analytic answers to that question, but there are some empirical studies that examine the practical import. We emphasize those treating complex stratified, multistage cluster samples and models using least squares estimators of regression model parameters. At the end of the discussion we mention other analytic techniques. We concentrate on sample designs that are equal probability selections.

In addition to the philosophical debate reviewed above, the field of variance estimation on complex statistics is complicated by the failure to identify most accurate estimators of sampling variance. There are three major candidates: (1) balanced repeated replications, a technique well-suited to paired selection designs, (2) jackknife estimation, another replication technique, and (3) a Taylor's series expansion estimator, using numerically derived estimates of partial derivatives. In addition to that there are bootstrapping estimators (McCarthy and Snowden, 1985). There are several empirical comparisons of these estimators (e.g., Kish and Frankel, 1974). These should be reviewed for insights into features of the alternative estimators. Our concern here is the magnitudes of design effects for analytic statistics found in stratified multistage samples.

Kish and Frankel (1974) performed the first large empirical studies of design effects on complex statistics in complex samples. These were based on repeated subsamples drawn from a large master sample, based on the Current Population Survey. Two hundred to 300 samples were drawn from this master sample, and the same statistics were calculated on each of the samples. This produced an empirical sampling distribution for every statistic. In addition, sampling error calculations reflecting the sample design were performed on each sample. The statistics examined included means on the total sample and differences of subclass means. Table 6.7 shows data from that example and from three other surveys for which estimates of sampling variance for means and for complex statistics are available. In every case the design effects for means on the total sample are larger than design effects for regression coefficients for models involving those same variables. Nevertheless, the design effects of regression coefficients remain larger than 1.0. That is, clustering effects appear to influence statistics measuring both relationships and simple means.

Table 6.7 Square Roots of Design Effects (deft's) and Sample Information for Ratio Means and Complex Statistics from Four Surveys

	1967 CPS	Health Exam Survey	Election Survey	Health & Nutrition Exam Survey
Topic	Employment issues	Health, demographic variables	Voting	Blood pressure, caloric intake
Means on total sample	1.443 (8)	1.80 (19)	1.106 (5)	2.22 (3)
Correlation coefficients	1.413 (12)	1.26 (51)	1.096 (4)	
Regression coefficients	1.108 (8)	1.29 (48)	1.015 (4)	1.57 (2)
Partial correlations	1.449 (6)	1.40 (48)	1.041 (4)	
Average sample size	846.5	3091	1111	20,749
Average cluster size	14.1	73.6	11.8	319.2
Number of strata	30	NA	47	35

Source: L. Kish and M. Frankel, "Balanced Repeated Replications for Standard Errors," *Journal of the American Statistical Association*, Vol. 65, No. 331, September 1970; M. Frankel, *Inference from Survey Samples: An Empirical Investigation*, Survey Research Center, 1971; W. Harlan et al., *Dietary Intake and Cardiovascular Risk Factors, Part II. Serum Urate, Serum Cholesterol, and Correlates*, Series 11, No. 227, National Center for Health Statistics, 1983.

6.9 SUMMARY

To many readers of survey results, sampling error is erroneously interpreted as a summary measure of all the errors in the estimates. Other difficulties in the data collection process, however, can reduce the quality of survey statistics. Sampling error represents only one component of total error. Treating standard errors as reflections of total error ignores biases in statistics that can arise from coverage, nonresponse, and measurement errors. In addition, such a view ignores the variability in survey results due to these same error sources.

However, as we noted in Chapter 1, because sampling error is routinely *measurable* (with probability sampling) and *measured* (by both analysts of the data and designers of future studies), it is often subjected to disproportionate attention among all the other errors of survey estimates. That which is measurable permits evaluation of reduction efforts. These efforts, although expensive (through the cost of stratification or increased sample sizes), reduce the measures of error that analysts use to qualify their conclusions from survey statistics. With such visible rewards for spending scarce research dollars, it is understandable that researchers often think first about purchasing a large, efficient sample, then later about how to reduce more complex nonsampling errors. Those nonsampling errors unfortunately yield themselves less easily to empirical measurement and control exerted through design features.

Probability sampling offers its users unbiased estimates of a variety of descriptive statistics and estimates of their sampling error. Without known probabilities of selection, the analyst is forced to posit some model of the process by which the data were generated. Thus, users of quota samples assume homogeneity within quota cells for estimates of population means. These samples may provide better information about the larger target population than do probability samples, but they lack the model-free estimates of sampling error which probability samples offer. Well-designed nonprobability samples, like those analyzed by Stephenson (1969) (see Section 6.3.2), resemble full probability samples on many of their attributes. When they are compared to full probability samples they yield small differences.

Advocates of probability samples have traditionally believed that the strength of unbiasedness and measurable sampling errors were sufficient causes to avoid other selection methods. This position is well founded if sampling error were the only concern. Especially when errors of nonresponse are also considered, these unique attractions disappear. A national probability sample survey with a 50 percent response rate in the large urban areas of the country and 90 percent in other areas is unlikely to have the same bias and variance properties as one with 100 percent response everywhere. The Smith and the Stephenson analyses in Section 6.3.2 demonstrate this. The competition between probability and nonprobability sampling is thus related to the differences in perspectives on survey error described in Chapter 1.

We reviewed in this chapter, however, another debate relevant to sampling error—that between describers and modelers, between design-based and model-based estimation. Here the conflict concerns whether the selection method has effects on how survey statistics and their sampling errors should be estimated. This is a debate common to environments in which survey data are used to estimate causal models of social processes. The debate centers around whether the model under investigation is well specified and whether all inference must be conditional on the specification of some model, implicit or explicit. This debate is likely to continue.

Finally, because sampling errors are measurable and reducible under a wide variety of designs, they have yielded themselves to explicit cost and error modeling. Optimal sampling designs under fixed research resources can be identified. These minimize sampling error relative to all other designs. This use of cost and error modeling is the foundation of attempts in this book to acknowledge cost constraints as limits on efforts to reduce nonsampling errors. The reader will note that these attempts are only partially successful and are critically affected by the weak measures available for some of the nonsampling errors.

CHAPTER 7

EMPIRICAL ESTIMATION OF SURVEY MEASUREMENT ERROR

A man with one watch always knows what time it is; a man with two watches always searches to identify the correct one; a man with ten watches is always reminded of the difficulty of measuring time.

Anonymous

To err is human, to forgive, divine—but to include errors in your design is statistical.

Leslie Kish, *"Chance, Statistics, and Statisticians"*, Journal of American Statistical Association, *Vol. 73, No. 361, 1978, p. 2*

7.1 A FIRST EXAMINATION OF OBSERVATIONAL ERRORS VERSUS ERRORS OF NONOBSERVATION

Thus far the discussion has focused exclusively on errors of nonobservation, failures of a survey to measure some portion of the population, and on the costs involved in reducing those errors. Beginning with this chapter the attention turns to errors in data actually obtained by the survey, termed "measurement errors." Although there are many possible ways to classify these errors, they are seen as arising from four principal design features in the measurement process: the interviewer, the respondent, the survey questionnaire, and the mode of communication.

As with coverage, nonresponse, and sampling error, we are first interested in assessing the magnitude of measurement errors in real survey data. The reader may profit from reviewing the various definitions of measurement error that we introduced in Chapter 1 (Sections 1.4 to 1.7). In one sense, this chapter examines how those concepts of error are operationalized in data collection designs. It will present alternative research designs used to estimate measurement errors. Once we review these alternative estimation approaches in this chapter, we search for evidence for the causes of measurement error and illustrate survey designs intended to reduce the four sources of measurement error. This is done in Chapters 8 to 11.

In this chapter we find the following:

1. Different designs are used to estimate variable measurement errors and biases.

2. Researchers who conceptualize an error as a bias often seek to identify the *cause* of the error and eliminate it; those who conceptualize an error as a source of increased variance in survey statistics often attempt to measure it and adjust their conclusions to its presence.

3. Most past work studies only one of the design features (the interviewer, the respondent, the questionnaire, or the mode) as a source of measurement error.

4. There are different approaches used by those building causal models (studying relationships among variables) and those calculating descriptive statistics.

Many of the approaches to measurement error estimation are based on formally specified models of the measurement process. Just as we saw in Chapter 1, the various models label error terms in different ways. In addition, the standard notation used for the different models across the fields will differ. In our attempt to integrate these various perspectives we have chosen a unifying notation. To those readers familiar with one or another of these different approaches, the chosen common notation will not be familiar and will slow their comprehension of the material. We hope that the ease of comparing the models, given a unified notation, is worth that price.

Finally, although this chapter is devoted to presenting research designs and estimators of measurement errors, it also contains examples of those designs, to aid comprehension. We do not discuss them in full here. For that reason, many will be examined again in later chapters devoted to different sources of measurement error.

7.2 LABORATORY EXPERIMENTS RESEMBLING THE SURVEY INTERVIEW

Research on survey measurement error is sometimes difficult to perform in the uncontrolled setting of the survey interview. For example, if unusual additional data must be collected to study measurement errors, the very collection of those data may interfere with the survey interview. Although this is not the typical design, some studies use nonsurvey settings to learn about components of error.

For example, Rustemeyer (1977) had interviewers conduct "mock interviews" with supervisory personnel in an office, while a tape recording was made of the interaction. This resembled an experimental setting where several sources of variability typical of real survey settings were removed: (1) the setting was constant over different interviews done by an interviewer and (2) the answers given by the "respondent" (the supervisor) were the same for each interviewer. Stripping the measurement process of irrelevant sources of variability allowed the research to concentrate on errors associated with interviewer behavior.

As with all experimental work that is removed from the setting to which inferences will be made, such research is subject to some questions. Did the tape recording affect the interviewer behavior in a manner to increase or decrease questioning errors? Did the use of supervisory personnel as "respondents" affect these errors? Does the office setting produce more or fewer errors than natural settings? Such questions cannot usually be answered by the experimental designs themselves. For that reason, controlled laboratory designs are often used as precursors to survey measures themselves.

Another set of experiments relevant to measurement error uses traditional psychology laboratories. One line of research focused on the medium of data collection as a source of measurement error, comparing audio and visual contact between persons with contact in an audio medium only (see Short et al., 1976, for a comprehensive summary). Most of these experiments were motivated by interest in the process of communication in business meetings in face to face settings and by telephone. One experiment (Champness and Reid, 1970) asked 36 subjects to communicate the contents of a 200 word letter to another subject. Three forms of the letter varied the positive or negative content (e.g., using words like "possible" versus "impossible") and certainty or uncertainty (e.g., "will be possible" or "may be possible"). The subjects communicated either face to face, by telephone, or in a situation without visual contact but with direct audio communication. Subjects were given as much time as desired to communicate the contents. The subjects receiving the information were asked several questions about the content.

There were no significant differences among the media in the accuracy of receivers' reports of the letter content. Here, the goal of the researcher is to understand the influence of the medium, stripped away from all the other characteristics often related to the choice of medium in surveys (e.g., the sampling frame, the centralization of interviewers). Applying results to surveys requires replication in real survey settings.

There are many other laboratory experiments comparing face to face communication with telephone communication when the task involves conflict resolution or affective perceptions. Clearly, some of these have relevance to the communication and cognitive task presented to interviewers and respondents in a survey situation, but the setting and tasks are somewhat different from those of the survey interview. In the Champness and Reid experiment, for example, the subjects were given an unlimited amount of time to communicate the letter. In survey settings, however, communication is limited by time constraints generally imposed by respondents anxious to return to their normal activities.

The use of laboratory experiments to study survey measurement error properties is currently increasing because of the interest in possible insights from cognitive psychological theories into the causes of respondent errors in surveys. Cognitive theories have by and large been tested and refined in controlled laboratory settings, using student subject pools. The survey response task resembles some of those studied in the laboratory tests, and has led to speculation about the direct applicability of the cognitive theories to the survey setting. Tests of these speculations are using both laboratory settings (Lessler et al., 1985) and more traditional surveys. For example, one technique used in the cognitive research laboratories is protocol analysis or "think aloud" techniques (see Royston et al., 1986). This asks the respondent to verbalize his/her thoughts while answering the question. This attempt to observe the cognitions "as they happen" is an effort to learn what tasks the questions present to the respondent and the methods used to perform them. Respondents are sometimes asked to paraphrase the survey question in order to provide insight into their comprehension of the question's intent. Respondents are asked to assign levels of confidence to the answers they provide. The number of seconds passing between the delivery of the question and the response (a "response latency" measure) is recorded as an indicator of the depth of cognitive processing required to obtain an answer.

Most of the laboratory studies do not specify a formal measurement error model. However, the experimental situations often permit the assessment of the true value of measures taken on the subjects. In those cases the actual error in response is determined. Most often the implicit measurement model seems to focus on measurement bias, assuming that the experimental variables being manipulated produce constant effects over replications for all subjects. When the true values of measures are not easily observed, experimental treatment groups are compared. For example, alternative wordings of the questions might be compared to measure the effects of wording on the response task. These comparisons are used to estimate whether differences in statistics are to be expected under the different conditions. Few of the experiments attempt to measure variation in these differences over replications of the measurement. They thus exclude measurement error variability from their perspective.

Such laboratory experiments are clearly inappropriate tools for estimating survey measurement error directly. However, they may be valuable as a way to identify *causes* of survey measurement error. They may be able to strip from the survey interviewing situation extraneous and confounding characteristics of the response behavior, retain the main influences, manipulate variables that have direct impact on the quality of recorded survey data, and reduce mistakes in communication of question, respondent cognitive processes, or interviewer receipt and recording of answers. If so, survey researchers could then attempt to implement new procedures in the more uncontrolled survey setting to reduce measurement errors.

7.3 MEASURES EXTERNAL TO THE SURVEY

One way to estimate measurement error directly is to compare the result of a survey measurement to another indicator of the same characteristic, one likely to be more accurate. There are two methods that are often used to do this: one obtains external data on individual persons in the survey, while another compares external population parameter values with survey-based estimates of the same quantity.

Perhaps the most common method of assessing measurement error is the use of a "record check study" or "validation study." Such a study generally assumes that information contained in the records is without error, that is, the records contain the true values on the survey variables. There are three kinds of record check study designs:

1. **The reverse record check study** (or retrospective design), in which the entire sample is drawn from the record file for persons with a trait under study, interviews are taken containing questions about information also contained on the records, and a comparison of survey data with record data is performed to estimate measurement error. The phrase "reverse record check" presumably is used because after the survey

is over the researcher goes back to the records (which were the source of the sample) to check the survey responses.

2. **The forward record check study**, in which the sample is drawn from a separate frame (e.g., area frames, telephone list frames) and after survey responses are obtained, relevant record systems containing information on respondents are searched. In some studies survey questions must be asked about the location of records containing similar information on the sample person, a request is made to the sample person to obtain the record file, the record is acquired, and a comparison of survey values is made to record values. The word "forward" implies that the researcher moves from the survey data to new sources of record data for the validity checks.

3. **The full design record check study**, promoted by Marquis (1978), which combines features of both designs, obtains a sample from a frame covering all persons of the population, and seeks records from all sources relevant to those persons. Survey errors associated both with failure to report an event as well as erroneously reporting an event are measured by comparison to all records corresponding to the respondent.

Reverse record check surveys were the basis of assessing error in measures of victimization for the National Crime Survey (e.g., Law Enforcement Assistance Administration, 1972). In these surveys police department records of reported crimes were sampled, and the victim whose name appeared on the record was sought out for an interview. The interview asked the person to report crimes of a certain type that occurred during a fixed period of time. The interview data were compared to the police record data, and tables like that in Table 7.1 were presented. This shows that, for example, 62.6 percent of the violent crimes sampled from the police records were reported by the victims, when those victims were successfully located and interviewed. Matching rates were lower for violent crimes than for property crimes.

It is important to note that reverse record check studies by themselves fail to measure errors of overreporting. For example, there are no estimates in Table 7.1 about what proportion of the incidents reported in the survey but not matched to the police records were correctly reported (i.e., eligible incidents occurring within the given reference period). These can occur because the police record system does not contain all victimizations that occurred. Instead, reverse record checks can measure what portion of the records sampled correspond to events reported in the survey. They can also measure whether the characteristics of those events are the same on the record as in the survey report.

Table 7.1 Cases Sampled from Police Records by Whether Crime Was Reported in Survey "Within Past 12 Months" by Type of Crime

Type of Crime	Total Police Cases Interviewed	Percentage Reported to Interviewer as "Within Past 12 Months"
All crimes	394	74.1%
Violent crimes	206	62.6
Assault	81	48.1
Rape	45	66.7
Robbery	80	76.3
Property crimes	188	86.2
Burglary	104	90.3
Larceny	84	81.0

Source: Law Enforcement Assistance Administration, *San Jose Methods Test of Known Crime Victims*, 1972, Table C, p. 6.

The only record data accessed in a reverse record check are the sample records, and the design can only record what proportion of the events in the sample records are reported. In a sample of records of individual crime incidents as above, this failure may be severe. In a sample of physician records for persons, if the validation study is estimating measurement error for reporting of doctor visits, the loss is limited to visits to other physicians not covered by the record system.

Forward record check studies appear to be less common. They generally entail contacts with several different record-keeping agencies and (in the United States) explicit permission of the respondent to seek record information from the agency holding it. For example, Andersen and Anderson (1967) asked a national cross-section household sample of 7803 persons about hospital stays of overnight or longer. The hospitals corresponding to the over 1000 hospital stays were asked to list all stays reported for the respondent during the 12 month reference period for the survey. For the responding hospitals there were about 100 survey reports for which no record evidence could be found and 10 record reports for which no survey report was made (by respondents reporting some other stay). This yields a net overreporting finding; respondents appear to report more stays than are on the records. The strength of forward record check designs is their ability to measure overreports in the survey.

In contrast to reverse record checks, forward record checks are limited in their measurement of underreporting. They learn about the failure to report events only when mention of those events appear on records corresponding to other events which *are* reported. Records are not searched for those respondents who fail to report any event. Hence, if their nonreporting is in error, these underreports escape notice by the forward record check design.

The full design record check avoids the weaknesses of the reverse or forward design but requires a record data base that covers all persons in the target population and all events corresponding to those persons. There are some examples of this in countries with good population registers (Kormendi, 1988) and in certain restricted population cases in the United States. For example, medical care studies of members of health maintenance organizations can be full design record check studies with the assumption that all medical care is provided by the organization and the inference is limited to the population of members of the organization.

All the record check designs share three other limitations. First, most contain the explicit assumption that the record systems are without errors of coverage, nonresponse, or missing data. For example, in a police record-keeping system the assumption is that all crimes reported to the police are indeed filed in the record system. This assumption is challenged by the fact that some reports are "unfounded" by police officers, who, after taking the report of the victim, judge that the event is not worthy of completion and filing. Another example of coverage error is the misplacement of a hospital record in the filing system.

The second limitation common to record check studies results from problems of matching the respondent survey records with their administrative records. Neter et al. (1965) demonstrate that matching errors can have important effects on the estimation of measurement errors. They examine the case in which within different subgroups of the population records might be mismatched with interview data. For example, in a reverse record check of bank accounts in a survey on assets, mismatches are most likely among persons living in the same household and sharing a surname. *If mismatches occur at random within the subsets,* the expected mean difference between interview responses and mismatched records will be equal to that of the expected mean difference between interview responses and correctly matched records. However, even under such restrictive assumptions, the variance in response errors will be overestimated with the possibility of mismatching and the regression of measured response error on the matched record value will have a smaller slope than that of correct response error on correct record value.

The bases of these observations are the practical problems of record check study administration. If the respondent is "Harold Thompson," living at "123 Main Street," but if the record closest to a match is "Harold

Thomas" at "123 Main Street," what should the researcher do? What if "Harold Thomas" lives at "123 Maynard Street"?

Miller and Groves (1985), in a reverse record check study for victimization reporting, describe mismatching problems not on the level of the respondent but on the event being described. A ratio was used to measure the degree of match between survey reports and police records. In the numerator was a count of all survey reported incidents that fulfilled the match rule. In the denominator were all sampled police reported incidents. It was possible for more than one survey report to match the same police report. The ratio is found to vary between 0.136 and 1.149 depending on the strictness of criteria concerning the time, location, and type of crime reported. For example, the respondent reports someone stealing the hubcaps from his car in April 1984, and the police record is for the stealing of hubcaps and breaking an antenna in May 1984. Has the respondent misdated the event or reported another event?

The third limitation is that the record itself is assumed to be without measurement error. "Measurement error" in this context means that the contents of the record apply to the person in whose file they are placed, that the characteristics of the event (e.g., date, duration) appear without error, and that no characteristics of the event have been omitted. If errors on the records are allowed within the research design, then differences in statistics calculated on the records and those on the survey instruments are estimates of "relative difference in measurement error." Furthermore, errors on the record may be correlated with the errors in the survey reports. For example, victims of crime are sometimes offenders as well. Drug users are disproportionately victims of assaults. Both survey responses and police records for these events are subject to great distortion in reports. This is an example of positive correlation among errors in the two measurement systems.

Using the terminology introduced in Chapter 1 (Section 1.4), one would say that record check studies are usually employed to estimate measurement errors that are conceptualized as biases.[1] They often explicitly compare measurement procedures (e.g., different questioning strategies) to see which will match the record data most closely. That one doing so is judged to have the lowest "measurement bias." Generally omitted from this view is the possibility of measurement variance differing between the two approaches.

---

[1] This is clearly not a limitation of record check study design. Most could be used to estimate variable measurement errors, but most investigators who use record check studies conceptualize the errors as fixed over replications of the survey.

An alternative use of external data to validate survey statistics does not obtain matching reports on a person level. Instead, survey-based estimates of population parameters are compared to estimates of the same parameters based on another methodology. For example, surveys of political behavior routinely ask persons whether they voted in the last election and compare the survey estimate with the published number of votes. Groves and Kahn (1979) compare their household sample survey estimate of 69.1 percent voting in the 1972 presidential election with the 57.1 percent reported by cumulating actual vote counts precinct by precinct. As estimates of measurement error, the comparisons are subject to coverage errors between the full population of registered voters and the household population, nonresponse errors in the survey, vote fraud, and arithmetic errors in the voting statistics. Furthermore, although these comparisons are useful on the aggregate level, they provide no data on the measurement errors of different subgroups in the population. For example, we obtain estimates of "net" measurement error but are not able to estimate separately errors of reported voting among nonvoters and reported nonvoting among voters. Thus, the potential correlation of errors and true values cannot be investigated. This is the advantage of matching survey report to records on a person level.

The largest weakness with the above designs that utilize external data to estimate survey measurement errors is that only a limited number of variables are measured in administrative record systems. These involve voting records, hospital records, criminal records, school records, business records, financial records, and several other sets of "objective phenomena." The designs therefore provide no assistance to those interested in the measurement error properties of attitudes, beliefs, or activities that do not involve contact with an organization. Furthermore, the difficulties in obtaining access to a record base, assessing its quality, and matching survey records to the administrative records, all produce large costs in record check studies. This book continually reminds us that reduction in survey errors often requires increase in survey costs. Record check studies exemplify how the quality of estimation of measurement error also is related to the costs of the methodological study.

7.4 RANDOMIZED ASSIGNMENT OF MEASUREMENT PROCEDURES TO SAMPLE PERSONS

The randomized experiment is perhaps the most frequently used method to estimate survey measurement error. This tool is given different labels among practitioners of surveys. Mahalanobis (1946) described the method of "interpenetrated" samples, in which subsamples of identical design were given different treatments (e.g., measurement techniques, interviewers assigned). The phrases "split sample experiment," "random half-sample" experiments, or "split ballot" experiments are most often used when only two or at least a small number of alternative design features are compared (e.g., two wordings of a question). "Split ballot" appears to be the label of choice for social scientists.

Randomization is a tool used to study both measurement errors conceptualized to vary over replications of a survey and those that are fixed features of the design. When randomization is used to study variable errors, the investigator attempts to employ and compare many different entities viewed as the source of the error (e.g., many different interviewers, for interviewer variance estimates). When errors conceptualized as biases are studied, generally only two or three alternative designs are compared (e.g., two different wordings of a question). These differences commonly result from the fact that the precision of estimates of variance is a function of the number of error-producing units (e.g., interviewers) used, while estimates of bias are based on the assumption that one method being compared is errorless or at least the preferred method.

Randomized assignment of interviewers to subsamples of respondents is introduced in order to measure "correlated response variance," arising from the fact that response errors of persons interviewed by the same interviewer may be correlated. This is sometimes called the "interviewer variance" component and is based on the model that the set of interviewers would vary over replications of the survey, and hence the measurement error of the survey statistics will vary over replications. On first encounter such a model may seem unattractive to some analysts, but the model is the basis of a large literature in the estimation of interviewer effects on survey data (see Chapter 8 for a full discussion).

The model used for sampling error asks the analyst to conceptualize variation in the values of statistics over repeated sampling of persons from the population, even though only one sample is actually drawn. This model probably has more intuitive appeal because survey samples generally form only a small portion of the frame population. In contrast, the selection of interviewers to work on a survey often relies on the identification of the best applicants (sometimes among a very small number) for the job. The "population" of interviewers from which the successful applicants are chosen (the frame population) is often only vaguely defined, and the possibility that different interviewers might be selected in another trial of the survey is not so obvious as with random selections of persons at the sampling step.

From another perspective, however, the concept of interviewer variance can be an important part of the variability in results. If different interviewers do obtain different answers from the same respondent, the analyst would expect the survey results to depend on which interviewers were used. Conditioning inference from his/her survey to only those surveys that use the same interviewers would be seen by most as severely restricting possible conclusions from the study. Indeed, if the values obtained would not be expected with another set of interviewers, the results have much less value to most analysts.

Alternatively, the researcher can view the set of interviewers actually used on the survey as the result of a stochastic, random selection process. Despite attempts to apply deterministic selection criteria to interviewer applicants, and despite applications of the same training procedures to all interviewers, each set of interviewers obtained exhibits variation in skills. The particular set of interviewers used on any survey is a collection of outcomes from this stochastic process of selection and training and thus will induce the levels of correlated response errors expected from the selection and training design. Measuring the internal variation among interviewers actually used can provide estimates of variation to be expected in replications of the survey design due to different sets of interviewers. The stability of the interviewer variance estimates is a function of the number of interviewers used in the design (and to a lesser degree the number of cases they complete).

Randomization is used differently when measurement errors are seen to arise from the survey question. In contrast to the variable measurement errors above, the form of the survey question is often viewed as a fixed property of the design. It is productive of "measurement bias" not "measurement variance." Split sample experiments are used to compare survey statistics based on two different versions of the question. Achieved differences between the two values of the statistic are estimates of relative bias in the two question forms. A similar viewpoint is generally taken for experiments with interviewer training, interviewer behavior, and mode of data collection.

Why do researchers in split sample experiments generally conceive of the errors as biases rather than variable errors? One answer is that with only two or three different forms, it is simple to display the values for the statistic of interest from each one separately. With scores of interviewers such display of individual results becomes less informative. Another answer is that there is less prior belief on the investigator's part that the two forms of the question are equivalent than that the various interviewers are substitutable for each other. However, that answer would seem to disappear upon careful inspection of individual differences among interviewers. A third answer is that the researcher can conceive of using only one form of the question over all possible trials of the survey. Thus, any measurement errors connected with it would be present in all replications of the survey. In contrast, the conditioning of all inference for all statistics on the particular set of interviewers used is not as attractive. Finally, a fourth answer is that the purpose of split sample experiments is explicitly the estimation of differences between techniques. *Explanations* or *causes* of differences are being sought. In contrast, variable errors are measured in order to place qualifications on inferences from the survey data; the causes of the variability in errors are not necessarily of any interest to the investigator.

7.4.1 Split Sample Experiments Concerning Measurement Error

There are large literatures in survey methodology based on comparisons of results in random subsamples of a survey. For example, measurement errors associated with question wording are usually investigated this way. Schuman and Presser (1981) compared two forms of many different survey questions intended to measure the same construct by asking one question of one random half-sample and another of the complement half-sample. Survey estimates based on the two different questions were compared to measure the effect of question wording. For example, in a random half-sample of the households of a personal interview survey, the selected adult respondent was asked:

> "Do you think the United States should forbid public speeches in favor of communism?"

In the complement half-sample, an alternative form was asked:

> "Do you think the United States should allow public speeches in favor of communism?"

The interest in these two forms stems from the hypothesis that the questions have exact opposite meanings. Everyone answering "yes" to the first should answer "no" to the second; every person answering "no" to the first should answer "yes" to the second. The goal of this research is to test this hypothesis. To do so requires some controlled comparison of results. Although this might be obtained by asking both questions of the same respondent, this option is rejected by most researchers. The experience of answering the first question is thought to affect the cognitive activities in answering the second. Instead, each form is administered to random half-samples, and the overall results are compared, as in Table 7.2. A smaller percentage of people (39.3 percent) support "forbidding" such speeches than say they should not be allowed (56.3 percent).

Table 7.2 Percentage of Respondents Forbidding and Not Allowing Speeches in Favor of Communism

Do you think the United States should forbid speeches in favor of communism?		Do you think the United States should allow speeches in favor of communism?	
Yes (forbid)	39.3%	No (not allow)	56.3%
n	(409)	n	(432)

Source: H. Schuman and S. Presser, *Questions and Answers in Attitude Surveys*, Academic Press, Orlando, 1981, Table 11.2, p. 281.

Such an experiment is used to suggest that the expected results from the two forms over all samples are different. The difference (39.3 − 56.3 = −17.0 percent) could be used as an estimate of *difference* in measurement bias between the two question wordings. Such a formal labeling of the difference as related to measurement bias, however, is not used by Schuman and Presser, nor by most other researchers in this tradition. Instead, the result prompts the use of social psychological and psycholinguistic arguments to explain the difference in results. They note different connotations of "forbid" and "allow." They call attention to the different grammatical structures in the questions. The arguments appear to imply that the questions measure different concepts which, although closely related to one another, are subtly different.

This line of research is quite distinctive from the others reviewed in this chapter. Throughout most of the work the researchers do not explicitly address whether the two forms of the question are viewed as measures of the same underlying concept, subject to different response biases, or whether they are really measures of two different concepts. Instead, the focus of the research is on learning what features of the questions produce the differences. By not explicitly addressing the underlying concept to be measured by the questions, notions of measurement error are not key to the research.

Split sample experiments are also used to study measurement biases associated with interviewer behavior. In a large set of experimental studies Cannell and his colleagues (see Cannell et al., 1981, for a summary) have explored the use of controlled interviewer behavior during the survey interaction. For an "experimental" random half-sample, the designs ask for a commitment of the respondent to give correct answers to the interviewer's questions, provide instructions to the respondent regarding the intent of individual questions, and give standardized feedback to the respondents regarding the quality of their performance.

This set of interviewer behaviors was compared to another (called the "control" procedure) which allowed more flexibility but was the natural result of traditional interviewer training. In terms of the survey error structure we are using in this book, the work could be classified as attempting to estimate measurement bias associated with training and interviewing procedures. In most of the studies by Cannell, all interviewers used both methods, alternating between the "experimental" method and the "control" method.

Most of the studies show that the new interviewing procedures yield increased reporting of events believed to be underreported in most surveys. For example, Table 7.3 shows that the use of commitment, controlled feedback, and instructions leads to a higher percentage of respondents reporting at least 1 day in bed in the last 2 weeks because of illness (10.0 percent) relative to the typical interviewing procedures (7.3 percent). Similar results are found for other health events. Such results are sometimes interpreted as estimates of reduction in measurement bias. This conclusion is based on (1) the premise that such health events tend to be underreported and (2) evidence of higher reporting in the experimental condition. The premise is an assumption that can be evaluated only by external data (e.g., see Madow, 1967) and not with the experimental data themselves. In contrast to the question wording literature in attitudinal areas, however, this research generally has the implicit assumption that the two procedures measure the same underlying concepts, and thus differences between procedures are often related directly to changes in measurement biases.

It is noteworthy, however, that most of the measurement error literature employing split sample methods (1) does use explicit measurement models in their work and (2) does include variable measurement errors in the approach. If it were to be specified, a formal measurement error model applicable to this work probably resembles

Response = True Value + Method Effect + Random Error

$$y_{ij} = X_i + M_{ij} + \varepsilon_{ij},$$

where y_{ij} = response obtained for the ith person using the jth method or form;

X_i = true value of the characteristic for the ith person;

M_{ij} = effect on the response of the ith person of using the jth method;

Table 7.3 Percentage Reporting at Least One Health Event by Interviewing Procedure for Various Types of Events

Statistic	Percentage Reporting at Least One Event in 2 Weeks by Interviewing Procedure	
	Control	Experimental
Days in bed because of illness	7.3%	10.0%[a]
Loss of work day because of illness	6.3	8.8[a]
Days of reduced activity	8.4	11.5[a]
Dental visits	6.8	7.4
Doctor visits	17.4	17.5
Acute conditions	14.9	17.7[a]
Average n	4217	3993

[a]Difference between control and experimental procedures statistically significant at the 0.05 level or less.

Source: C. Cannell et al., *An Experimental Comparison of Personal Interview and Telephone Health Surveys*, National Center for Health Statistics, 1987, Table M. Reference period for all questions was the 2 week period preceding the interview.

ε_{ij} = deviation for the ith person from the average effect of the jth method.

We use the term "random error" above, which is given many meanings across the different literatures concerned with survey error. In this context we use it to mean that its value is uncorrelated with any other term in the measurement model and its expected value over replications is zero for each person in the population. This meaning is used consistently throughout this chapter.

Although there is no formal model employed, several features of the measurement do seem obvious from the literature. First, the method effect is generally not viewed to be constant over all subgroups (M_{ij} probably varies over different persons, i). *This is an important distinction between this literature and those we shall encounter later in this chapter.* A favorite correlate is educational attainment, most often used as a proxy indicator for cognitive facility with the question and answering task. Second, partially because of this, across persons the method effect is correlated to the true values of the concept being measured. Thus, in contrast to the models studied below, the covariance of the true values and the method effects are not assumed to be zero, $\text{Cov}(X_i, M_{ij}) \neq 0$. For example, those who have actually experienced some health event are the only group susceptible to the change in interviewer behavior studied by Cannell.

Research using split sample methods sometimes acknowledges variable measurement errors but rarely does it attempt to measure them. (Indeed, the split sample design itself is poorly suited to such estimation.) That is, using the model above, for example, there is little empirical work in this field concerning the variance of ε_{ij} over replications for the same person. There is no empirical investigation about whether the reliability (in the psychometric sense of the term; see Chapter 1, Section 1.5) of one form of measurement is higher than another. That is, over repeated administrations of one question wording does the stability of answers differ from that of another wording? Instead, the emphasis is on error that is viewed to be constant over replications of the design.

This omission of attention to variable errors leads to the possibility that while measurement bias is reduced with one wording or one interviewer behavior or another, measurement error variance (variation over administrations) is increased. If that were true, the technique preferred on bias terms alone may not be preferred in terms of total error.

7.4.2 Interpenetration for Estimation of Variable Measurement Error

Sometimes the investigator is faced with several possible measurement procedures that are viewed as essentially equivalent ways of obtaining the needed data. By "essentially equivalent" is meant that any differences in results arise from an unknown or unmeasured cause. There are two commonly used examples of this: interviewers as a source of variable measurement error in surveys and the ordering of questions or response categories as a source of measurement error. In both cases, there will typically be no prior guidance for the choice of best procedure (i.e., the best interviewer or the best ordering). This is the question of a "reducer" interested in eliminating error from surveys. Instead, efforts will be made to measure the variability in results possible from these sources. That is, "measurers" versus "reducers" contribute to these literatures.

To estimate the component of variance in survey statistics attributable to some survey design feature, interpenetration of the design is used. For example, random subsets of the sample are given different orderings of the questions or response categories.

In one sense, we view this research as related to the split sample methods, when the number of methods is very large (instead of two forms of a question wording, we are interested in hundreds of possible orderings of the questions). In another sense, however, the approach is very different from that using split sample methods. This is because the researcher is generally interested not in measuring bias associated with each of the methods but rather variation in results due to method. An effort is made to (1) include all possible *but equivalent* methods in each trial of the survey, (2) induce in each survey statistic the variability inherent in the measurement process, and (3) measure the variance component in the survey statistics corresponding to the different methods. Thus, the goal is direct estimation of variable measurement error, not measurement bias.

The measurement models for interviewer effects are identical to those applicable to the split sample investigations. Typically, however, additional constraining assumptions are placed on terms in the model. These are generally introduced not because they enhance the realism of the model but rather because they aid the empirical estimation of terms in the model. The Hansen–Hurwitz–Bershad (1961) model, sometimes referred to as the U.S. Census response error model, is based on

Response = True Value + Interviewer Effect + Random Error

$$y_{ijt} = X_i + M_{ij} + \varepsilon_{ijt},$$

where y_{ijt} = observed response on a survey item for the ith person, the jth interviewer on the tth trial;

X_i = the true value for the variable for the ith person;

M_{ij} = the effect on the response of the ith person of the jth interviewer;

ε_{ijt} = a deviation of the ith person by the jth interviewer on the tth trial from the true value and the average interviewer effect.

First of all, it should be noted how similar this model is to the one implicitly used by those employing split sample methods. The "interviewer effect" here replaces the form or method effect in the prior discussion. The difference between the two approaches lies in sets of assumptions about the terms.

The model does not focus on possible biasing effects of interviewers (M_{ij} terms) or of the individual errors (ε_{ijt} terms). That is, over all interviewers and persons, the expected value of the interviewer effect terms is zero: $E(M_{ij}) = 0$. Similarly, the expected value of the ε_{ijt} terms is zero: $E(\varepsilon_{ijt}) = 0$. These assumptions are not too consequential for the estimation of interviewer variance, but they do affect estimates of total mean square error. That is, if the entire set of interviewers tend to induce overreports or underreports of the X_i, then the sample mean will be biased regardless of which interviewers are chosen for the study. This possibility is eliminated by the assumption of $E(M_{ij}) = 0$. Finally, the random error terms, ε_{ijt}, are viewed to be completely nonsystematic disturbances to the obtained response, whose values are unrelated to which interviewer is assigned or the true value for the respondent being examined. That is, $\text{Cov}(X_i, \varepsilon_{ijt}) = 0$ and $\text{Cov}(M_{ij}, \varepsilon_{ijt}) = 0$. To any researcher who has observed interviewers in action, this assumption usually causes some concern. Many vivid examples of how interviewers adjust their behavior to idiosyncrasies of respondents are available. These cases violate this assumption to the extent that different interviewers adjust their behavior in different ways *and* different measurement errors result from these differences.

These assumptions are desirable not because they enhance the realism of the model but because they aid in the estimation of error terms. Given these assumptions the variance of a single observed value (y_{ijt}) over persons, interviewers assigned, and trials can be expressed as

$$\text{Var}(y_{ijt}) = \text{Var}(X_i) + \text{Var}(M_{ij}) + \text{Var}(\varepsilon_{ijt}) + 2\,\text{Cov}(X_i, M_{ij}) .$$

These variances are dispersions of values of the three terms over (1) different persons in the population, who are interviewed by (2) different interviewers, over (3) all possible trials. However, given the restrictive assumptions above, the variance of the true values (X_i) need be measured only over all persons. The variance of the interviewer effect terms (M_{ij}) need be measured over all combinations of interviewers and persons. Only the variance of the random effect term, ε_{ijt}, needs to be measured over persons, interviewers, and trials. Without the assumptions made above, just as a covariance term between true values and interviewer effects exists, $\text{Cov}(X_i, M_{ij})$, so too would covariance terms among random errors (ε_{ijt}), true values (X_i), and interviewer effects (M_{ij}). Most of these could not be empirically estimated without further replication and interpenetration. Such designs would be very expensive; hence, we encounter another example of costs constraining attempts to estimate survey errors.

The covariance term, $\text{Cov}(X_i, M_{ij})$, reflects the possibility that interviewers will influence persons with high values on the survey characteristic differently from those with low values. That is, there will be a correlation between the interviewer effect and the true value of the variable. There are some examples of this in the social science literature. Schuman and Converse (1971) show that white interviewers appear to influence more moderate racial attitudes among black respondents who would give more radical answers to the same questions asked by a black interviewer. Note that this kind of covariance is of prime interest to those investigators using split sample methods to compare a small number of measurement procedures.

Sometimes this covariance is assumed to be zero—that the effects interviewers have on responses are constant across different persons they interview. In practical terms this excludes the possibility that the interviewers affect different respondents in different ways. If these assumptions are made then the variance of a survey statistic can be described more simply; for example, a single observation has a variance due to each of the three factors

$$\text{Var}(y_{ijt}) = \text{Var}(X_i) + \text{Var}(M_j) + \text{Var}(\varepsilon_{ijt}) \ .$$

That is, the variance of a sum of independent random variables is simply the sum of their individual variances.

The real focus of the model is describing the effects of interviewers on descriptive sample statistics, like the sample mean. If a simple random sample of persons of size n were selected, partitioned at random into J subsets of equal size m, so that each of J interviewers would have an equal workload, then the variance of the sample mean would be

$$\text{Var}(\bar{y}_t) = \left(\frac{N-n}{N-1} \right) \frac{\text{Var}(X_i)}{n}$$

$$+ \frac{\text{Var}(M_j) + \text{Var}(\varepsilon_{ijt})}{n} \left(1 + \rho(m-1) \right),$$

where

$$\rho = \frac{\text{Var}(M_j)}{\text{Var}(M_j) + \text{Var}(\varepsilon_{ijt})}$$

is an intraclass correlation coefficient associated with interviewers. This correlation measures the extent to which response errors ($y_{ijt} - X_i = M_j + \varepsilon_{ijt}$) made by respondents of the same interviewer are correlated. Do interviewers influence respondents to make response errors of a similar type? To the extent that this correlation is large and positive, they do. This is the origin of the term "correlated response variance."

The expression above has two components. The first term is the sampling variance of the mean, reflecting the fact that different samples of persons will produce different mean values. The second is a measurement variance or response variance component. The reader will note that the correlation of measurement errors within interviewer workloads acts to inflate the response variance component. It is a multiplier effect on the other term,

$$\frac{\text{Var}(M_j) + \text{Var}(\varepsilon_{ijt})}{n},$$

sometimes called the "simple response variance."

If each interviewer were assigned only one respondent, there would be no correlations among measurement errors across respondents. In that case only the simple response variance would exist. This is the case in which $m = 1$, so the multiplier effect of the interviewer influence does not exist. This observation also leads to the prescription to use more interviewers on a survey, each completing smaller numbers of interviews, in order to reduce the measurement variance associated with interviewers. Some confusion arises among most practitioners when that prescription is given. How can interviewer effects be eliminated by having each interviewer complete only one interview? Two comments are in order. First, there is no claim that interviewer effects are eliminated. The model merely states that the correlated component of the error is eliminated. This merely means that, since each interviewer is collecting data from only one person, there can be no correlations in measurement errors across persons because of interviewers. What happens, if the model is true, is that all measurement errors are captured in the "simple response variance" term. The response deviations will include both the interviewer effect and the respondent effect on deviations from the true value. Measurement error variance *is* reduced, but only to the extent that positive correlations existed in response deviations among respondents of the same interviewer under the other case ($m > 1$).

There is a second negative reaction to the notion that interviewer effects are reduced when one interview is taken by each interviewer. This stems from the belief that interviewers become better with practice on the same questionnaire. The first interviews taken by the interviewer have lower quality than later ones. If each interviewer did only one interview, the worst data quality would result (and probably the worst response rates would be obtained). These beliefs are best viewed as a criticism of the model assumptions. They probably include the observation that measurement *bias* would result from the $m = 1$ case. That is, such inexperienced interviewers are likely to commit errors in a constant direction with similar results over replications (e.g., failure to probe incomplete answers). In addition to bias, however, the observations note that variability among inexperienced interviewers is greater than variability among experienced interviewers. This essentially says that the M_{ij} term for a particular interviewer will vary throughout the survey period, changing with greater experience. The effect that any interviewer has on a response is dependent on how many times the interviewer has administered the questionnaire. The model as stated above does not permit this possibility. If the experience factor does affect interviewer variance, then the estimates of interviewer variance or ρ must be conditioned on the particular mix of experiences interviewers bring to the study on the first day of the survey period. The use of one study to estimate the likely correlated response variance component if interviewer workload would change is not permitted.

Kish (1962) uses a slightly different approach than Hansen et al. (1961). He combines the true value X_i and the random error term, ε_{ijt}, into one term that can vary both because of sampling variance and variability over trials and interviewers, thus

$$y_{ijt} = X_{ijt} + M_j,$$

where $X_{ijt} = X_i + \varepsilon_{ijt}$, in the model above, and all other terms are defined as above.

With this perspective, the variance of the sample mean for simple random samples with relatively small sampling fractions is

$$\text{Var}(\bar{y}_t) = \left(\frac{1}{n}\right)(\text{Var}(X_{ijt}) + \text{Var}(M_j))[1 + \rho_{int}(m - 1)],$$

where

$$\rho_{int} = \frac{\text{Var}(M_j)}{\text{Var}(X_{ijt}) + \text{Var}(M_j)}.$$

Note that Kish's $\rho_{int} \neq \rho$ under the Hansen-Hurwitz-Bershad model. Indeed, $\rho_{int} < \rho$, because ρ_{int} includes in its denominator variation in true values *and* variation in error terms. The ρ includes in the denominator only variation in the error terms. These are merely definitional differences in most applications of the models. There are no strong arguments to recommend one approach over the other.

The requirements to estimate the interviewer variance component are random assignment of sample cases to interviewers, giving each sample case an equal chance of assignment to each interviewer selected for the survey trial. With this assignment scheme, under the Hansen-Hurwitz-Bershad model and the Kish model, a simple analysis of variance can be run, estimating between- and within-interviewer components of the variance. That is, the sums of squared deviations and mean squares presented in Table 7.4 are calculated. From these, the Hansen-Hurwitz-Bershad model derives a sample-based estimate of

$$\rho\big(\text{Var}(M_j) + \text{Var}(\varepsilon_{ijt})\big) = \left(\frac{1}{m}\right)(V_a - V_b),$$

where V_a and V_b are defined as in Table 7.4. Note that a direct estimate of ρ cannot be obtained with a one-time survey with interpenetration, even with the assumption that the interviewer effects are not correlated with true values. Some researchers call $\rho\big(\text{Var}(M_j) + \text{Var}(\varepsilon_{ijt})\big)$ the correlated response variance.

Table 7.4 Analysis of Variance Table for Measurement of Interviewer Variance Components

Source of Variation	Degrees of Freedom	Sum of Squares	Mean Square
Between interviewers	$J-1$	$SS(b) = m\sum_{j=1}^{J}(\bar{y}_{.jt} - \bar{y}_{..t})^2$	$V_a = \frac{SS(b)}{J-1}$
Within interviewers	$n-J$	$SS(w) = \sum_{j=1}^{J}\sum_{i=1}^{m}(y_{ijt} - \bar{y}_{.jt})^2$	$V_b = \frac{SS(w)}{n-J}$

Those following the Kish model get a direct estimate of ρ_{int} by

$$\rho_{int} = \frac{\left(\dfrac{V_a - V_b}{m}\right)}{\left(\dfrac{V_a - V_b}{m}\right) + V_b}.$$

The direct estimate of ρ_{int} offers a statistic that measures interviewer effects on variable measurement error that is unit free. The estimator used in the Hansen–Hurwitz–Bershad approach is measured in squared units of the variable itself. Thus, it cannot be directly compared across different statistics measured in different units (e.g., mean income in dollars and mean number of cars owned by the household). The ρ_{int} values in the population are bounded by $-1/(m-1)$ and $+1$, similar to any intraclass correlation. Thus, the researcher can compare ρ_{int} values across different statistics to check on differential sensitivity to interviewer effects.

By far the most popular estimator for interviewer variance in practice is based on the Kish model. The studies using the model interpenetrate the sample across the interviewers and use the analysis of variance model (usually with appropriate adjustments for unequal numbers of completed interviews among interviewers). They often compute ρ_{int} for many different statistics in the survey, usually means and proportions on the total sample. Table 7.5 presents a summary of several studies using personal interviews. Average values of ρ_{int} will appear very small to readers accustomed to correlation coefficients among substantive variables. The statistics presented for the World Fertility Surveys are those exhibiting the largest interviewer effects among those examined. They have mean values of ρ_{int} between 0.05 and 0.10. A more common value of mean ρ_{int} is between 0.01 and 0.02.

With a value of 0.01, following the model above, the inflation in variance for the sample mean from interviewer variation is $[1 + 0.01(m - 1)]$, where m is the number of interviews completed by each interviewer. If each interviewer completes 10 interviews, then the variance is increased by the factor 1.09, a 9 percent increase. (This is approximately a 5 percent increase in standard errors.) If workloads of 40 are used (as is often the case in telephone surveys), the factor becomes 1.39 for the variance, about an 18 percent increase for standard errors. Thus, although 0.01 is a small value for a correlation coefficient, as an intraclass correlation it can have large effects on the total variance of an estimated population mean.

However, increases for subclass statistics (e.g., a mean for particular age groups within the full population sampled) will be proportionately smaller, because m in the expression above would be based on the number of subclass members interviewed by each interviewer. Note that this use of ρ_{int} is subject to the same criticisms lodged above, if interviewer effects are a function of the number of cases they complete.

Table 7.5 Summary of ρ_{int} Values from Various Personal Interview Surveys

| | | Values of ρ_{int} | |
Study Description	Number of Statistics	Minimum, Maximum	Mean Value
World Fertility Survey (O'Muircheartaigh and Marckwardt, 1980)			
Peru—main survey	5	0.020, 0.100	0.050
Peru—reinterview	5	0.000, 0.150	0.058
Lesotho—main survey	5	0.050, 0.190	0.102
Canadian Census, 1961 (Fellegi, 1964)	8	0.000, 0.026[a]	0.008
Study of Blue Collar Workers (Kish, 1962)			
Study 1	46	−0.031, 0.092	0.020
Study 2	25	−0.005, 0.044	0.014
Canadian Health Survey (Feather, 1973)	39	−0.007, 0.033	0.006
Consumer Attitude Survey (Collins and Butcher, 1982)	122	−0.039, 0.199	0.013
Study of Mental Retardation (Freeman and Butler, 1976)	17	−0.296, 0.216	0.036
Interviewer Training Project (Fowler and Mangione, 1985)	130	0.001[b], 0.048	0.005[b]

[a] ρ_{int} was calculated from Fellegi's data, adjusted for parameters obtained after reinterview survey.
[b] The Fowler and Mangione study assigned the value 0.001 to all negative ρ_{int}'s.

The above discussion is only a short introduction to what is a very large field of empirical investigation. It emphasizes those attributes of the field that are useful to compare to other approaches of measurement error estimation. Other estimators are presented by Bailey et al. (1978) and by Fellegi (1964). The complications of estimating correlated response variance in the presence of a complex sample design is also discussed by Bailey et al. (1978). The complications of variable workload sizes is addressed by Kish (1962). Estimates of sampling variance for the estimates of interviewer effects are described in Bailey et al. (1978) and Groves and Magilavy (1986). Chai (1971) presents an extension to this model to estimates of interviewer variance on regression coefficients. Anderson and Aitken (1985) discuss the complication of estimating interviewer effects on dichotomous variables.

7.4.3 Practical Problems with Randomization as a Procedure for Estimating Measurement Error

Randomization internal to survey design is a powerful tool to estimating measurement error. However, all forms of interpenetration (both split sample methods and randomization over many units) require some assumptions about the nature of the measurement process:

1. **Other nonsampling errors are constant over assignment groups; that is, there exist equivalent groups receiving the different data collection procedures.**[2]

At sample selection this equivalency is normally introduced by randomization, giving each person (or household) in the sampling frame an equal chance of being assigned to each of the experimental groups. During data collection, however, this equivalency is often destroyed through nonresponse. In a random half-sample experiment, if the two experimental groups do not have nonrespondent groups that are equivalent, then estimates of measurement error differences between the two experimental groups will be confounded with nonresponse error differences. For example, if interviewers using one interviewing technique are less successful in gaining cooperation, the effort to estimate measurement error differences purely is threatened. Similarly, if

---

[2] The notion of "equivalent groups" means most strictly that all moments of univariate and multivariate distributions on the survey measurements are the same for all groups of persons.

interviewers obtain different levels of nonresponse error among their assigned groups due to differences in their ability to do the job, then estimates of correlated response variance will be confounded with nonresponse error differences across interviewers. These are two examples of nonresponse error affecting the estimates of measurement error.

There are also examples of different sources of measurement error causing nonequivalent groups. Many of the split ballot experiments do not randomize interviewer assignments jointly with randomizing measurement technique to sample person. As a result some interviewers may conduct many more interviews with one form than another. If interviewers vary in the form differences they obtain, the overall form difference is dependent on the mix of interviews done by form over the different interviewers. In our terms, the estimates of variance of form differences are too low; they do not reflect the variance from interviewers as well as from sample persons. The two experimental groups are nonequivalent because the set of interviewers and assignment patterns of interviewers to respondents are not equivalent.

2. There are no "Hawthorne–like" effects that challenge the external validity of the experiment.

That is, the behavior in each experimental group is assumed to be replicated outside the experimental setting. There are three types of challenge to this assumption. The first is most applicable to experimental manipulations of interviewer behavior. The knowledge of interviewers that they are participants in an experiment observing their own behavior may alter their behavior. If told to perform the same tasks using identical instructions, outside the experimental context, they may behave differently.

The second problem concerns contamination of measurement techniques. This is most obvious in question wording experiments. If interviewers are aware of the two forms of questioning, they may deliver each with a different style than if they were unaware of the other measurement form.

The third problem concerns human adaptation to routinized procedures. Many survey experiments compare techniques that are novel to the survey administrators and interviewers. When comparing the results obtained from first efforts to apply the new techniques, no evidence is obtained concerning whether repeated applications of the same techniques by the same staffs would yield similar results. This observation leads to the belief that replications of experiments are often needed in measurement error studies. This is true of many of the experimental comparisons of mode of data collection, in which experienced interviewing staffs are used for the personal mode and inexperienced for the telephone mode.

Many of these issues arise in other experimental sciences. Often these concerns in experimental design lead to the use of various "blinding" procedures. "Blinding" refers to the deliberate withholding of knowledge about the experimental conditions from those implementing the experimental design. For question wording experiments, blinding would involve the deliberate ignorance of the survey administrators of the experiment. The "survey administrators" should include any person whose activities can affect the recorded data. This would probably include the trainers, supervisors, and interviewers. This ignorance implies that different interviewers would do interviewing for different question forms, if these were being manipulated. It would mean that the randomization of assignment would not be revealed to interviewers, although the impact of knowledge of this on measurement errors is probably small. The burden of blinding is that, other things being equal, a smaller number of interviewers would administer each form of the question and separate staffs must be supervised.

3. **The magnitude of the measurement error associated with some unit in the design is independent of the number of units used in the design.**

Any variable error arising from some component of the survey (e.g., interviewers) can be measured with more precision when many units are used in the design. The more interviewers who are used on a project, the more stable are the estimates of interviewer variance. However, use of a larger number of interviewers than is typical merely to measure interviewer variance more precisely may be problematic. Such problems arise if the causes of interviewer effects are changed with larger numbers of interviewers.

One such possibility is that the difficulty of training large numbers of interviewers on the study may adversely affect the quality of training of all interviewers. With poorer training, interviewer variability in the performance of their tasks may increase, and the true values of interviewer variance are higher. Hence, the implicit model is that the level of correlated response variance associated with interviewers is independent of the number of interviewers used.[3]

7.5 REPEATED MEASUREMENTS OF THE SAME PERSONS

There is a large body of methodological literature in the social sciences which attempts to learn about measurement error by taking several measurements on the same person. In some designs these measurements are viewed as being fully equivalent to one another, as replicates which have the same expected values. Thus, they are used to investigate variability in response behavior over trials or replications (i.e., "measurement variance," "response variance," "reliability"). Sometimes reinterviews of the same person are conducted, asking the questions a second time. Sometimes multiple measures of the same concept are used in the same survey. Other designs attempt to understand measurement errors in one item by introducing other measures in the survey. These include direct questioning of the respondent about their understanding of a survey question, use of other questions to test comprehension of a question, recording of interviewer observations about the respondent behavior during the questioning, use of supervisory monitoring or taping of interviews, and questioning of other persons about the respondent.

These various techniques differ radically in the kind of information provided to the researcher about measurement error. The designs with repeated measures of the same quantity (or related quantities) on a person generally produce quantitative estimates of measurement error components; the designs with auxiliary measures about the questioning or response process generally provide indicators about the conformity of the measurement to the specified survey design (e.g., training guidelines for interviewer behavior, requisite respondent understanding of the questions). In this section we describe these different procedures and comment on their relative strengths and weaknesses.

---

[3] Survey statisticians using these models often incorporate this assumption into a statement that they are focusing on variable errors over replications of a survey *under the same essential survey conditions*. Most would probably agree that this means that the quality of training for interviewers is independent of the number of interviewers used on the survey.

7.5.1 Reinterviews to Estimate Measurement Error

Any reinterview design that produces estimates of response variance or reliability is based on a model of the measurement process (sometimes explicitly stated; other times only implicit in the procedures). A key premise is the stability of true values over trials. Although the concept of repeated trials does not require time to pass between trials, practical applications of the design do require repeated measurement over time. Thus, the researcher at some point is required to specify that the true value of the measure for a respondent is the same at the time of the first measurement and the second. That is,

Response at Trial 1 = True Value + Random Error

$$y_{i1} = X_i + \varepsilon_{i1} \,,$$

Response at Trial 2 = True Value + Random Error

$$y_{i2} = X_i + \varepsilon_{i2} \,,$$

where y_{i1} = obtained response for the ith person at time (trial) 1;

y_{i2} = obtained response for the ith person at time (trial) 2;

X_i = true value of the item for the ith person;

ε_{i1} = response deviation from true value for the ith person at time (trial) 1;

ε_{i2} = response deviation from true value for the ith person at time (trial) 2.

Note that relative to the prior models that include a method or interviewer effect on response, these simple models do not incorporate such an effect. The two models assume that $E(\varepsilon_{i1}) = E(\varepsilon_{i2}) = 0$; that is, the expected values of the two measures are the same quantity, X_i, a value that is constant over time. This assumption really has two components: (1) the assumption that the measurement on the second trial is an exact replicate of that on the first trial (e.g., same question, same context to the question, same interviewing procedures, same medium of data collection), and (2) the underlying characteristic that is being measured has not changed between the first and second trial for the person being measured.

With such a model, the difference between the two answers,

$$y_{i1} - y_{i2} = (X_i - X_i) + (\varepsilon_{i1} - \varepsilon_{i2})$$

$$= \varepsilon_{i1} - \varepsilon_{i2},$$

is merely the difference over trials in measurement error.

One added assumption permits an estimation of the variance in errors over trials. That assumption is that the errors are uncorrelated over trials, that $Cov(\varepsilon_{i1}, \varepsilon_{i2}) = 0$. With this additional constraint, the variance in errors over trials can be measured by

$$E(y_{i1} - y_{i2})^2 = E(\varepsilon_{i1} - \varepsilon_{i2})^2,$$

that is, the squared difference of observations can yield estimates of error variance.

Let us examine the plausibility of the various assumptions. Most can be violated through the respondent's memory of the first trial answer. That memory may prompt the respondent to give exactly the same answer (to appear consistent) or to give a different answer (to appear open-minded, flexible, especially on attitudinal items). Under this possibility, $E(\varepsilon_{i1}) = 0$, but once the ε_{i1} is committed, then ε_{i2} is determined to have a similar or dissimilar value. In this case, there is no guarantee that $E(\varepsilon_{i2}) = 0$. Furthermore, the errors are correlated; $Cov(\varepsilon_{i1}, \varepsilon_{i2}) \neq 0$. Thus, on the respondent side, memory effects can violate the assumptions. On the interviewer side, knowledge of the prior answer can induce correlation of errors across trials. Note well that the violation of the assumption may yield an overestimate or underestimate of the error variance (the direction depends on the sign of the correlation of errors). For these reasons, reinterview studies to estimate measurement error variance must be used carefully.

The simplest design of repeated measurement is applied on a routine basis in most U.S. Census Bureau surveys through their reinterview programs. A sample of households at which completed interviews were obtained are visited after the survey period. As described for the period of the early 1970s for the National Health Interview Survey, the reinterviews are conducted by supervisors (mostly male versus the required female NHIS interviewers), at least a week after the interview (Koons, 1973). The reinterview questionnaire asked about health events in the same time period as did the original interview. Thus, the time interval between the event to be recalled and the interview date was longer for the reinterview. In some of the reinterview households (about 80 percent) the reinterviewer was given the answers obtained on the first interview; in others, he was not. In those with knowledge of the interview results, sometimes the reinterviewer asked the assistance of the respondent to determine the correct answer for discrepancies between trials (about 64 percent of all reinterview households). A measure of reliability in response is provided by the "index of inconsistency" which is defined by

$$I = \frac{E(\varepsilon_{i1} - \varepsilon_{i2})^2/n}{\sigma_y^2},$$

where I = index of inconsistency;

n = number of reinterviewed cases;

σ_y^2 = variance over persons (i) of the observed values on y, the survey measure.

Thus, the index of inconsistency is the ratio of the variance of response errors to the total variance of the measure. Given the assumptions of the measurement model, the index obtains values between 0.0 and +1.0.

With a dichotomous variable, for example, whether the respondent suffered from one or more chronic conditions in the past 12 months, the components of the index of inconsistency simplify to functions of cells in a contingency table (see Table 7.6). It becomes

$$I = \frac{(b + c)}{\left(\dfrac{(a + c)(b + d) + (a + b)(c + d)}{n}\right)}.$$

Different results are typically obtained when the reinterviewer is privy to the first responses obtained. The index of inconsistency is lower when reconciliation is used. For example, Table 7.7 compares three different reinterview procedures, yielding very different estimates of the index of inconsistency. In the first, the reinterviewer is not made aware of the answers obtained in the initial interview. In the second, the reinterviewer is given these answers but is instructed not to attempt a reconciliation of discrepant answers before recording the reinterview response. In the third, the reinterviewer is both made aware of the initial response *and* asked to resolve inconsistencies. Table 7.7 shows that the knowledge of the prior response typically reduces the inconsistencies in responses, whether or not the reconciliation step is formally conducted. For example, when the reinterviewer is unaware of the first answer, response variance between trials represents about 31 percent of observed variance for reports of chronic conditions (by definition, given the model, all other variance is attributable to sampling variability, variation across persons in true values). When he is made aware of the first answer, response variance forms only 22 percent of total observed variance. When reconciliation is added, it drops to 17 percent of observed variation in responses. This is direct evidence of the effect of making the two measurements dependent on one another. It implies that such dependency tends to induce a positive correlation between the first recorded response and the second.

Table 7.6 Response Outcomes for Interviews and Reinterviews

Results of Reinterview	Results of First Interview		
	$x_{i1} = 1$ Has a chronic condition	$x_{i1} = 0$ Does not have a chronic condition	Total
$x_{i2} = 1$ Has a chronic condition	a	b	$a + b$
$x_{i2} = 0$ Does not have a chronic condition	c	d	$c + d$
Total	$a + c$	$b + d$	$n = a+b+c+d$

Table 7.7 Indexes of Inconsistency for Three Reinterview Procedures in the Health Interview Survey

Measure, Persons with One or More	Knowledge and Use of First Interview Result		
	No Knowledge	Knowledge, No Reconciliation	Knowledge, Reconciliation
Chronic conditions in past 12 months	30.9	22.2	17.4
Hospital episodes in past 12 months	7.6	7.0	6.0
Restricted activity days in past 2 weeks	44.5	28.6	18.3
Bed days in past 2 weeks	41.1	26.6	15.8
Time-lost days in past 2 weeks	37.6	32.9	21.4
Hospital days in past 2 weeks	12.8	19.5	19.1

Source: D. A. Koons, *Quality Control and Measurement of Nonsampling Error in the Health Interview Survey*, National Center for Health Statistics, March 1973, Table 11, p. 20.

In the psychometric literature "reliability" is sometimes assessed using replicated measures, in a test–retest design. The same measurement models are used. An added assumption is that $\text{Var}(\varepsilon_{i1}) = \text{Var}(\varepsilon_{i2})$. "Reliability" is assessed by the correlation of the two observed scores,

$$\rho_{y_{i1}, y_{i2}} = \frac{\text{Cov}(y_{i1}, y_{i2})}{\sqrt{\text{Var}(y_{i1})\text{Var}(y_{i2})}}.$$

Readers more familiar with the psychometric definition of "reliability" will note that the index of inconsistency, I, is closely related to the index of reliability, $\rho_{y_{i1}, y_{i2}}$. Indeed,

$$\rho_{y_{i1}, y_{i2}} = 1 - I.$$

Thus, the two disciplines have very similar forms of estimation for variable measurement errors.[4]

7.5.2 Multiple Indicators of the Same Characteristics Administered in the Same Questionnaire

Another approach also uses replicated measures to estimate measurement error, but it uses multiple measurements of the same characteristic in a single survey. The approach is connected with a radically different set of estimation procedures, firmly based in the psychometric literature. In this approach measurement error associated with a particular method of data collection and/or a particular question can be assessed. "Measurement error" here is defined as a component of variance in the observed values of indicators, not corresponding to variability in the true values of the underlying measures. In the terminology of this book, it corresponds to variable errors of measurement only. "Method" has been used to mean the mode of data collection (personal, telephone, or self-administered), the format of the question (5 point scale, 10 point scale, open questions), the respondent rule (self-response, proxy response), or various other characteristics of the measurement.

The most striking difference between this approach and those above is that an explicit model of correspondence between two or more different survey questions is posited. We introduce this notion by dealing with the simplest (and most constraining model), **parallel measures**, when two questions measure the same underlying characteristic with the same degree of precision. Later we remove the restrictive assumptions required for parallel measures. In notation, the simplest set of parallel measures is

$$y_{ikm} = X_{ik} + \varepsilon_{im} \text{ and } y_{ikm'} = X_{ik} + \varepsilon_{im'},$$

---

[4] This example of the index of inconsistency and the reliability coefficient is typical of other comparisons of terms between survey statistics and psychometric approaches to measurement error. The psychometric approaches typically measure the positive side— validity and reliability; the survey statistics models measure the negative side— measurement bias and response variability.

where $y_{ikm}, y_{ikm'}$ = indicators (m and m') of the kth underlying characteristic for the ith person;

X_{ik} = true value of the kth characteristic for the ith person;

$\varepsilon_{im}, \varepsilon_{im'}$ = random error terms for the two indicators.

Each of these measurements has the simple classical true score assumptions (reviewed in Chapter 1). That is, $E(\varepsilon_{im}) = E(\varepsilon_{im'}) = 0$, the indicators have the same expected value, because they are measuring exactly the same thing. Furthermore, they do it equally well; $\mathrm{Var}(\varepsilon_{im}) = \mathrm{Var}(\varepsilon_{im'})$. And the error terms are independent of the values of the characteristic; $\mathrm{Cov}(X_{ik}, \varepsilon_{im}) = \mathrm{Cov}(X_{ik}, \varepsilon_{im'}) = 0$. There is nothing new here so far.

Another assumption gives a boost to assessing the reliability of y_{ikm} as an indicator of X_{ik}. Just as psychometricians were forced to assume independence between measurement error in a test–retest situation, a similar assumption is required here in order to get an unbiased estimate of the correlation of scores over replications. So in this situation, we assume that the error committed on one indicator is not correlated to the error on the other; that is, $\mathrm{Cov}(\varepsilon_{im}, \varepsilon_{im'}) = 0$. This makes the existence of two parallel measures even better than the test–retest situation, because there is no threat of change in true values (X_i) with parallel measures administered. Conditional on the two being parallel measures and errors independent, the reliability of y_{ikm} or $y_{ikm'}$ is

$$\rho_{y_{ikm}, y_{ikm'}} = \frac{\mathrm{Cov}(y_{ikm}, y_{ikm'})}{\sqrt{\mathrm{Var}(y_{ikm})\mathrm{Var}(y_{ikm'})}}.$$

Parallel measures require the most extreme assumptions permitting reliability assessment. Other models of correspondence between two indicators permit more complex assessment of reliability. Alwin and Jackson (1980) review different models.

The key difference between this technique and those reviewed above is that a substantive theory is the basis on which the expected values of the two questions are equated. Typically, there is no empirical data external to the survey to support this theory. Rather, the substantive theory asserts the equivalency, and measurement errors are estimated *given the theory.*

When several characteristics are measured, each with the same set of different methods, the multitrait multimethod matrix results, originally proposed by Campbell and Fiske (1959). Implicit is a measurement model,[5]

$$\text{Response} = \text{Population Mean} + \text{Influence of True Value} + \text{Method Effect} + \text{Random Error}$$

$$y_{ijkm} = \mu_k + \beta_{km} X_{ik} + \alpha_{jm} M_{ij} + \varepsilon_{im},$$

where y_{ijkm} = observed value of the ith person using jth method to measure the kth characteristic using the mth indicator;

μ_k = mean value of the kth characteristic for the population studied;

β_{km} = "validity" coefficient for the mth indicator of the kth underlying characteristic;

X_{ik} = for the ith person, the true value of the kth characteristic;

α_{jm} = "method effect" coefficient for the mth indicator of the jth method;

M_{ij} = for the ith person, the common effect of using the jth method;

ε_{im} = a random deviation for the ith person on the mth indicator.

This is sometimes referred to as the common factor model (Alwin, 1974) because multiple indicators, each subject to "specific" error (method effect) and random error, are used for an underlying trait (factor). To the

---

[5] There seems to be no uniform notation to present the multitrait multimethod model. The notation used here was designed to emphasize the similarities between models in the various fields.

extent that the influence of the true value (X_{ik}) on the response is large, the indicator is said to have high validity. With such a model and appropriate numbers of indicators, traits, and measurement methods, empirical estimation of some components of measurement error is permitted. Specifically, the researcher can estimate what portion of the variance in observed indicators corresponds to variation in true scores (in X_{ik}), what portion corresponds to variation in method effects (in M_{ij}), and what portion corresponds to random error (in ε_{im}). The method effects concern the error-producing features of the differences among methods. The random errors contain all other sources of deviation from the true value.

As with several of the techniques above, the estimates of measurement error are interpretable *only* if certain model-based assumptions are true. For example, a multitrait multimethod matrix based on nine indicators of three traits using three distinct methods is the most general case. Here, the full set of models are those in Table 7.8. Note that each of the three traits (X_{ik}) and each of the three methods (M_{ij}) are associated with three of the nine survey questions (y_{ijkm}). (The three by three design is used to get sufficient information to estimate the various parameters in the models.) Each indicator (y_{ijkm}) is viewed as a measure of one trait (X_{ik}) and a representative of one method (M_{ij}).

These equations need further explanation. It seems relatively straightforward to consider the respondent's answer to a question to be a simple linear function of his/her true value on the trait. Thus, the common factor model, with multiple indicators for each trait, has some intuitive appeal. In this context, however, what is the meaning of the value of the method effect? Following the model, there is some reaction of the respondent to a particular method of measurement that affects his/ her answers to items measuring each of the substantive traits. Different items may be differentially affected by this "method reaction" but the source of the method effect for that person stems from this same reaction.

Before examining the models, some empirical examples should be presented. Bohrnstedt (1983) describes a case where the methods are telephone, personal, and self-administered versions of the same measurements but does not explicate the meaning of the method effects. Presumably, the values of the method factor for each person consist of all the response errors common to the method. Thus, for example, if a person reacts to a telephone survey by giving less attention to the respondent task, whatever effects are shared by items of any trait using that method will be the value of the underlying method factor. The shared effect may not correspond merely to lack of attention; instead, another respondent may be both inattentive and hostile because of being disturbed at home and thus tend to answer all questions by agreeing or

Table 7.8 Nine Measurement Models for a Multitrait Multimethod Analysis with Three Methods and Three Traits, Estimated by Nine Indicators

Method 1	Trait 1	Item 1	$y_{i111} = \mu_1 + \beta_{11}X_{i1} + \alpha_{11}M_{i1} + \varepsilon_{i1}$
Method 1	Trait 2	Item 2	$y_{i122} = \mu_2 + \beta_{22}X_{i2} + \alpha_{12}M_{i1} + \varepsilon_{i2}$
Method 1	Trait 3	Item 3	$y_{i133} = \mu_3 + \beta_{33}X_{i3} + \alpha_{13}M_{i1} + \varepsilon_{i3}$
Method 2	Trait 1	Item 4	$y_{i214} = \mu_1 + \beta_{14}X_{i1} + \alpha_{24}M_{i2} + \varepsilon_{i4}$
Method 2	Trait 2	Item 5	$y_{i225} = \mu_2 + \beta_{25}X_{i2} + \alpha_{25}M_{i2} + \varepsilon_{i5}$
Method 2	Trait 3	Item 6	$y_{i236} = \mu_3 + \beta_{36}X_{i3} + \alpha_{26}M_{i2} + \varepsilon_{i6}$
Method 3	Trait 1	Item 7	$y_{i317} = \mu_1 + \beta_{17}X_{i1} + \alpha_{37}M_{i3} + \varepsilon_{i7}$
Method 3	Trait 2	Item 8	$y_{i328} = \mu_2 + \beta_{28}X_{i2} + \alpha_{38}M_{i3} + \varepsilon_{i8}$
Method 3	Trait 3	Item 9	$y_{i339} = \mu_3 + \beta_{39}X_{i3} + \alpha_{39}M_{i3} + \varepsilon_{i9}$

disagreeing, by answering yes or answering no, or by giving short answers. Andrews (1984, pp. 412-413) presents models where the shared methods of the indicators are question wording properties:

> Imagine a survey item that asks respondents to evaluate their own health by picking one of several answer categories ranging from "very good" to "very bad." The answers will vary—partly because people differ in the way they perceive their own health (valid variance). In addition, the answers may vary because people interpret the answer categories differently (e.g., "very good" may mean something more positive to some respondents than to others). This is measurement error attributable to the method (methods variance).

> Now, if a second survey item using the same response scale is included in an analysis with this item on health, and if each respondent is consistent in the way he or she interprets the meaning of these categories (as could be expected), the measurement errors attributable to the method would be the same in both items. Respondents who tended to be "too high" on the first item—because of the way they interpreted the answer categories—would also tend to be "too high" on the second item—because it used the same categories.

Andrews probably overstates the need for similar effects on two traits measured by the same method. All that is necessary under the model is that respondents make a response error for all items using the scale, not that the direction of the error is constant over all items.

There are several assumptions that are necessarily made about the nature of the terms in the model. Whenever survey estimates are based on models, it is important to conceptualize what the assumptions really mean. Sometimes it helps to give examples of violation of the assumptions:

1. $E(\varepsilon_{im}) = 0$; that is, except for random variation, the models perfectly determine the value of the nine items for each person. (This assumption could be violated if some of the indicators also share some other source of influence; for example, if two of the methods were subject to more effects of social desirability than the third. Thus the models omit another term reflecting social desirability effects.)

2. $\text{Cov}(\varepsilon_{im}, \varepsilon_{i'm}) = 0$, for all $i \neq i'$; that is, between persons in the population, deviations from one measurement are uncorrelated with the deviations from another. (This assumption could be violated with the same case as above, a misspecification of the measurement model to omit another shared influence, such as the interviewer.)

3. $\text{Cov}(X_{ik}, \varepsilon_{im}) = 0$, for all k, m; that is, over persons in the population, the deviations from the expected values under the model are uncorrelated with the true values of the trait. (This can be violated if the nature of response errors is such that persons with high values on the trait tend to make different kinds of errors than those with low values; for example, if income is the underlying constraint, persons with high incomes may underestimate incomes and persons with low values may overestimate, regardless of what measurement method is employed.)

4. $\text{Cov}(M_{ij}, \varepsilon_{im}) = 0$, for all i, j, m; that is, over persons in the population (i), the deviations from expected values under the model are uncorrelated with the magnitudes of the effects of the measurement method on the person's response to an item. [An example of a violation of this assumption concerns the use of different scale labels (e.g., "very good" to "very bad" versus "excellent" to "poor"). The assumption is violated in the case that for some traits being measured, persons who interpret "very good" to mean something more positive than others tend to become subject to more social desirability effects than those who interpret it in the normal manner.]

The models do permit the true values of the traits to be correlated with the method factors over different persons. That is, persons with high values on the underlying traits may be more or less affected by the method of measurement than those with low values on the trait. (This flexibility makes the approach very attractive since it frees the investigator from one of the restrictive characteristics of most of the models in the interpenetrated designs for interviewer variance.)

Instead of presenting separate equations for each of the questions, multiple indicator models are generally depicted graphically, following the conventions of path analysis. For example, Figure 7.1 presents a graphical representation of the nine-indicator, three-method, three-trait measurement discussed above. The graph uses squares to refer to the nine different indicators. Circles are used to indicate the underlying traits and methods, with traits on the top of the graph and methods on the bottom. Straight lines with directional arrows represent causal influences. That is, the answers to survey questions are viewed as being caused by the true values on the underlying trait and a reaction to a method. Curved arrows represent covariances or correlations that do not reflect direct causal relations. The small straight lined arrows pointing into the boxes (indicators) correspond to the random error terms affecting survey responses (the ε_{im}). Next to the curved and straight lines appear the coefficients corresponding to the relationship indicated. Note that the graph, by omitting various lines, asserts the assumptions reviewed above.

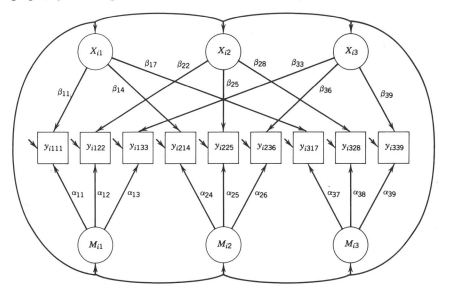

Figure 7.1 Path diagram for multitrait multimethod model, using boxes to denote questions and circles to denote underlying constructs.

In a study that is discussed in more detail in Chapter 10, Andrews (1984) estimated coefficients in 125 different multitrait multimethod models, using several surveys and different substantive topics. The models resembled that in Figure 7.1 except that three further constraining assumptions were made. The first was that the method factors were uncorrelated among themselves, $\mathrm{Cov}(M_{ij}, M_{ij'}) = 0$, for all $j \neq j'$. The second was that the method factors were uncorrelated with the substantive traits, $\mathrm{Cov}(X_{ik}, M_{ij}) = 0$, for all i, j, and k. The third was that the effects of a method were constrained to be equal for all indicators using the method. Using standardized variables allowed use of the estimated β coefficients for X_{ik} in the models to be interpreted as validity measures, as correlations between the underlying trait and the indicator. Similarly, estimated α coefficients for the method factor were interpreted as method effect parameters. Finally, the random variance parameters were estimated. This yielded measures on the same scale (0 to 1) estimating the components of error. Table 7.9 presents summary values over the 125 different models. It shows that the mean validity coefficient was 0.81 and that the method effect was 0.16. The squares of the coefficients provide estimates of the components of variance of survey measures. These estimates imply that 66 percent of the observed values' variance reflects variation in true scores, 3 percent reflects variance due to method, and about 28 percent reflects random variance.

Although the multitrait multimethod approach above is designed to estimate measurement errors, the estimation approach has been generalized to permit estimation of error properties in structural equation models. The method is implemented using the LISREL program (Joreskog and Sorbom, 1984) in a confirmatory factor analysis. Here the researcher is allowed to specify a particular structural model, using substantive (exogenous) variables to predict observed scores on another variable (endogenous, dependent), but to include in the structural model a specification of the behavior of measurement errors, similar to those in the multitrait multimethod approach. Using the terminology of econometrics, the approach permits implementation of the "errors in variables" and "errors in equation" models, with correlated errors. The method is limited only by the identifiability of the system of equations and the constraints posited on them. This software offers analysts interested in structural equation parameters a powerful way to include errors in their models. Interested readers should consult the growing literature in the social sciences on this approach (Long, 1983).

Table 7.9 Measurement Quality Estimates Based on 125 Multitrait Multimethod Analyses

	Validity	Method Effect	Random Error
Mean	0.81	0.16	0.53
Median	0.81	0.16	0.55
Standard deviation	0.10	0.11	0.16
Number of estimates	2115	2115	2115

Source: F. M. Andrews, "Construct Validity and Error Components of Survey Measures: A Structural Modeling Approach," *Public Opinion Quarterly*, Vol. 48, 1984, pp. 409–442, Table 2.

7.5.3 Measuring Characteristics of the Interview Itself

Many times investigators are not comfortable with the assumptions required in the models above, or alternatively, they believe that multiple measures or internal randomization of procedures is too expensive to implement. Despite this, they would like to obtain some information about measurement errors. In these cases, additional observations are sometimes taken by the interviewers or by an observer of the interview situation. In contrast to the approaches above, these are not attempts to replicate the measurement process with an equivalent procedure, nor to compare different procedures. Instead, most of these techniques provide only indications of whether the measurement process was implemented as designed.

Sometimes these techniques are used to measure characteristics of the respondent presumed to be related to measurement error. Many questionnaires include "interviewer observations," completed by the interviewer after the interviewing session is completed. Figure 7.2 presents an example of these interviewer observations. These observations generally refer not to single questions in the instrument but rather to respondent behavior or attitudinal states (as judged by the interviewer) that persisted during extended periods of the interview. They might also attempt to measure characteristics of the interviewing situation (e.g., whether or not others were present during the interview).

Section T

COMPLETE THE FOLLOWING QUESTIONS AFTER THE INTERVIEW.

T1. SEX OF RESPONDENT:

| 1. MALE | | 5. FEMALE |

T2. WAS THERE A LANGUAGE PROBLEM THAT MADE IT DIFFICULT FOR YOU TO INTERVIEW
THIS RESPONDENT?

| 1. YES, MAJOR PROBLEM | | 3. YES, MINOR PROBLEM | | 5. NO, NO PROBLEM |

GO TO T3

    T2a. (EXPLAIN)_____

    _____

T3. WERE THERE ANY OTHER PROBLEMS THAT MADE IT DIFFICULT FOR YOU TO INTERVIEW
THIS RESPONDENT?

| 1. YES, MAJOR PROBLEMS | | 3. YES, MINOR PROBLEMS | | 5. NO, NO PROBLEMS |

GO TO T4

    T3a. (EXPLAIN_____

    _____

T4. IN GENERAL, THE RESPONDENT'S UNDERSTANDING OF THE QUESTION WAS:

| 1. EXCELLENT | | 2. GOOD | | 3. FAIR | | 4. POOR |

**Figure 7.2 Illustrative interviewer observation form completed by the interviewer
after an interview.**

T5. PLEASE DESCRIBE THE RESPONDENT'S ABILITY TO EXPRESS (HIMSELF/HERSELF) USING THE SCALE BELOW.

VERY ARTICULATE   EXPRESSES SELF WITH
  EXCELLENT   GREAT DIFFICULTY;
  VOCABULARY   LIMITED VOCABULARY

T6. WAS RESPONDENT SUSPICIOUS ABOUT THE STUDY BEFORE THE INTERVIEW?

T7. OVERALL, HOW GREAT WAS RESPONDENT'S INTEREST IN THE INTERVIEW?

T8. DID RESPONDENT EVER SEEM TO RUSH (HIS/HER) ANSWERS, HURRYING TO GET THE INTERVIEW OVER?

T9. DURING THE INTERVIEW DID RESPONDENT EVER ASK HOW MUCH LONGER THE INTERVIEW WOULD TAKE?

T10. OTHER PERSONS PRESENT AT INTERVIEW WERE: (CHECK MORE THAN ONE BOX IF APPROPRIATE.)

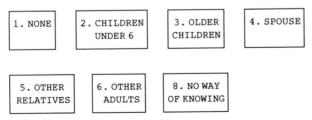

Figure 7.2 (Continued)

T11. HOW MANY TIMES WAS THE INTERVIEW INTERRUPTED? (ANY TIME THE INTERVIEWER WAS PREVENTED FROM CONTINUING TO ASK QUESTIONS OR RESPONDENT TO RESPOND TO THEM)

_____ TIMES

(IF INTERRUPTIONS)

T11a. APPROXIMATELY HOW MANY MINUTES WERE TAKEN UP BY INTERRUPTIONS?

_____ TIMES

T11b. WHAT WHERE THE CAUSES OF THESE INTERRUPTIONS?
(CHECK AS MANY AS APPLY.)

[ ]   1. CHILD(REN) FROM THE HOUSEHOLD

[ ]   2. OTHER ADULT(S) FROM THE HOUSEHOLD

[ ]   3. PHONE CALL(S)

[ ]   4. VISITOR(S) TO THE HOUSEHOLD

[ ]   7. OTHER (SPECIFY) _____

[ ]   8. NO WAY OF KNOWING

Figure 7.2 (Continued)

These questions are not often used in estimates of measurement error. If they were to be used, the appropriate measurement model would have to include them as direct and/or indirect influences to response behavior. For example, because a respondent asked many times for questions to be repeated, they might not have fully comprehended the intent of the questions. Hence, their responses might be more subject to prior beliefs about the purpose of the survey, to influences from nonverbal behavior of the interviewer, or to single words in questions.

Another set of techniques involving observations of the interview situation utilizes an observer to the question and answer process (some third person in addition to the interviewer and the respondent). In personal interview situations, this is sometimes another interviewer (or supervisor) who records observations during the questioning. Another technique is to tape record the interview. With that approach no observer is present, but after the interview is complete a supervisor reviews the tape (see Cannell et al., 1975). With telephone interviews in centralized facilities, a monitor listens to the interview, usually with an unobtrusive method of tapping into the telephone line (see Mathiowetz and Cannell, 1980). All these procedures require that a third person evaluate the measurement process through the exclusive use of audio and/or visual observation, not by asking the respondent any questions directly. With this restriction they most often focus on measuring adherence to training guidelines. Did the interviewer read the question as written, was approved probing used, did the recorded answer capture the response? To be useful for understanding measurement error, such documentation requires an assumption that the prescribed behavior for interviewers is that which minimizes measurement errors and that deviations from those prescriptions will induce more error. Rarely is this assumption stated formally, and there is little empirical evidence in the literature to support or refute this assumption.

7.5.4 Direct Measures of Cognitive Steps in Response Formation

Another line of research in measurement errors is more typical of clinical psychology: the questioning of the respondent about his/her reactions to the survey question. For example, Belson (1981) asked respondents several questions that included the term "weekday," for example,

> When the advertisements come on between two television programmes on a weekday evening, do you usually watch them?

Later, he asked them to tell the interviewer what they understood was meant by the term "weekday." Table 7.10 shows the distribution of answers to that question. Surprisingly, only a little more than half of the cases (53.4 percent) used the word "weekday" to refer to a time period from Monday through Friday. The largest misinterpretation was to include all days of the week or not to use the word as a meaningful qualifier in any way.

Belson asserts that measurement error results from differences in interpretations of the answer. The technique, in contrast to those above, however, does not produce estimates of measurement error directly. Implicit in the work is the assumption that if two respondents mean different things by the words in the question, their answers to the question will have different amounts of error in them for measuring the intended concept.

A similar technique asks the respondent more direct questions relevant to the measurement error properties of an item. Ferber (1956) asked the respondent for his/her opinion on the issue of a guaranteed annual wage, as a policy of the federal government:

Table 7.10 Percentage of Respondent Interpretations of the Word "Weekday"

Interpretation of "Weekday"	Percentage of Cases
From Monday till Friday (5 days)	53.4%
From Monday till Saturday (6 days)	0.9
From Monday till Friday, plus Sunday	0.9
From Sunday till Saturday (7 days)	33.2
Any day of the week	0.5
Not clear which days were considered	4.6
Reference apparently not considered at all	6.5
n	(217)

Source: W.A. Belson, *The Design and Understanding of Survey Questions*, Gower, London, 1981, Table 8.6, p. 361.

What is your attitude toward allowing labor to have a guaranteed annual wage?

> For___
> Against___
> Neutral___
> Don't Know___
> No Answer___

Then they followed the question with another:

(If an attitude is expressed) Why?

And finally,

As you interpret it, what do the unions mean by a guaranteed annual wage?

The answers to the last question were coded according to whether the respondent seemed to know what the phrase meant. The results were that many people who gave answers to the attitudinal question did not have an accurate understanding of "guaranteed annual wage." Almost half of those who gave no answers to the meaning of the term (47.0 percent) expressed an attitude on the policy issue. The vast majority of those who gave incorrect definitions of the term (83.6 percent) expressed an attitude. These two percentages should be compared to the 90.6 percent of those who correctly identified the term, who expressed an attitude.

This finding is linked to the notion of "nonattitudes," discussed more fully in Chapter 9. The author cites this as evidence that answers of those uninformed about the various issues suffer more completely from measurement error. It is important to note the assumptions implicit in this conclusion. First, the knowledge question is assumed to be measured without error. Some persons giving the correct answer in the implementation of the procedure may not always do so over repeated trials. Second, the attitudes expressed by those who provided incorrect answers about the meaning of "guaranteed annual wage" have larger errors attached to them. This notion needs more clarification. It is not clear that over trials their answers are less stable than those who provided correct answers, nor is it clear that their answers are less predictive of behaviors or other attitudes related to the measurement.

The most direct and intrusive use of questioning the respondent about measurement error attributes are debriefing interviews like that used by Martin (1986). At the end of a questionnaire on criminal victimization, the interviewer asked the questions in Figure 7.3. These seek reports of psychological states during the interview but also ask the respondent to reveal whether he/she violated the prescriptions of the interview by not reporting a victimization eligible for reporting under the survey rules. In one form of the questionnaire about 9 percent of the respondents noted that they failed to mention a crime. This provides a direct assessment of the underreporting of victimization but does not reveal the type of crime missed. The assumption in the use of this technique is that these debriefing questions are themselves not subject to measurement errors.

7.5.5 Evaluating Repeated Measurements as a Procedure for Estimating Measurement Error

Under the heading of "repeated measurements" we have treated radically different approaches. Two of them attempt to produce empirical estimates of error parameters (reinterviews, multiple indicator models); two of them are used most often to observe cognition or behavior that *may* be related to measurement error (observations about interviews and questioning respondents). The latter methods generally do not posit a direct link between the behaviors observed and measurement errors in the data. For example, if interviewers change the wording of a question when delivering it, the investigators do not study empirically how that appears to change the responses obtained. Rather, the investigations appear to use the departures from specified procedures as *prima facie* evidence of measurement error. Empirical estimation of the missing link in this argument would provide a valuable assist to understanding survey measurement error.

The reinterview and multiple indicator model approaches presented above resemble one another in that formal models of equivalence in measurements are made. The reinterview approach posits that the two trials yield independent observations of the same characteristic and the two have the same error features. The multiple indicator models assemble questions that are posited to be indicators of the same characteristic. *Only when these models are true*, do the validity coefficients and method effect coefficients provide meaningful estimates of errors.

To illustrate this point, we might confront the multitrait multimethod approach with "social desirability" effects. Social desirability is the hypothesized cause of respondents failing to report personal characteristics that they believe the society negatively evaluates. These beliefs are usually viewed to be the cause of measurement bias. That is,

X1. We are finished with all the questions about you and about crimes that happened to you. Now I'd like to ask a few questions about your reactions to the interview. Our purpose is to discover the problems people have with our questionnaire so we can improve it.

At the beginning of the interview I told you we were interested in crime incidents that happened to you during a period of time. Do you happen to remember what that time period was?

    1. LAST SIX MONTHS, SINCE (DATE)
    2. OTHER
    8. DON'T KNOW

X1a. Do you think the time frame was intended to be exact or approximate?

    1. EXACT
    2. APPROXIMATE
    3. DON'T KNOW

X3. At the beginning, before I asked you any questions, did you think you would have any crimes to report?

    1. YES
    2. NO
    3. DON'T KNOW

X4. I (asked questions/gave examples) to help you remember crimes that might have happened to you. You told me about one incident. Did you find you were still thinking about that incident when we went back to the examples/questions?

    1. YES
    2. NO

X5. While I was asking you the (questions about/examples of) crimes that might have happened to you, did you lose track or have a hard time concentrating?

    1. YES
    2. NO

Figure 7.3 Illustrative debriefing questionnaire administered to the respondent by the interviewer after the main interview has been completed.

Did you feel bored or impatient?

1. YES
2. NO

Were you reminded of types of crimes that you hadn't already thought of on your own?

1. YES
2. NO

X6. Was there an incident you thought of that you didn't mention during the interview? I don't need details.

1. YES
2. NO (GO TO X7)

X6a. Were you unsure whether it was the type of crime covered by the survey?

1. YES
2. NO

Was the incident a sensitive or embarrassing one?

1. YES
2. NO

Did it happen to you?

1. YES
2. NO

X7. How certain are you that you've remembered every crime that happened to you during the last six months — very certain, pretty certain, or not certain at all?

1. VERY CERTAIN
2. PRETTY CERTAIN
3. NOT CERTAIN AT ALL

Figure 7.3 (Continued)

the respondents are expected to adjust their responses in the same way over trials regardless of any other design feature.

Suppose in the context of a multitrait multimethod analysis that all three of the methods suffer from some degree of social desirability effects. Imagine, for example, the trait being measured is "suffering from hemorrhoids." To make the case most general, allow the degree of social desirability effects to vary across persons (γ_i), and view each as deviation from a base social desirability bias δ_{im}, so that

$$y_{i111} = \mu_1 + \beta_{11}X_{i1} + \alpha_{11}M_{i1} + \delta_{i1} + \gamma_i + \varepsilon_{i1},$$

$$y_{i214} = \mu_1 + \beta_{14}X_{i1} + \alpha_{24}M_{i2} + \delta_{i4} + \gamma_i + \varepsilon_{i4},$$

$$y_{i317} = \mu_1 + \beta_{17}X_{i1} + \alpha_{37}M_{i3} + \delta_{i7} + \gamma_i + \varepsilon_{i7}.$$

In constructing these $(\gamma_i + \varepsilon_{im})$ terms we need to specify the real nature of their effects on respondent behavior. Most discussions of social desirability effects posit that those respondents who have an attribute that is seen to be socially undesirable will report that they have a more social desirable value on the attribute. This social psychological effect is therefore inherently correlated to the true value of the respondent on the attribute; that is, the covariance between X_{i1} and the sum $\delta_{im} + \gamma_i$ is nonzero. The specification of both δ_{im} and γ_i reflects the belief that part of the social desirability effect is a function of the respondent (γ_i) and part of it is a function of the specific measure used as an indicator of the trait, the δ_{im} term. Furthermore, we would specify $\gamma_i = f(X_{i1})$, where $f(\cdot)$ is some function not necessarily fully specified. To illustrate our point, we specify $\gamma_i = f(X_{i1}) = bX_{i1}$, where b is some real-valued number fixed across persons. Then the entire set of nine equations becomes

$$y_{i111} = \mu_1 + (\beta_{11} + b)X_{i1} + \alpha_{11}M_{i1} + (\delta_{i1} + \varepsilon_{i1}),$$

$$y_{i122} = \mu_2 + \beta_{22}X_{i2} + \alpha_{12}M_{i1} + \varepsilon_{i2},$$

$$y_{i133} = \mu_3 + \beta_{33}X_{i3} + \alpha_{13}M_{i1} + \varepsilon_{i3},$$

$$y_{i214} = \mu_1 + (\beta_{14} + b)X_{i1} + \alpha_{24}M_{i2} + (\delta_{i4} + \varepsilon_{i4}),$$

$$y_{i225} = \mu_2 + \beta_{25}X_{i2} + \alpha_{25}M_{i2} + \varepsilon_{i5},$$

$$y_{i236} = \mu_3 + \beta_{36}X_{i3} + \alpha_{26}M_{i2} + \varepsilon_{i6},$$

$$y_{i317} = \mu_1 + (\beta_{17} + b)X_{i1} + \alpha_{37}M_{i3} + (\delta_{i7} + \varepsilon_{i7}),$$

$$y_{i328} = \mu_2 + \beta_{28}X_{i2} + \alpha_{38}M_{i3} + \varepsilon_{i8} \,,$$

$$y_{i339} = \mu_3 + \beta_{39}X_{i3} + \alpha_{39}M_{i3} + \varepsilon_{i9} \,.$$

In this case the "validity coefficients" reflect how closely the indicator is affected by an underlying characteristic that is not purely X_{i1} but rather X_{i1} adjusted by response errors associated with social desirability. Similar comments can be made about the random error variance terms. This problem is implicit to the common factor model on which the approach is based. The estimated validity coefficients represent shared variance among indicators, regardless of whether some of that shared variance is variation in errors common to the indicators. Note further that a variety of question effects could be used as examples of this problem: the degree of filtering with the don't know option, the existence or absence of a middle alternative, and so on.

If the above is correct, one would presume that β and α terms would be sensitive to the presence of different combinations of traits and methods. That is, alternatively substituting traits that were or were not targets of the response bias should change the validity coefficients, correlations among the unobserved, and the residual terms. A sensitivity analysis could be a weak test of model misspecification.

Indeed, substituting terms in the model that should have no effect on estimated parameters should be a basic tool in investigating the quality of the model specification in the multitrait multimethod approach. This could be done with data that have more than the required number of indicators, but no hypothesized error specifications that would prevent estimation of parts of the model with fewer indicators. Checking the assumptions of the measurement models *in addition to estimating errors* appears not to have been practiced by most users of the method.

7.6 SUMMARY OF INDIVIDUAL TECHNIQUES OF MEASUREMENT ERROR ESTIMATION

Although this chapter has presented a large number of different techniques used by researchers to study measurement error, they can be classified into a small number of categories:

1.   Replication of identical measures over trials.

2.   Replication of indicators of the same concept within trials.

3.   Randomization of measurement procedures to different persons.

4.   Collection of correlates of measurement error.

Even these four categories could be collapsed into three, given the correctness of the model often used in category 2. If the indicators are viewed to be parallel measures, with identical expected values and measurement variances, then multiple indicator approaches are similar to replication of identical measures over trials. Categories 1 to 3 are used to provide direct empirical estimates of variable measurement errors. These are the index of inconsistency, I, for reinterview studies, the ρ_{int} statistics for interviewer variance, and the validity coefficients of the multiple indicator models. Category 3, in the split sample version in which a small number of techniques are randomly assigned to subsets of the sample, is generally used to gather evidence of differential measurement bias. Only differences in measurement bias can be obtained, not empirical estimates of absolute measurement bias. Finally, category 4 does not provide direct empirical estimates of variable or fixed measurement errors. Rather, it is used to gather data on whether the survey is being implemented as designed—whether interviewers are performing their tasks according to training guidelines and whether respondents fully understand their duties in the interview.

All the single procedures used in this chapter require assumptions about the nature of the measurements in order to obtain empirical estimates of error. For this reason, it is preferable to view these as "model-based" estimates of measurement error. The survey methodological literature is filled with implementations of these various designs. Rarely do the researchers include checks on the assumptions of their models. For example, record check studies rarely evaluate the quality of the records and rarely examine effects of different matching rules between survey reports and records. Reinterview studies typically do not examine differences among reinterviewers in measured response differences. Interviewer variance studies do not examine changes in the correlated component of response variance with different workloads. Multiple indicator model studies often do not test the sensitivity of validity coefficients to different indicators.

These observations do not imply that the estimates of error from these procedures are consistently flawed. They imply that the reader cannot sometimes evaluate whether error is overestimated or underestimated. Only through replication of the analyses, direct efforts to challenge the assumptions, and tests of the robustness of the findings to slight changes in the models will the correctness of the measurement models on which the error estimates are based be assessed.

7.7 COMBINATIONS OF DESIGN FEATURES

There is little empirical work that jointly employs two or more of the techniques described above. This is probably because different academic disciplines contributed to the various methodologies and because the techniques are best suited either to measurement bias estimation or measurement error variance estimation, not to both at the same time. We have already noted that most work in survey methodology has traditionally concentrated on one or another of these errors.

Although it is difficult to find examples of empirical work that employs multiple perspectives, some can be located and are reviewed below. Other combinations of techniques can be described and evaluated, even if no examples of them yet exist.

7.7.1 Interpenetration for Measurement Bias Estimation and Interpenetration for Variable Interviewer Effect Estimation

There are several examples of survey experiments that simultaneously compared two methods of survey measurement *and* interpenetrated the assignments of sample cases to interviewers. With this combination, two different questions can be addressed:

1. What is the likelihood that the two methods would yield different values on the survey statistic (i.e., there are measurement bias differences) in replications of the design, allowing both for the selection of another sample *and* the use of a different set of interviewers?

2. In addition to possible differences in measurement bias in the two procedures, is there evidence that one is more susceptible to greater variable measurement error due to interviewer effects?

The first question is answered by including the interviewer component of variance in estimates of the variance of the statistics from the two forms. Statistical tests of hypotheses or confidence intervals about the statistics can thus reflect both variable sampling error and variable measurement error. For example, Groves and Kahn (1979) used split samples to compare two different formats for a life satisfaction scale (this technique is reviewed in more detail in Chapter 11). One used a seven point scale, numbered 1 to 7, with 1 labeled as "completely dissatisfied," 4 as "neutral," and 7 as "completely satisfied." The respondent was asked to give a number between 1 and 7 to describe, for example, his feelings about his "life as a whole." The other-half sample employed an unfolding technique that classified respondents on a seven-point scale based on their choosing labels:

1. Terrible
2. Unhappy
3. Mostly Dissatisfied
4. Mixed
5. Mostly Satisfied
6. Pleased
7. Delighted

The percentage of respondents giving answers 5, 6, or 7 on the first scale was 83 percent. On the second scale, 80 percent gave the answers "mostly satisfied," "pleased," or "delighted." The estimates of interviewer effects using ρ_{int} were -0.005 for the percentage based on the terrible to delighted scale and .01 for that from the dissatisfied to satisfied scale. Incorporating just sampling and interviewer error into an estimate of variance leads to an estimate of the standard error for the 3 percentage point difference between the two forms of 3.21 percentage points. If only sampling error were considered, the comparable number would be 3.17. In this case interviewer variance increases the standard error only slightly; in other cases the increase can be much greater.

Similar joint use of split sample methods and interpenetrated assignment of interviewers appeared in Cannell et al. (1987) for tests of the effect of CATI and the use of controlled interviewer behavior.

7.7.2 Interpenetration of Interviewer Assignments and Reinterviews

Fellegi (1964) extended the basic design proposed by Hansen et al. (1961) to include both interpenetration and replication of the measurement on each sample person. The design, actually implemented in the 1961 Canadian Census, requires the division of the sample at random into $2j$ distinct subsamples, two subsamples for each of the j interviewers working on the survey. Each interviewer is assigned to two of the subsamples. Thus, each of the subsamples is assigned to two interviewers, one to do the interviewing on the first wave and another on the reinterview wave. On the first wave of the survey the interviewer completes initial interviews with one of the subsamples assigned (while at the same time another interviewer is interviewing her subsample assigned for reinterview). On the reinterview survey, the interviewer completes second interviews with the other assigned subsample. By the end of this design, all respondents have been interviewed twice by two different interviewers, each assigned at random to them.

The bonus to the researcher from this complicated design is that new measurement error components can be estimated. Fellegi presents various estimators for these parameters. The most interesting component is the correlation between response deviations and true values on the variable being measured. This is a term that was assumed to be zero in single interview designs, but that assumption was forced on the researcher because of insufficient data to estimate it.

7.7.3 Interpenetration of Interviewer Assignments and Multiple Indicators

A conceptual link between the multiple indicator approach and the errors studied by those who build models of interviewer effects is sorely needed. The interviewer effects model is generally a simple one:

$$y_{ij} = X_i + M_j + \varepsilon_{ij},$$

where y_{ij} = recorded response of the ith respondent interviewed by the jth interviewer;

X_i = true value for the ith respondent;

M_j = the effect of the jth interviewer;

ε_{ij} = a random error term.[6]

The typical assumptions are that $E(M_j) = E(\varepsilon_{ij}) = 0$, and that $\mathrm{Cov}(M_j, \varepsilon_{ij}) = \mathrm{Cov}(X_i, \varepsilon_{ij}) = \mathrm{Cov}(X_i, M_j) = 0$.

This should be contrasted to the multitrait multimethod model:

$$y_{ijkm} = \mu_k + \beta_{km} X_{ik} + \alpha_{jm} M_{ij} + \varepsilon_{im}.$$

One conceptualization of how interviewer effects might be incorporated into the multiple indicator model is that they affect (in a variable way over

---

[6]This model differs from those presented in Section 4.2 by the omission of the trial subscript (t). This is because the empirical estimation procedures proposed in this section are implemented with an interpenetrated design imbedded in a single trial survey.

interviewers) how the trait, method, and residual terms blend to produce
the response. That is, some interviewers may increase the sensitivity of
the respondent to the method effect, others to the underlying trait. Taken
from another perspective, the interviewers that happen to be selected for
the study produce as a group a set of β's for validity coefficients (as well as
the other parameters estimated in the model). If another set of
interviewers would have been used for the study, another set of β's might
have been produced. Therefore, the model is

$$y_{ijj'km} = \mu_k + (\beta_{km} + \varepsilon'_{j'}) X_{ik} + (\alpha_{jm} + \varepsilon''_{j'}) M_{ij} + (\varepsilon_{im} + \varepsilon'''_{j'})$$

where $y_{ijj'km}$ = response obtained from the ith person to a
question using the jth method, administered
by the j'th interviewer for the kth underlying
characteristic using the mth indicator;

$\varepsilon'_{j'}, \varepsilon''_{j'}, \varepsilon'''_{j'}$ = deviations from the expected coefficients
associated with using the j'th interviewer, with
the expectation of all ε equal to zero.

Thus, interviewers are a source of variability over replications of a
multiple indicator analysis in the estimated coefficients. This source of
variability has not been reflected in prior estimates of coefficients in
multitrait multimethod models.

How can we measure this source of variability? One option is a
jackknife estimator (Wolter, 1985) that would group respondents
interviewed by the same interviewer together and sequentially drop one
interviewer's group from the data file. To do this, an interpenetrated
design for interviewer assignment should have been introduced so that
there is some assurance that the expected values of the β's and α's are
equal for all interviewer groups, in the absence of interviewer effects.

7.8 SUMMARY

Chapter 1 introduced the notion of the roles of "measurer" and "reducer"
with regard to survey error. Research on measurement error is easily
classified into two groups corresponding to these two roles. The reducer
camp could contain most of the laboratory studies of interviewer behavior
and respondent cognitions (Section 7.2), the use of external data for
validation (Section 7.3), and the split sample experiments comparing two
question wordings (Section 7.4.1), and direct questioning of respondents

about the survey process (Section 7.5.4). If not immediate to the purposes of this research, then indirectly so is the improvement in survey data quality by the reduction of measurement error. This research is productive of guidelines for survey design—in questionnaire construction, in interviewer behavior, and in mode of data collection.

In contrast to these efforts are those devoted to quantitative estimates of measurement error or at least proxy indicators of measurement error. These would include interpenetration of the sample for interviewer variance estimates (Section 7.4.2), reinterviews for measures of response variance (Section 7.5.1), multiple indicators of the same concept for estimating error variance due to question type (section 7.5.2), and monitoring and interviewer observational data on the interview (section 7.5.3). All these inform the analysts about the real or potential measurement errors in the data they are analyzing. Sometimes these techniques provide estimates of error which can supplement the estimates of sampling error on survey statistics, to clarify how guarded the inference should be from the survey estimates to the target population (e.g., estimates of interviewer variance). Other times, they provide quantitative information which cannot be formally integrated with other estimates of error but form a separate caution to the inference (e.g., frequencies of respondent queries about the meaning of a survey question).

The reader will note that no mention of survey costs has appeared in this chapter. Unfortunately, the costs related to reducing measurement errors have not been much explored. There are two possible reasons for this. First, much recent attention to measurement error estimation (e.g., uses of structural equation modeling techniques) was developed for use by secondary analysts of data and was not primarily devoted to design decisions seeking minimal measurement error for available resources. Second, that research of a more prescriptive character for survey designers (e.g., some work in question wording) has by its nature tended to examine improvements in survey design that cause little increase in cost. Despite these reasons for ignoring cost implications of measurement error reduction, however, cost considerations do arise with regard to measurement error reduction. We discuss two of these (interviewer variance reduction and multiple indicators to increase reliability) in the coming chapters.

This chapter attempted to lay the groundwork necessary to consume a large literature in survey measurement error. The next four chapters of the book present the findings of these two approaches to exploring survey measurement error (that of the "measurers" and that of the "reducers"). Those chapters give the results of methodological research into the magnitudes and causes of measurement error. This research uses one or more of the designs presented in this chapter. At various points in those later chapters, the reader may find cause to review sections of this chapter in order to critically evaluate the findings presented.

CHAPTER 8

THE INTERVIEWER AS A SOURCE OF SURVEY MEASUREMENT ERROR

Interview techniques smuggle outmoded preconceptions out of the realm of conscious theory and into that of methodology.

Charles L. Briggs, Learning How to Ask, *1986, p. 3*

8.1 ALTERNATIVE VIEWS ON THE ROLE OF THE OBSERVER

The survey interviewer is one source of measurement error which has been studied both by those researchers interested in *reducing* measurement error and those interested in *estimating* the size of those errors. This error source is of interest to those who conceptualize the interviewer effects as inducing constant errors in survey statistics over replications of the survey and to those who focus on variability across interviewers in errors they create in the data. Indeed, the attention given by survey methodologists to the data collector as a source of error makes the methodology distinctive among alternative data collection strategies.

In some sciences in which observations are taken in order to test specific hypotheses (e.g., physics), efforts *are* made to seek replication of findings by having another laboratory make the same observations under similar circumstances. That is, they incorporate into their data collection practice consideration of variable errors across laboratories. The different replications are considered only "similar," not completely equivalent. This is because it is acknowledged that the complex of features of the laboratory (e.g., the investigator, the physical equipment) may interact in ways not specified in the theory being tested to affect the phenomena under study. This perspective brings the investigator and the observational environment into the set of design features that can affect research results. Replication is desirable in order to estimate the effects that the measurement environment can have on results or, at the very least, observe whether the results are robust to changes in them.

In psychological laboratory experiments, there is also a concern about the effect on subject performance of experimenter characteristics, situational factors in the laboratory, and experimenter expectations of likely subject behavior (Rosenthal, 1966). However, most psychological experiments continue to have designs with single experimenters and without formal plans to replicate using different environments and different personnel. Despite the absence of designs that seek to measure the variability in results due to the investigators, however, there are efforts to reduce these effects. Hence, there are recommendations to blind the experimenter and subjects to the treatments, to minimize experimenter contact with subjects, and to design control groups to measure experimenter effects (Rosenthal, 1966, pp. 402-403).

These two approaches to observer effects—replication for measurement and reduction efforts—are both represented in survey research. One set of research efforts resemble the latter—methods to assure that the interviewers are "neutral" collectors of the data, acting to obtain the most accurate information from respondents. This research is dominated by attention to training interviewers. The goal is nothing less than the elimination of the interviewer as a source of measurement error. Most of the research is devoted not to the routine measurement of interviewer effects but to evaluating different ways to reduce them. Another set of research examines the effect of interviewers on data by replication of measurements using different interviewers and comparisons of results across individual interviewers. That approach is more interested in measurement of error than in direct reduction of error. Sometimes the replication method is implemented by interviewing the same respondent twice with two different interviewers. Other times different interviewers are assigned random subsamples, and each of their workloads is viewed as a replication of the entire survey process.

Both of these efforts have brought interviewer effects into the concerns of the investigator, but only the latter acknowledges the interviewer at the time of estimation and inference. The failure of most research to study both reduction and measurement of interviewer effects is a serious weakness in this field. For example, there are too few studies of interviewer variance that simultaneously experiment with different ways of reducing those errors. Such experiments could lead to optimal designs of interviewer activities.

8.2 THE ROLE OF THE SURVEY INTERVIEWER

To understand how interviewers can contribute to survey measurement error, it is useful to review their responsibilities before, during, and after the interview. The interviewers locate and contact the sample household. They describe the study to the household member who answers the telephone or the door. They define the purpose of the study and the sampling procedures. They persuade respondents to cooperate. Depending on the respondent rule used (e.g., self-response of one randomly selected adult), they select respondents. When the interview begins they deliver questions, answer the respondent's queries, and probe to clarify questions. They record answers of the respondents. Finally, they edit the interview prior to submitting it for further processing. Because interviewers are the agents who implement the entire survey design, they can contribute to several sources of error in the design. Separating their effects on survey measurement error from their effects on coverage and nonresponse error can rarely be done.

Interviewer effects on survey measurement error occur in four ways. First, social psychologists sometimes view the survey interview as a structured social interaction (e.g. Kahn and Cannell, 1957), subject to many of the social influences present in other interactions. Thus, interviewer demographic and socioeconomic characteristics can affect the behavior of respondents. Second, interviewers can administer the questionnaire in different ways. Despite guidelines to the contrary they can reword questions, fail to ask some questions, or record the wrong answers. Attention to this problem is integral to a psycholinguistic view or a cognitive science view of the survey interview. The stimuli (the survey questions) to the respondent vary across interviewers and different respondent answers result. Third, even if the questions are read exactly as written, interviewers can emphasize different words or use different intonation in delivering the words. Fourth, in reaction to respondent difficulties with the questionnaire (e.g., failure to understand a question), interviewers assist the respondent in different ways, use different probing techniques, and thus yield variation in responses. All four factors are reasons for *variable* measurement errors across interviewers. It is difficult to find examples of measurement errors associated with interviewers that are fixed (i.e., replicable) and of constant magnitude (i.e., every interviewer making the same error). Such errors could arise if none of the interviewers was trained to probe when they receive inadequate answers.

This chapter reviews efforts both to measure and to reduce interviewer effects on survey data. It begins with the focus on *measurement* of interviewer effects, as evidenced by differences among interviewers in the responses they obtain. It posits that the amount of survey error in the recorded answer of each survey question varies by which interviewer was assigned to the respondent. This variation exists because different interviewers produce different amounts of measurement error in respondents they interview. We review both measures of interviewer variance components and techniques for measuring compliance with interviewer training guidelines. Next, the chapter discusses research in interviewer training procedures and experimental alterations of traditional interviewer behavior. These are primarily attempts to minimize bias, error associated with actions that interviewers would naturally perform if not instructed to do otherwise. The implicit assumption is that such instruction or quality control procedures can eliminate behaviors that lead to some measurement errors. In the absence of such training every replication of a survey is subject to those errors.

8.3 DESIGNS FOR MEASURING INTERVIEWER VARIANCE

Despite the thousands of surveys conducted every year there exist only a handful of empirical estimates of interviewer variance components. All designs for estimating interviewer variance share the feature that different interviewers are assigned to equivalent respondent groups. Using this equivalency, differences in results obtained by different interviewers are then attributed to effects of the interviewers themselves.

In designs with "replication" two interviewers are randomly assigned to the same respondent, one as an initial interviewer and the other as a reinterviewer. Comparisons of results between the two interviewers of each pair are used to measure the impact that the interviewer has on the survey responses. In this design the two answers to a single question might be related to one another both because the interviewers share some attribute or because the respondents remember and feel some commitment to their first answer. Conversely, in remembering their prior answers, respondents might tend to give somewhat different responses (especially on attitudinal questions) in an attempt to appear more open-minded. For this reason differences between responses obtained by the interviewer and reinterviewer do not only measure interviewer variance but also effects of memory of the first response.

"Interpenetrated" interviewer assignments avoid multiple interviews with the same respondent. Instead, they estimate interviewer variance by assigning to each interviewer different but equivalent sets of respondents, ones who have the same attributes on the survey variables. In practice this equivalency is assured through a randomization step. That is, the sample is partitioned into subsets at random, each therefore having the same attributes *in expectation* as the others, and then each interviewer works on a different subset. With this design each interviewer conducts a small survey with all the essential attributes of the large survey, except its size.

Interpenetrated interviewer assignments take a different form in personal interview surveys compared to centralized telephone surveys. In personal interview survey designs, interviewer assignments are often geographically defined. The assigned areas have sizes sufficient for one interviewer's workload. Pairs of assignment areas are identified and randomly assigned to pairs of interviewers. Within each assignment area each interviewer of the pair is assigned a random half of the sample housing units. Thus, each interviewer completes interviews in two assignment areas, and each assignment area is handled by two different interviewers.

The design consists of one experiment (a comparison of results of two interviewers in each of two assignment areas) replicated as many times as there are pairs of interviewers. Hanson and Marks (1958), Hansen et al. (1961), and Bailey et al. (1978) describe estimation procedures that can be used with this design.

Why are geographical clusters randomly assigned to interviewers? Why not assign each interviewer a random subsample of the housing units in the sample, without regard to geography? There are two reasons to implement randomization only within geographically defined assignment areas. First is the desire to avoid the costs of interviewers traveling long distances to interview each respondent. Second and equally important is the desire to randomize assignments among interviewers who in any likely replication of the survey would be given interviewing duties in the same areas.

For example, in most national face to face surveys, interviewing in New York City is performed by interviewers residing in New York City. An interpenetrated design which would assign some of these cases to interviewers who live in Los Angeles or Chicago or Franklin, Nebraska, would reflect interviewer variability not observable under realistic design options (i.e., the effects of using interviewers from another region). For both those reasons, randomization among interviewers *within* their "usual" assignment areas is desirable.

In centralized telephone surveys, there are no cost savings enjoyed by restricting the randomization of cases to geographical areas. Furthermore, most facilities allow assignment of any sample case to any interviewer working in the facility. In such cases the entire sample that is active at any one moment in the survey can be partitioned at random and assigned to the interviewers working on the project. Thus, for example, in paper questionnaire surveys, the cover sheets (forms listing the sample number and providing a call record on the case) can be shuffled, divided into groups, and assigned at random to interviewers as they arrive at the facility.[1] In a computer-assisted telephone interviewing (CATI) system, the sample file can be sorted in a random order prior to access by interviewers. In addition, call scheduling routines can be designed to continue the randomization throughout the survey period.

Although the marginal cost increase of interpenetration in centralized telephone surveys is lower than in most personal interview surveys, there remain some difficulties. Centralized telephone interviewing is typically organized into shifts (of about 4 to 6 hours). Each interviewer is typically assigned to work 4 to 8 shifts a week, at times that are mutually desirable for the interviewer and the facility supervisors. Because of this, individual interviewers are prevented from dialing a sample number at all the possible times that it should be called. For example, some households might be absent from the housing unit for most hours of the day, being available only on a restricted number of hours on the weekend. If the interviewer assigned the case does not work a weekend shift, this sample case could never be reached and would not be interviewed. Hence, interpenetration with fixed subsample assignments to interviewers is not normally done in centralized facilities.

Instead, some variant of the plan described by Stokes (1986) is used, which treats each shift of interviewers as a new group, eligible for assignment at random to the remaining cases that are active in the sample. In some sense, the sample is constantly being rerandomized for assignment to those interviewers working at a given moment during the survey period. This process of interviewer assignment and survey production can be thought of as a randomization within shift, yielding equivalent sample groups for all interviewers in the same shift.[2] Thus, comparisons of results obtained by different interviewers are conducted within shifts. However, some shifts use many interviewers; others use only a few. Thus, the stability of the within-shift, between-interviewer comparisons is highly variable over the shifts, and some collapsing of shifts is normally done to obtain more stable estimates of interviewer variance. Stokes (1986) found that estimates of intraclass correlation

---

[1] An unanticipated benefit of randomized assignment of cases to interviewers occurs in a centralized facility. Interviewers express a preference for such an allocation procedure because they are assured that their performance will not be harmed by the supervisor giving them only the difficult cases to dial (e.g., sample numbers in large metropolitan areas). Randomization enforces equity.

[2] A "shift" is most conveniently defined as any unique collection of interviewers, all of whom are engaged in interviewing at the centralized facility at the same time. The length in time of a shift is thus dependent on the administrative decisions regarding the scheduling of different interviewers.

associated with interviewers were smaller with full collapsing over shift than with within-shift comparisons. Groves and Magilavy (1986) found no important differences among shifts in the magnitudes of the correlations. The estimation problems for interviewer effects in centralized telephone facilities will no doubt be a topic of some attention in the coming years, as the number of such facilities increases.

8.4 INTERVIEWER EFFECTS IN PERSONAL INTERVIEW SURVEYS

In this section we review the magnitudes of interviewer effects found in personal interview studies and comment on their impacts on the total variance of estimates. When studying the magnitudes of interviewer effects found in past studies, one is confronted with a variety of statistics (see Chapter 7, Section 7.4.2). This complicates assessment of how large these effects are, relative to other sources of error. When raw variance component statistics are cited, it is difficult to compare the results of different studies. Such comparisons might be useful if one could identify those aspects of the designs that appeared to be associated with high or low interviewer effects. Unfortunately, most of the studies provide little documentation on characteristics of the interviewers, interviewer training guidelines, supervision of the interviewing process, and evaluation procedures for interviewers. These, among other design aspects, would form the **essential survey conditions** within which interviewer variability is being studied. Differences in results obtained among interviewers might be a function of the labor market for interviewers, what they were told to do, and what persons they encounter in their job. Although little can be done to eliminate ignorance about design features of past studies, the comparability of interviewer effects across studies can be enhanced by using intraclass correlation estimates with similar statistical properties. The Kish ρ_{int} is well suited for this because it is a unit-free measure, a ratio of variance between interviewers to the total variance of a measure. This measure is discussed in more detail in Chapter 7, Section 7.4.2, and is defined as

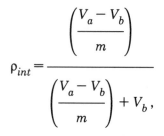

$$\rho_{int} = \frac{\left(\dfrac{V_a - V_b}{m}\right)}{\left(\dfrac{V_a - V_b}{m}\right) + V_b},$$

where V_a = between mean square in a one-way analysis of variance
with interviewers as the factor;

V_b = within mean square in the analysis of variance;

m = total number of interviews conducted by an interviewer.

Table 7.5 (reproduced here from Chapter 7) presents a brief summary of the results from 10 personal interview surveys. Overall, the values of ρ_{int} will appear negligibly small to those accustomed to correlations between substantive variables. For example, the results from the World Fertility Survey apply to those survey questions with the highest ρ_{int} values found, but even these lie below 0.15. The Study of Mental Retardation has unusually high values of ρ_{int} (mean = 0.036) that might be attributable to the use of nonprofessional interviewers. Most other surveys have average ρ_{int} values lying between 0.01 and 0.02. The overall average for all studies in the table is 0.031. Summarizing over all the studies, one would conclude that ρ_{int} values below 0.02 are most common, that demographic variables are less affected by interviewer behavior, and that ρ_{int} values higher than 0.02 are possible in surveys on sensitive topics, those with questions that respondents find difficult to understand or those employing less-well-trained interviewers.

A common model of how interviewer variation inflates the variance of the sample mean is $\text{deff}_{int} = 1 + \rho_{int}(m - 1)$, where m is the interviewer workload (see Chapter 7). A ρ_{int} of 0.02 with a workload of 10 interviews each yields a deff_{int} of 1.18, an 18 percent increase in variance. A workload of 25 yields a deff_{int} of 1.48, a 48 percent increase in variance. Thus, even apparently small interviewer intraclass correlations can produce important losses in the precision of survey statistics.

Table 7.5 Summary of ρ_{int} Values from Various Personal Interview Surveys

| | | Values of ρ_{int} | |
Study Description	Number of Statistics	Minimum, Maximum	Mean Value
World Fertility Survey (O'Muircheartaigh and Marckwardt, 1980)			
Peru—main survey	5	0.020, 0.100	0.050
Peru—reinterview	5	0.000, 0.150	0.058
Lesotho—main survey	5	0.050, 0.190	0.102
Canadian Census, 1961 (Fellegi, 1964)	8	0.000, 0.026[a]	0.008
Study of Blue Collar Workers (Kish, 1962)			
Study 1	46	−0.031, 0.092	0.020
Study 2	25	−0.005, 0.044	0.014
Canadian Health Survey (Feather, 1973)	39	−0.007, 0.033	0.006
Consumer Attitude Survey (Collins and Butcher, 1982)	122	−0.039, 0.199	0.013
Study of Mental Retardation (Freeman and Butler, 1976)	17	−0.296, 0.216	0.036
Interviewer Training Project (Fowler and Mangione, 1985)	130	0.001[b], 0.048	0.005[b]

[a] ρ_{int} was calculated from Fellegi's data, adjusted for parameters obtained after reinterview survey.

[b] The Fowler and Mangione study assigned the value 0.001 to all negative ρ_{int}'s. Average value presented is median ρ_{int}.

8.5 INTERVIEWER EFFECTS IN CENTRALIZED TELEPHONE SURVEYS

One of the attractions of centralized telephone interviewing, among early proponents of the methodology, was the possibility of constant supervision of interviewers. This feature was felt to promise reduction of interviewer variance through enforcement of interviewing guidelines. In addition, it was believed that the ability of interviewers to hear one another administering the questionnaire would lead to an exchange of techniques over time so that interviewers would tend to resemble one another in their behaviors.

Tucker's work (1983) in measuring interviewer intraclass correlations in a centralized setting was the first to investigate these possibilities. He used results from 11 national polls conducted by CBS News and *The New York Times*. Each study analyzed from 10 to 18 variables, and the mean values of ρ_{int} for the 11 studies ranged from -0.003 to 0.008 with an overall average of 0.004. These figures were substantially below those from face to face interview surveys.

Nine centralized telephone surveys conducted at the Survey Research Center were examined by Groves and Magilavy (1986) for magnitudes of interviewer effects. Eight of the surveys involved a two–stage stratified national sample of randomly generated telephone numbers following Waksberg (1978). The ninth survey, the 1980 Post Election Study, was a telephone reinterview with respondents chosen earlier in an area probability sample.

Several aspects of the study designs are summarized in Table 8.1. Response rates ranged from 59 to 87 percent, with unanswered numbers included in the denominators of these rates.[3] Row 4 lists the number of interviews obtained in the studies. In each case, this indicates the interviews generated by the random assignment of telephone numbers to interviewers. Interviews resulting from nonrandom assignments (e.g., refusal conversions, select appointments) were excluded from both this number and the analysis. In all surveys one adult was randomly selected in each sample household. Also included in Table 8.1 is the number of interviewers employed for each study and the average workload (i.e., number of interviews) per interviewer. Not only does average workload vary across studies, 12 for the November 1981 Consumer Attitude Survey to 58 for the Health in America Survey, but workloads across interviewers within studies exhibit a similar degree of variation.

All the studies collected standard demographic data in addition to data on the specific topics listed in Table 8.1, and there is a common set of substantive measures across surveys. The 1980 Post Election Study provides a repetition of several questions which appeared in the Study of Telephone Methodology. Some of the Telephone Methodology economic questions were repeated in the monthly Consumer Attitude Surveys.

---

[3] The lowest response rate, 59 percent, reported for the Study of Telephone Methodology reflects an exceptionally large proportion of unanswered numbers. In later studies, improved methods of identifying nonworking numbers have caused a decrease in this proportion and consequently, a more accurate response rate. The high response rate (87 percent) observed in the 1980 Post Election Survey reflects the fact that this was a reinterview study.

Table 8.1 Study Descriptions and Summary of Overall Findings

	Study of Telephone Methodology	Health and Television Viewing	Health in America	1980 Post Election Study	Monthly Consumer Attitude Survey				
					November 1981	December 1981	January 1982	February 1982	March 1982
Field period	April–May 1976	March–April 1979	October–December 1979	November–December 1980	November 1981	December 1981	January 1982	February 1982	March 1982
Response rate	59%	67%	80%	87.3%	73.5%	72.8%	73.3%	76.9%	73.8%
Number of interviewers	37	30	33	22	31	28	21	25	26
Number of interviews	1529	954	1918	697	370	350	386	379	366
Average workload	41.3	31.8	58.1	31.7	11.9	12.5	18.4	15.2	14.1
Major topics	Political, Economic, and social issues	Health and TV viewing	Health	Political issues	Economic issues	Economic issues	Economic issues	Economic issues	Economic issues
Number of variables analyzed	25	55	25	42	30	30	30	30	30
Range of ρ_{int}	−0.0080 0.0560	−0.0150 0.1650	−0.0070 0.0097	−0.0154 0.1710	−0.0217 0.0895	−0.0373 0.0546	−0.0221 0.0916	−0.0419 0.0657	−0.0356 0.0729
Mean value of ρ_{int} Mean value of $deff_{int}$	0.0089 1.36	0.0074 1.23	0.0018 1.10	0.0086 1.26	0.0184 1.20	0.0057 1.07	0.0163 1.28	0.0090 1.13	0.0067 1.09

Source: R.M. Groves and L.J. Magilavy, "Measuring and Explaining Interviewer Effects in Centralized Telephone Surveys," *Public Opinion Quarterly,* Vol. 50, 1986, pp. 251–266.

Health issues were covered by both the Health and Television Viewing Study and the Health in America Survey. The 30 variables analyzed for the Consumer Attitude Survey are identical in each of the five months. These various repetitions enhance the utility of the investigation of interviewer variability by question type. The number of variables analyzed for each study ranges from 25 to 55.

8.5.1 Magnitudes and Precision of Estimated Interviewer Effects

Table 8.1 presents the range of estimates obtained in each of the SRC studies, along with the mean value of ρ_{int}.[4] The average ρ_{int} values seem to vary around 0.01 over the nine surveys and to exhibit large variability over the different statistics examined. The highest average ρ_{int} (0.0184) is found for one of the Consumer Attitude Surveys, but comparison over the other months for the same statistics shows a range of 0.0057 to 0.0184, with most lying nearer 0.01. The statistics examined for these surveys are mainly attitudinal in nature. The lowest average ρ_{int} (0.0018) is found for a mainly factual survey on health conditions, Health in America. This survey also had the largest interviewer workload.

8.5.2 Stability of Interviewer Effect Estimates

Since estimates of ρ_{int} are based on samples of respondents and interviewers, they will vary over replications of the survey. That is, these estimates of interviewer variance are themselves subject to variance. Unfortunately, few previous studies present estimates of how stable the ρ_{int} values are. Table 8.2 illustrates the importance of that failing by showing estimates of individual ρ_{int} values and their standard errors for 30 statistics, 10 from each of three SRC telephone studies.[5] This table

---

[4] Some of the statistics are means computed on continuous variables. Other questions required the respondent to answer on a four- or five-point scale (e.g., like very much, ..., dislike very much). To obtain estimates of interviewer effect for these variables the proportion choosing the modal response category was analyzed. It is possible for estimates of ρ_{int} to be negative (Bailey et al., 1978). In previous studies these negative values have often been presented as zeros. Here, the actual value is presented to give the reader evidence of the instability of the estimates. The negative values are also included in all summary calculations.

[5] In order to estimate the sampling variance of the ρ_{int} values, a jackknife estimator was used by dividing the sample into the k subsamples associated with each interviewer. That is, we omit elements from one interviewer's (i) workload and compute $\rho_{int_i} = S_{a_i}^2 / (S_{a_i}^2 + S_{b_i}^2)$

illustrates the range and magnitude of ρ_{int} values estimated throughout the nine surveys. The standard errors for the measures of interviewer effects are often larger in size than the estimates themselves, leading one to conclude that the values of ρ_{int} presented here could have been much larger *or much smaller* had different sets of interviewers been chosen to work on the surveys. These standard errors are estimates of the dispersion of the distribution of ρ_{int} values over replications of the surveys, with variation due to the selection of different interviewers from the population of all interviewers and from sampling variability of the respondents.

In order to relate these results to the past research, these standard error estimates can be used to illustrate the largest ρ_{int} values likely to be encountered with these designs. If we assume that the ρ_{int} are distributed roughly according to a t-distribution, we can estimate the value beyond which only 10 percent of the ρ_{int} and deff_{int} values lie. These are called the "maximum" values of ρ_{int} and deff_{int} in Table 8.2. These maximum values of ρ_{int} are in the general range of point estimates presented in the literature for personal interview surveys. Such values would then imply that the centralization of interviewing has little effect on between–interviewer differences. Indeed, with the larger workloads often found in centralized telephone interviewing, the possible impact on the total variance is much higher than would be true in personal interviews.

It has been assumed for some time that the measurement of interviewer effects would be enhanced by centralized telephone interviewing facilities. The ability to incorporate an interpenetrated design at a relatively small cost makes this an attractive setting for the evaluation of interviewers. However, it remains difficult to draw conclusions from the results presented here because of the imprecision of these estimates. The instability of the estimates is mainly due to the small number of interviewers employed for telephone surveys. Since the past literature in the area has generally failed to address the problem of imprecision, it is also possible that many of the effects judged to be important in those studies would not be found in replications.

The researcher faced with unstable estimates of interviewer effects has the alternatives of abandoning their measurement, altering the survey

---

on $(k-1)$ interviewers' cases. There are k possible ρ_{int_i} values and a jackknife variance estimator of ρ_{int} is $\text{Var}(\rho_{int}) = \frac{1}{k(k-1)} \sum_{i=1}^{k} (\rho_{int_i} - \rho_{int})^2$. Although the properties of the jackknife estimator are known only in the case of a linear statistic, it has also performed well empirically for a variety of nonlinear estimators including regression and correlation coefficients (Kish and Frankel, 1974). These empirical results suggested its use in this case.

Table 8.2 Illustrative Values and Jackknife Estimates of Standard Errors of ρ_{int}

Study: Statistic	ρ_{int}	Standard Error of ρ_{int}	"Maximum" Value of ρ_{int}	"Maximum" Value of deff_{int}
Health and Television Viewing:				
Proportion ever having had arthritis or rheumatism	0.0227	0.0132	0.0400	2.23
Proportion ever having had diabetes	−0.0059	0.0053	0.0010	1.03
Frequency of drinking alcohol in the previous month				
Beer: not at all	0.0040	0.0091	0.0159	1.49
Hard liquor: not at all	0.0152	0.0117	0.0305	1.94
Time elapsed since last doctor's visit: less than 6 months	0.0266	0.0144	0.0455	2.40
Proportion able to report date of most recent doctor's visit, precise within a week	0.0310	0.0152	0.0509	2.57
Number of ameliorative health behaviors reported	0.0285	0.0141	0.0470	2.45
Number of ways mentioned that TV is good or bad for children	0.0157	0.0129	0.0326	2.00
Number of "bad reactions" to medicine reported	0.0076	0.0133	0.0250	1.77
Number of health symptoms reported	−0.0032	0.0062	0.0049	1.15
Health in America:				
Number of physician visits during the past 2 weeks	0.0092	0.0076	0.0192	2.10
Number of hospitalizations during the past year	0.0040	0.0041	0.0094	1.54
Proportion reporting health status as excellent	0.0085	0.0077	0.0186	2.06
Proportion reporting 2-4 physician visits during the past year	0.0002	0.0035	0.0048	1.27
Mean number of chronic conditions reported	0.0096	0.0063	0.0179	2.02
Proportion reporting no bed days during the past year	−0.0070	0.0029	−0.0032	0.82
Proportion reporting no operations during the past year	−0.0031	0.0030	0.0008	1.05

Table 8.2 (Continued)

Study: Statistic	ρ_{int}	Standard Error of ρ_{int}	"Maximum" Value of ρ_{int}	"Maximum" Value of deff_{int}
Proportion reporting no bed days during the past 2 weeks	−0.0057	0.0025	−0.0024	0.86
Proportion reporting 2 weeks–6 months since last physician visit	0.0003	0.0042	0.0058	1.33
Proportion reporting 2 weeks–6 months since last dental visit	0.0017	0.0048	0.0080	1.46
Consumer Attitude Survey – December 1982:				
Proportion reporting that they are better off financially than they were 1 year ago	−0.0232	0.0171	−0.0008	0.99
Proportion reporting an increase in the price of things during the last 12 months	0.0101	0.0181	0.0338	1.39
Percent change in prices during the last 12 months	−0.0325	0.0178	−0.0092	0.89
Proportion reporting it is now a bad time to buy a house	0.0112	0.0239	0.0425	1.49
Proportion reporting it is now a good time to buy major household items	0.0242	0.0258	0.0580	1.67
Proportion reporting that they expect a decrease in interest rates during the next 12 months	0.0089	0.0271	0.0444	1.51
Proportion reporting that they expect an increase in unemployment during the next 12 months	−0.0210	0.0151	−0.0012	0.99
Proportion reporting that they expect no change in their financial situation during the next 12 months	0.0308	0.0269	0.0660	1.76
Proportion reporting that they expect their family income to increase 15% or less during the next 12 months	0.0546	0.0385	.1050	2.21
Proportion believing that the current government's economic policy is fair	0.0508	0.0361	0.0981	2.13

Source: R.M. Groves and L.J. Magilavy, "Measuring and Explaining Interviewer Effects in Centralized Telephone Surveys," *Public Opinion Quarterly*, Vol. 50, 1986, pp. 251-256.

design to improve their precision (chiefly by increasing the number of interviewers used), or increasing their precision through replication of surveys. Ceasing attempts to measure this source of survey error seems premature; the other two options deserve more thorough comment.

Increasing the number of interviewers is not recommended in this case, because the hiring and training of larger numbers of interviewers may force a change in essential survey conditions that help to determine the magnitudes of interviewer effects in the survey data. If, for example, the number of interviewers were increased greatly, the quality of training or amount of supervisory attention given to each interviewer might be altered (ignoring the higher costs incurred with such a design). With departures from training guidelines interviewer variance might increase, as well as other nonsampling errors (e.g., nonresponse). Thus, we face the Heisenberg-like result that to measure a particular survey error well *in a single survey*, we may change its value.

Another way to obtain useful knowledge about the average levels of interviewer variability is repeated measurement of interviewer effects over surveys. By cumulating results over surveys, we hope to take advantage of the fact that averages of the same survey statistic (in our case, interviewer variance estimates) over replications have greater stability than a single measure of it does. Ideally, averages would be constructed from repeated measures of the same population, using the same interviewer selection and staffing rules, and the same statistics. For example, four replications of a survey might reduce the standard errors of average ρ_{int} values by half, relative to those from a single survey.[6] This is a possible strategy in an ongoing survey. Rarely, however, are exactly equivalent measures obtained in several repeated studies. Lacking that, cumulation of ρ_{int} estimates over statistics and across surveys might be attempted within the same survey organization. Although this summarization technique would mask variation in ρ_{int} over different *statistics*, it would provide information about typical interviewer effects to be expected in similar surveys. This is similar to the construction of generalized sampling variance estimates in complex samples, used to provide the analyst with average design effects for clustering in the sample (see U.S. Bureau of the Census, 1974; Kish et al., 1976; Wolter, 1985). Given the severe weaknesses of the other approaches, the construction of generalized interviewer effect estimates through averaging over surveys is attractive.

In Figure 8.1 the cumulative distribution of ρ_{int} values is plotted for each study over all statistics for which ρ_{int} was calculated. It may be

---

[6] This assumes independence in the selection of interviewers across the replications and no change in true values of interviewer effects over replications.

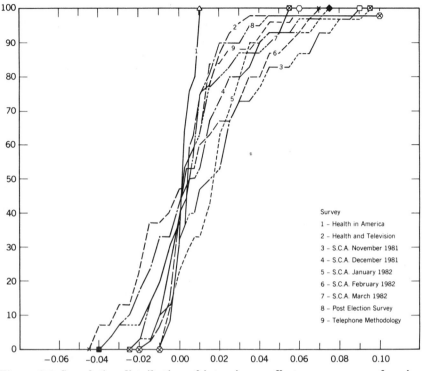

Figure 8.1 Cumulative distribution of interviewer effect measures ρ_{int} for nine SRC surveys.

observed that in general the *shapes* of the distributions are related to sample size and numbers of interviewers, with the range of values of ρ_{int} decreasing as these increase in number. This mirrors the estimates of standard errors of ρ_{int}. We see that while each study yielded some examples of high values of ρ_{int}, the majority of values cluster around 0.01. The mean values of ρ_{int} and the ranges are given in Table 8.1.

Cumulation of estimates of interviewer effects in figures such as Figure 8.1, together with other summary values of ρ_{int} over surveys and statistics, can demonstrate likely magnitudes of ρ_{int} to be expected given a set of interviewer procedures. If average values are higher than desired, attention needs to be devoted to implementing questionnaire changes or training changes to reduce them. Furthermore, as estimates of individual statistics are replicated, similar figures can be constructed for subsets of statistics (e.g., figures for questions on particular domains of attitudes, open questions versus closed questions).

The overall mean ρ_{int} is 0.009 for the nine surveys and approximately 300 statistics examined. Viewing this mean as an average of 300 independent statistics would imply a standard error about it of *approximately* 0.001, using the guidance of Table 8.2.[7] Such an average implies an increase in variance of estimates by roughly 10 percent for each 10 interviews added to the interviewers' workloads (relative to the hiring of other interviewers to complete the study). Such results might guide design decisions (following Kish, 1962) to minimize variance of statistics, given fixed resources.

8.6 EXPLAINING THE MAGNITUDE OF INTERVIEWER EFFECTS

In addition to lack of concern about stability of estimates, another weakness in the interviewer effects literature is the failure to seek *explanation* for interviewer effects. Instead, they have been viewed as a component of total variance of statistics without suggestions about designs to reduce them. The explanations for effects might lie within the nature of interviewer training guidelines, of the interviewer task (e.g., probing instructions, mode of interviewing), and of the survey instrument itself (e.g., question type). The research reviewed in this section sought explanations for interviewer effects as part of their design.

Groves and Magilavy (1986) and others have examined three design features: the form of the question, the type of interviewer behavior used, and the use of computer assistance in the data collection. In addition to these design features they also monitored certain aspects of interviewer behavior and examined both characteristics of the interviewers and the respondents as correlates of interviewer effects.

8.6.1 The Influence of Question Form on Interviewer Effects

Previous face to face surveys have examined question type in an effort to identify the correlates of interviewer variability. Different question types may present to the interviewer communication tasks that vary in complexity. For example, factual questions (those for which there is a knowable, verifiable answer) might be thought to be less subject to effects of different wording or inflection in their delivery than would be attitudinal questions (e.g., "How many times in the last 2 weeks did you

---

[7] This is based on a weighted average sampling variance of 0.0004575 on individual ρ_{int} and an assumption of independence among the 297 ρ_{int}'s. The lack of independence of the various ρ_{int}'s stems from the use of the same interviewers and respondent pairs on multiple statistics of the same survey. It is expected that the correlation among such ρ_{int}'s would act to increase the standard error of the mean ρ_{int}.

visit a physician's office?" versus "How well do you like TV game shows?"). Comparisons of interviewer effects on factual and attitudinal measures have been a common exercise in past research, and the results over several studies are mixed, some of them finding attitudinal questions subject to higher interviewer variance. Kish (1962), for example, found no important differences in interviewer effects by question type. However, several studies (Hansen et al., 1961; Fellegi, 1964; Feather, 1973; O'Muircheartaigh, 1976; Collins and Butcher, 1982) have concluded that factual items are less susceptible to interviewer effects. Larger effects have been observed for attitudinal questions, especially those with open-ended responses (O'Muircheartaigh, 1976), emotionally charged questions (Fellegi, 1964), difficult items such as income and occupation (Hansen et al., 1961), and questions that lacked specification regarding the acceptable interviewing procedure (Feather, 1973).

Fowler and Mangione (1985) used a regression model predicting ρ_{int} values. The predictor variables were characteristics of the survey questions: the difficulty of the question task (assessed cognitive burden), the ambiguity of terms in the questions, the sensitivity of the topic, whether the question was a factual one or an attitudinal one, and whether it was an open or closed question. All these attributes were assigned to questions based on the judgment of the researchers. The regression model was built using 130 transformed ρ_{int} values as observations of the dependent variable.[8] In a model with direct effects only, the difficulty of the item was the only characteristic with large positive effects (achieving a p-value of 0.10). The interaction term in the model indicating difficult opinion items was the dominant predictor of ρ_{int}. Other coefficients had values counter to prior hypotheses.

In the SRC surveys all 297 statistics from the nine different surveys were classified as factual or attitudinal (155 factual questions and 142 attitudinal questions), and since some questions were used in several surveys, there was an opportunity for replicated comparisons. In examining these results there appears to be no evidence that factual items as a class are subject to any different interviewer effects than are attitudinal questions. Over all surveys, the average value of ρ_{int} was 0.0098 for factual items compared to 0.0085 for attitudinal items.

Another variation in question form concerns whether the respondent is supplied answer categories for his/her choice or is free to formulate his/her own answer—closed questions versus open questions. Again using data from eight of the nine different SRC surveys, ρ_{int} values were

---

[8] The transformation had two steps. First, negative ρ_{int} values were set to 0.001. Second, a logit transformation is taken—$\log [\,\rho_{int}/(1 - \rho_{int})\,]$.

compared for open and closed questions. Despite suggestions that open questions might be subject to larger interviewer effects, there was no tendency for this to be the case over the 63 open and 192 closed questions examined from the eight surveys. In fact, five of the eight surveys produced higher estimates of ρ_{int} for closed questions. Over all surveys, the average value of ρ_{int} for closed questions was 0.0082 compared to 0.0124 for open questions.

There were, however, aspects of interviewer behavior that did vary with open questions. For example, some open questions ask the respondent to mention several entities that belong to a certain set (e.g., "What do you think are the most important problems facing the country at this time?"); one statistic of interest to some analysts is the number of different items that are mentioned. This count is a function both of the respondent's ability to articulate problems and the interviewer's behavior in probing for more mentions (i.e., "Are there any others?"). The number of responses to a question appears to be subject to greater interviewer differences than the substantive response category into which the answers are coded. For example, for six questions asked in the Surveys of Consumer Attitudes, the ρ_{int} values for the percentage giving two mentions are higher each month than the ρ_{int} values for the modal response category of the first mention. Over all months, the average value of ρ_{int} for the percentage giving two mentions was 0.0162 compared to an average value of 0.0007 for the modal category of the first mention. The ρ_{int} for the substantive category of the second mention was 0.0120, close to the 0.0162 for the proportion with second mentions. This suggests that the differential behaviors that determine whether a second mention is given also might influence substantive responses on the second mention. This interpretation receives more support from the Fowler and Mangione (1985) study, which showed that the largest correlate of ρ_{int} among several interviewer behaviors was the failure to probe an open question.[9]

8.6.2 Effects of Interviewer Behavior Guidelines on Interviewer Variability

Fowler and Mangione (1985) report mixed results regarding the effect of length of training on ρ_{int} values. The highest values exist for the groups with the shortest and longest training periods. Although post hoc

---

[9] The other interviewer actions investigated were laughing, incorrect reading of the question, the use of a correct probe, the use of a directive probe, inappropriate feedback, inappropriate other behavior, and inaccurate verbatim recording of the answer.

explanations are provided (essentially an effect of overconfidence in the latter group), it is not clear what inferences are appropriate. Their results do suggest that the act of tape recording interviews may reduce interviewer variance, apparently by sharpening the interviewers' attention to the task.

In a series of experimental SRC surveys (of which Health in America and the study of Health and Television Viewing are two), interviewers were trained to limit their verbal interaction with the respondent to a specified set of probes or feedback, in addition to the survey questions (following Cannell et al., 1987). Most of these statements were entered into the questionnaire itself and were varied to fit the response behavior of the respondent. Interviewers were not permitted to engage in any other conversation with the respondent other than the delivery of the question and the use of specified feedback. Two experimental groups were used; one that limited interviewers to feedback written in a special version of the questionnaire (experimental group) and another that permitted them to use at their discretion (control group) only a small number of statements in response to answers given by respondents (e.g., "I see"). Although both restricted interviewer behavior to a set of specified actions in addition to question asking, the first technique was hypothesized to improve reporting (i.e., reduce response bias) over the second.

In both surveys the differences between the ρ_{int} values for the cases with specified feedback and those using feedback at the discretion of the interviewer were small relative to the standard errors of the estimates. Since one of these studies (Health in America) has unusually low ρ_{int} values overall, one might speculate that both procedures reduce interviewer effects relative to typical SRC training procedures. Even though the differences were minor, a tendency was observed for the estimates from the half-sample with specified feedback to be smaller. Combining 57 of the variables from both surveys, 58 percent of the estimates obtained from the experimental group were smaller (mean value of −0.0024) than those calculated for the control group (mean value of 0.0142). Such a finding merits attempts at replication over different substantive topics.

8.6.3 Effect of Computer–Assisted Telephone Interviewing (CATI) on Interviewer Variability

A CATI system might act on a different source of interviewer effects than the feedback experiments above, by using the software to eliminate certain types of interviewer error. One study, Health in America, randomly assigned cases to the use of a CATI system or to the use of a paper questionnaire (this is reported on in full in Mathiowetz and Groves, 1984). Each interviewer used both systems at different times throughout the study. The same questionnaire was used on both systems. Out of the 25 variables compared, the CATI cases have lower ρ_{int} values on 18 of the variables, with an average value of -0.0031 compared to an average value of 0.0048 for the paper questionnaire. The differences, however, were small relative to the standard errors of the ρ_{int} values. Attempts to replicate these results are needed.

8.6.4 Interviewer Characteristics and Interviewer Effects

Failure to read a question exactly as printed, inability to follow skip patterns correctly, and reading a question too fast are all believed to contribute to errors in recorded data (Marquis and Cannell, 1969; Lansing et al., 1971). The Health in America Survey incorporated a monitoring procedure into its design which evaluated these aspects of interviewer behavior to determine if they were related to the magnitude of interviewer variability. Two types of behavior, question reading and clarity and pace of question delivery, were analyzed in detail. Even with the special emphasis given to training interviewers for this study, interviewers showed significant variation in the proportion of questions read correctly for many of the variables. For 80 percent of the variables there was a significant difference between interviewers in the proportion of questions read well (i.e., correct pace, clear speech). However, these differences in interviewer delivery were not correlated with the magnitude of interviewer effects for statistics computed on those variables (e.g., the proportion choosing a particular category).

To search further for this relationship, scatterplots of the squared deviation of the individual interviewers' means from the overall study mean by scores from the monitoring data were constructed. A positive relationship between an interviewer's squared deviations and the proportion of "incorrect" behavior was expected. No such relationship was apparent. In both the Health in America and the Health and Television Viewing Surveys, other measures of interviewer performance were analyzed, such as their response rates, productivity, and supervisor evaluations. Again, no relationship between any one of these variables and interviewer deviations from survey averages were found. The so-called "better" interviewers on these characteristics did not deviate any more or any less from the overall mean than did the other interviewers.[10]

8.6.5 Respondent Characteristics and Interviewer Effects

For the SRC Health and Television Viewing Survey, Groves and Magilavy (1986) examined levels of interviewer variability by sex, education, and age of respondents. The predominance of female telephone interviewers and potential differences in their interaction with female and male respondents suggested the possibility of differences in interviewer effects by sex of respondent. There was no evidence that the gender of the respondent was related to interviewer effects. Although the mean value of ρ_{int} over all questions was slightly higher for female respondents, further investigation by question type found mixed results. For some question types the estimates of ρ_{int} were larger for males; for other question types the estimates were larger for females. In all cases the differences in the estimates were small.

Although past empirical results vary (Cannell et al., 1977; Schuman and Presser, 1977), it is reasonable to argue that poorly educated respondents might be more easily affected by the behavior and status of the interviewer than more highly educated respondents. Lower education groups may seek greater help from the interviewer in answering the questions or may use the inflection of the interviewer's voice as a cue for responses to questions they find difficult. Values of ρ_{int}, however, do not appear to be larger for respondents with less than 12 years of education in the Groves and Magilavy work. In a similar analysis Fowler and Mangione (1985) found the same result. Even when estimates of ρ_{int} are calculated by question type, differences by respondent education are not apparent.

Larger values of ρ_{int} are often expected for older age groups. There is some evidence that telephone surveys suffer greater nonresponse among

---

[10] The form of the distribution of mean values obtained by each interviewer in the Health in America Survey differed from those of past studies. Previous telephone data yielded distributions of interviewer means that contained few outliers and had relatively smooth patterns about the overall survey mean. These data, however, had many measures where one or two interviewers were extreme outliers to the distribution. Over different statistics the identity of the outliers varied. To evaluate the impact of these extreme values the interviewer variability analysis was performed for five statistics with high values of ρ_{int} (mean value of .0075), eliminating one or two outliers in each case. For each variable the new estimate of ρ_{int} was much smaller, with a mean value of $-.0004$ over the five variables. In other words, in that study one or two interviewers were responsible for most of the measured variability.

older persons (Groves and Kahn, 1979), and those who do respond might be more subject to influence by the interviewer because of their own suspicion about the nature of the survey, or greater tendency to fatigue during the 30 minute interview on the telephone. Sudman and Bradburn (1973) have noted that older respondents exhibit greater response errors on questions requiring recall of factual material. Collins (1982) found some evidence that ρ_{int} values were higher for older respondents. Fowler and Mangione (1985) found higher ρ_{int} values for the respondents less than 30 years of age and slightly higher than the average for those 60 or older. In the SRC telephone survey data the oldest age group tends to have larger ρ_{int} values than other age groups. Over all question types, the oldest age group had an average value of ρ_{int} equal to 0.0274 compared to an average value of 0.0012 for others.[11] As we noted earlier, ρ_{int} measures both variation in nonresponse and response errors over interviewers. These higher ρ_{int} values for the elderly may relate to variation in nonresponse error among the elderly across interviewers.

8.7 SUMMARY OF RESEARCH ON INTERVIEWER VARIANCE

It is clear that the instability of estimates from most single surveys is an impediment to understanding interviewer variability. Through cumulating results over many surveys subject to similar design and administration, generalized estimates of interviewer effects might be obtained. The analysis of the SRC telephone surveys showed that typical ρ_{int} values vary about 0.01, an average somewhat smaller than those reported from personal interview surveys (the average of those in Table 7.5 is closer to 0.02).

Going beyond the analysis of interviewer effects to aid decisions about interviewer workloads, there is a need to study correlates of interviewer effects. Some of these are design features, such as characteristics of the measurements or of interviewer behavior that might be altered to reduce interviewer effects. Others are characteristics of respondents that would portend different error levels for studies of different populations.

The results from these several examinations are that attitudinal questions are themselves not subject to greater interviewer effects than "factual" items. Open questions are not inherently subject to greater interviewer variability, but the number of answers obtained to an open

---

[11] There is also evidence that the 35–49 year old group experiences higher ρ_{int} values (mean 0.0099) but there is no theoretical justification for such a finding.

question appears to be sensitive to interviewer differences. These effects do not apparently influence the substance of the *first* mentioned answer but appear to be the result of variable probing behavior, which affects both the likelihood of the respondent giving a second mention *and* the substance of the second response given.

Interviewer effects may be able to be altered, both through use of controlled feedback and probing and of CATI systems that control the flow and questioning sequences of the interview. This is a promising result that deserves further research because it offers the designer real hope of reducing this source of survey error.

In contrast, there has been an unfortunate lack of success in using common indicators of performance to predict interviewer effects on survey data. Use of on-line monitoring of interviews did not collect information that seemed to be related to deviations from typical responses to the questions. Furthermore, the traditional indicators of interviewer quality (response rate, productivity, and supervisory evaluations) seemed to be unrelated to the errors measured by between-interviewer variability. This result adds further doubt to the relevance of these indicators for survey quality, although they retain their value for measurement of cost efficiency of surveys.

Finally, there has been some search for respondent level attributes that were related to interviewer effects. Gender and education of respondent are essentially uncorrelated to the effect, but age of the respondent seems to be related. Older respondents were found to exhibit slightly larger susceptibility to interviewer effects than younger respondents. Given past low response rates among the elderly in surveys, this result might reflect more variation in nonresponse bias across interviewers than in differential response errors. The finding deserves the attention of researchers especially interested in this subpopulation.

8.8 MEASUREMENT OF INTERVIEWER COMPLIANCE WITH TRAINING GUIDELINES

One reason for the existence of variability in results from different interviewers is that some interviewers follow the instructions given for administering the questionnaire, and others do not. Those who do not follow the training guidelines may increase response error in various ways. There are two sets of research efforts measuring how closely interviewers follow training guidelines in administering the questionnaire. The first is based on tape recorded interviews taken in face to face interview settings. After the interview is completed, the tapes are reviewed by a coder. The second set of research applies to centralized telephone interviews and is based on a monitor's listening in to an interview.

"Interaction coding" applied to surveys is designed to measure compliance with interviewer guidelines for identifying the proper questions to ask the respondent, for asking them in the manner prescribed, for clarifying the meaning of questions for respondents having difficulty, for probing to obtain complete and accurate data, and for recording the data as instructed (Cannell et al., 1975). With tape recordings of personal visit interviews, coders trained in the interaction coding system listen to the taped interviews and make judgments regarding the appropriateness of each behavior of the interview. This is a time-consuming process. Morton-Williams (1979) reports it taking "at least 45 minutes for a 30 minute interview and often longer" because coders often judge it necessary to listen to parts of the interview several times. In telephone interviews the monitor codes the interaction while the interview is ongoing.[12] This has led to a coding scheme that is simpler than that used for taped interviews.

Figure 8.2 presents the coding scheme of Mathiowetz and Cannell (1980), used for telephone interviews, but representing most of the major categories of behavior that are documented in taped interviews also. Some interaction coding in personal interviews has also coded the behavior of the respondent, as an aid in interpreting the reactive behavior of interviewers. The codes in Figure 8.2 tend to have four ordinal evaluative categories for each behavior type. The evaluation of question-asking measures the extent of agreement between the written question and the words used by the interviewer. "Repeating questions" concerns cases in which the interviewer judges that the respondent has not heard or understood the question and asks the question again. The deviations from training guidelines can consist of the interviewer repeating the question without need to do so or repeating incompletely or incorrectly. Sometimes the interviewer must use words not written in the questionnaire to clarify the meaning of a question ("defining/clarifying"). Whether instructed to do so or not, interviewers often acknowledge the answers of respondents by giving verbal feedback (e.g., "I see," "Thank you"). In the Mathiowetz and Cannell study this feedback was designed into the questionnaire and was of two types: short and long feedback. There are two sets of codes for feedback ("short feedback" and "long feedback"). The speed with which the question is delivered is another category of evaluation; for the Mathiowetz and Cannell work, a guideline

---

[12] Taping of interviews is illegal in many U.S. localities, without permission of the respondent and the use of a periodic beeping sound.

of two words per second was used. Monitors were trained in judging departures from this training guideline by listening to a tape recording of the questionnaire administered at the specified rate. Finally, there was a summary code for "overall clarity," which asks the monitor to record whether the interviewer had asked the question with improper inflection or with mispronunciations.

Interaction coding is a separate data collection activity, focusing on measuring weaknesses in the survey interviewing, but subject to errors itself. The interaction coding relies on the judgment of coders regarding departures from the training guidelines. It is to be expected that these judgments will vary. Not all the behavior categories can be used purely objectively. It is typical in this research that reliability estimates are made for coding performance. For example, Mathiowetz and Cannell (1980), find an average .85 agreement with a trainer of coders for the behaviors involving question wording, defining, and feedback, but .75 agreement for the pace and overall clarity codes. Brenner (1982), using tape recorded interviews, found a .98 agreement. It seems clear that the level of reliability across coders is maximized for those behaviors that minimize coder judgments.

There is a set of interaction codes that are consistent across six different studies, conducted in different organizations and among different interviewer types. These are whether the question was asked exactly as written, whether there was a slight change in wording, a major change, or whether the question was skipped erroneously. Table 8.3 presents findings from studies of mock interviewing situations with supervisors as respondents (Rustemeyer, 1977), experimental studies with real respondents (Marquis, 1971; Mathiowetz and Cannell, 1980), and ongoing production surveys (Oksenberg, 1981; Brenner, 1982; Prufer and Rexroth, 1985). Both telephone and personal interviewing modes are represented. The topics of the survey are varied.

The table presents estimates of the proportion of questions that were read exactly as written, with minor changes, with major changes, and those that were skipped entirely. Even a casual glance at the table will demonstrate that the proportions of questions read as they are written are highly variable, from about 30 to over 95 percent.

The Rustemeyer research with mock interviews studied over 200 interviewers. Some of the studies present their data in a way that variation over *interviewers* can be examined. Rustemeyer's study, for example, compares three groups of interviewers, as a way of studying the effect of training and experience on interviewer performance. "Experienced interviewers" are those with more than 3 months of field interviewing; "end of training" interviewers are those who had completed only two or three interviewing assignments; "new" interviewers had just

QUESTION-ASKING
11 Reads questions exactly as printed
12 Reads question incorrectly, minor changes
16 Reads question incorrectly, major changes
17 Fails to read a question

REPEATING QUESTIONS
21 Repeats question correctly
25 Repeats question unnecessarily
26 Repeats question incorrectly
27 Fails to repeat question

DEFINING/CLARIFYING
31 Defines or clarifies correctly
35 Defines or clarifies unnecessarily
36 Defines or clarifies incorrectly
37 Fails to define or clarify

SHORT FEEDBACK
41 Delivers short feedback correctly
45 Delivers short feedback inappropriately
46 Delivers short feedback incorrectly
47 Fails to deliver short feedback

LONG FEEDBACK
51 Delivers long feedback correctly
55 Delivers long feedback inappropriately
56 Delivers long feedback incorrectly
57 Fails to deliver long feedback

PACE/TIMING
65 Reads item too fast or too slow
66 Timing between items is too fast
67 Timing between items is too slow

OVERALL CLARITY
75 "Unnatural" manner of reading item (poor inflection,
    exaggerated or inadequate emphasis, "wooden" or
    monotone expression)
76 Mispronunciation leading to (possible) misinterpretation

Figure 8.2 Interaction codes for monitoring telephone survey interviewer behavior.

completed training but had no field experience. There are essentially no differences among the three groups in the percentage of questions they read exactly as written, despite the fact that the overall percentage is rather low (about 66 percent). The largest differences appear to occur for

Table 8.3 Results of Interaction Coding on Several Studies

Type of Question Delivery	Rustemeyer (1977) Interviewer Type			Mathiowetz and Cannell (1980) Question Type		Brenner (1982)	Oksenberg (1981) Question Type			Prufer and Rexroth (1985)	Marquis (1971)
	Experienced	End of Training	New	Open	Closed		Closed	Restricted Open	Open		
Exactly as written	66.9	66.4	66.9	95.8	95.4	57.3	56	51	30	93.6	62.4
With minor changes	22.5	17.9	19.9	1.9	3.7	6.4	36	44	46	6.4 }	19.1
With major changes	5.2	3.6	3.9	0.5	0.4	13.8	7	4	8		6.3
Not read	3.3	8.9	6.0	1.8	0.5	5.0	1	1	16	?	12.2
Number of questions	22,255	5,466	14,089	?	?	88,064	2,852	?	?	1,415	23,396
Number of interviewers	114	39	72	26	26	6	20	20	20	12	13
Number of interviews	456	117	288	208	208	72	105	105	105	35	164

Note: Researchers used somewhat different coding categories. Assignment to "Type of Question Delivery" categories was based on judgments of this author.

Source: A. Rustemeyer, "Measuring Interviewer Performance in Mock Interviews," *Proceedings of the American Statistical Association, Social Statistics Section, 1977*, pp. 341–346; N. Mathiowetz and C. Cannell, "Coding Interviewer Behavior as a Method of Evaluating Performance," *Proceedings of the American Statistical Association, Survey Research Methods Section, 1980*, pp. 525–528; M. Brenner, Chapter 5 in W. Dijkstra and J. von der Zouven, *Response Behaviour in the Survey–Interview*, Academic Press, London, 1982; L. Oksenberg, "Analysis of Monitored Telephone Interviews," Report to the U.S. Census Bureau; P. Prufer and M. Rexroth, "On the Use of the Interaction Coding Technique," *Zumanachrichten*, 17, November 1985; K. Marquis, Chapter 12 in J. Lansing et al., *Working Papers on Survey Research in Poverty Areas*, Survey Research Center, 1971.

385

the percentage of questions that are skipped, where the middle group
("end of training" interviewers) commit more errors (8.9 percent of all
questions versus 3.3 percent for the experienced interviewers and 6.0
percent for new interviewers).

Despite the fact that the training and experience factor does not
appear to be highly predictive of compliance with interviewing guidelines,
other studies show that the interviewers are highly variable on this
measure. For example, Oksenberg (1981) shows that the four best
interviewers on average delivered 11 percent of the questions with minor
changes and the four worst interviewers averaged 67 percent. This is a
large difference. Prufer and Rexroth (1985) find that rates of errors in
reading the question range from 0 percent for the three best interviewers
to 18.3 percent for the three worst. Brenner (1982) finds a range from 6 to
over 54 percent among the six interviewers studied.

Table 8.4 presents data from an investigation by Brenner in The
Netherlands. Face to face interviews were tape recorded and later
analyzed. The unit of analysis was the question; each interview contained
approximately 75 questions. Note the frequency of omitting a response
card (52.2 percent of questions) and the various errors regarding probing
(about 35 percent of the questions receiving some sort of directive
probing).

In addition to judging whether the interviewer has delivered the
question as written, interaction coders often make judgments about the
probing behavior of the interviewer. Probing is variously defined across
investigations, but it most often refers to verbal efforts by the interviewer
to convey the meaning of the question to the respondent. Sometimes
these efforts are initiated by a question from the respondent, other times
by a decision of the interviewer that the respondent is having difficulty
understanding the question. Probing in general is an activity that is not
specified for each question in the questionnaire but is guided by general
principles taught in interviewer training. The goal of probing is generally
believed to be the clarification of meaning of questions through the
definition of terms. Table 8.5 presents the results of interaction coding
research on probing behavior. Note again the large variability (44 to 90
percent correct) over the various studies.

The method of interaction coding is very different in telephone
surveys than in face to face surveys. The tape recording of personal
interviews permits the coder to study interactions between interviewer
and respondent carefully. In many states of the United States tape
recording of telephone interviews without obtrusive audio warning signals
is prohibited. Hence, interaction coding must be performed by a coder
listening to the survey interview. The coder (often called a "monitor" in
telephone surveys) cannot study the interaction before making a

Table 8.4 Descriptive Measures of Interviewer Actions (Dutch Personal Interviews)

Behavior	Percentage of Total Occurrences of Questions
Question asked as required	57.3%
Question asked with slight change	6.4
Question significantly altered	12.7
Question completely altered	1.1
Question asked directively	4.5
Response card omitted by mistake	52.2
Question omitted by mistake	5.0
Adequate probing	41.4
Repeating the question	2.9
Offering a leading probe	20.7
Directive probing based on respondent's answer	10.1
Directive probing based on interviewer's inference	4.4
Probing unrelated to task	20.5

Source: M. Brenner, "Response-effects of 'Role-restricted' Characteristics of the Interviewer," Chapter 5 in W. Dijkstra and J. van der Zouwen (eds.), *Response Behaviour in the Survey Interview*, Academic Press, London, 1982, Tables 5.2 and 5.3.

judgment about which kind of behavior code should be assigned. The coding is simultaneous with the interviewing. There are two effects of this: (1) reliability of judgments may be decreased and (2) the complexity of the coding procedures is often reduced on the telephone to make the coding feasible.

Unfortunately, although interaction coding can measure compliance with training standards, there have been few empirical links made between violating those standards and measurement error. Rewording of questions by the interviewer in itself does not imply that measurement error results. As is demonstrated in Chapter 10, it is true that answers to some questions are indeed affected by small wording changes, but answers to other questions seem rather immune to such alterations. Only the

Table 8.5 Percentage of Correct Probing Behaviors Among Those Observed

Evaluation of Probing Behavior	Rustemeyer (1977) Interviewer Type			Mathiowetz and Cannell (1980) Question Type		Brenner (1982)	Oksenberg (1981) Question Type			Prufer and Rexroth (1985)
	Experienced	End of Training	New	Open	Closed		Closed	Restricted Open	Open	
Percentage correct	80.7	86.0	80.3	79.2	85.6	44.3	75.6	56.9	89.7	62.4
Percentage incorrect	19.4	14.1	19.6	20.8	14.4	55.7	24.4	43.1	10.3	37.6

Source: See Table 8.3.

388

Rustemeyer study had direct measures of effect on recorded data. By examining the completed questionnaire, listening to the interview, and being aware of the true answers of the scripted respondent, the coder was able to judge whether the recorded answer correctly classified the respondent. Table 8.6 shows that from 10 to 14 percent of the items were coded incorrectly for the respondent's true characteristics (with less experienced interviewers making more errors). The Rustemeyer work developed a test score for the correctness of asking questions and finds that this test score is correlated at a 0.44 level with the number of errors made in recording answers. This level of correlation is lower than one might hope to establish the case that the interviewer behaviors in question produce measurement error. (Similarly, the score on probing behavior is correlated with recording errors at a 0.29 level.)

Interaction coding for face to face interviews using the techniques described is threatened by the high cost of having interviewers tape record interviews and by the laborious coding of the tapes. In centralized telephone surveys the marginal cost of the monitoring is much smaller, although the richness of the interaction coding is diminished. If techniques are developed to standardized monitoring systems, the method offers real promise for increasing information about the survey measurement process.

The empirical data from the interaction coding can be used in three distinct ways: (1) to give summary quality control evidence for the survey, for example, the percentage of all questions asked correctly; (2) to identify individual questions that cause interviewers and/or respondents difficulty (questions that were often subject to rewording, probing, respondent questions); and (3) to identify individual interviewers who tended to administer the questionnaire less well than others. The fact that the interaction coding data are empirical in nature permits summary in a variety of fashions.

What is largely missing from previous research, however, is evidence that the behaviors measured in interaction coding are themselves directly linked to measurement error in the survey data. This research should receive highest priority in the future.

8.9 EXPERIMENTS IN MANIPULATING INTERVIEWER BEHAVIOR

Concerns with interviewer effects on survey data led to early efforts by survey practitioners to reduce those effects. These efforts have focused most specifically on the methods used for training interviewers

Table 8.6 Errors in Recording of Respondent Answers as Judged by Coders of Tape Recorded Mock Interviews

Classification of Entry	Type of Interviewer		
	Experienced	End of Training	New
Consistent with response, incorrect due to interviewer error	90.6%	82.5%	86.1%
Correct answer, not given by respondent (interviewer guess)	4.6	8.4	8.7
Other error	?	?	?
Answer placed in wrong location	0.1	0.3	0.1
Correct omission of entry	NA	0.6	NA
Number of items	49,105	11,806	31,174

Source: A. Rustemeyer, "Measuring Interviewer Performance in Mock Interviews," *Proceedings of the American Statistical Association, Social Statistics Section, 1977*, pp. 341–346, Table A.

(pedagogical techniques) as well as the prescriptions for interviewer behavior (what the interviewers are told to do).

8.9.1 Effects of Interviewer Training

Interviewer training is the most direct attempt to alter the behavior of interviewers to assure uniform desired administration of the measurement procedures. Although all organizations use some amount of formal instruction, there is great heterogeneity in the amount and methods of training across organizations. There appear to be differences in the strictness of guidelines for reading every question exactly as worded. Organizations administering largely factual questions tend to be less strict on this feature. Similarly, some organizations emphasize probing in a nondirective way (i.e., they do not wish to do anything to alter the probability of one answer or another). Finally, organizations differ greatly in how many hours of training they give their interviewers prior to beginning production interviewing.

Fowler and Mangione (1985) addressed the question of how the *length* of the interviewer training period affected the behavior of interviewers and the quality of data. In this study 57 novice interviewers on a study of health behaviors and statuses in a suburban area of Boston, Massachusetts, were assigned at random to four training groups that varied on the number of days devoted to trainin⸺

1. The 1/2 day training group (this consisted of a 2 hour lecture on survey research, a demonstration interview, and some time before the session devoted to the trainees' reading of the interviewers' manual).

2. The 2 day training group (in addition to the material in the 1/2 day session, there was a discussion of interviewing principles, 2 hours of supervised role-playing interviews, use of tape recorded interviews as demonstrations, and exercises in interviewing activities).

3. The 5 day training group (in addition to the 2 day group activities, this group increased the amount of practice and role playing, with emphasis on probing, recording answers, and eliciting sample person cooperation).

4. The 10 day training group (in addition to the activities of the 5 day group, this group did a practice interview in a stranger's home in the presence of a supervisor, evaluated other interviewers' work—using interaction coding procedures, and read and discussed a survey methodological article).

In addition to manipulating experimentally the length of the training period, there was also experimental manipulation of the closeness of supervision. All interviewers were in contact with their supervisors at least once per week, but other attributes of the supervision varied depending on the assignment to one of three experimental groups:

1. A group receiving supervisory feedback on the number of hours worked, response rates, and costs per interview.

2. A group receiving, in addition, evaluation of a sample of their completed interviews on question flow, recording of open responses, and extent to which answers met question objectives.

3. A group that, in addition, tape recorded all their interviews, had a sample of these reviewed and evaluated on quality of interviewing and recording.

The three supervisory treatments were crossed by the four training procedures to produce $3 \times 4 = 12$ experimental cells to which interviewers were randomly assigned.

Using a six-call rule, a response rate of 67 percent was obtained, lower than a normal survey because of the absence of refusal conversion activities. There were no large (more than 6 percentage points) nor statistically significant differences across experimental groups in response rates. By using ρ_{int} as a measure of interviewer effects on response errors, there were no significant main effects of training or supervision. There was, however, a decreasing trend in average ρ_{int} values in conjunction with increased supervision (from a mean of .012 for the loosest supervision to .0008 for the closest supervision). There was a significant interaction effect on ρ_{int} for the training and supervision treatments, but it did not yield any logical interpretation. The authors conclude that the effects on tape recording on the quality of interviewing was the dominant source of improvement in the supervision treatment. In contrast to the lack of effects on ρ_{int} values, the results on interaction coding data were large (see Table 8.7). The groups more intensively trained consistently exhibited greater compliance with interviewing guidelines.

Billiet and Loosveldt (1988) describe a study in Belgium with a very similar experimental design as that of Fowler and Mangione (1985). Their study used 22 inexperienced female interviewers assigned to four experimental cells, defined by training and tape recording conditions. Two training regimens were compared: one 3 hour meeting versus five 3 hour sessions. In addition, half of the cases were tape recorded; half were not. The authors found lower missing data rates among the trained interviewers and that tape recording tended to reduce missing data rates among the untrained but not the trained interviewers. They also find that trained interviewers obtained different responses from untrained interviewers on questions requiring probing or clarification by the interviewers to obtain an adequate response. For other questions there were no differences.

8.9.2 Experiments with Interviewer Feedback to Respondents

Another set of research in manipulating interviewer behavior focuses on one of the activities frequently observed in survey interviews: providing feedback to respondents regarding their performance. In early examinations of tape recorded interviews, Cannell and his associates noted that interviewers varied greatly in the amount of utterances and comments given after the respondent had replied to a question. On the average, however, about one-quarter of the oral actions of the interviewer

Table 8.7 Measures of Interviewer Behavior from Taped Interviews by Training Group

Behavior	Length of Training Program			
	1/2 Day	2 Days	5 Days	10 Days
Mean number of questions read incorrectly	21	7	14	6
Mean number of directive probes	8	5	5	3
Mean number of failures to probe	8	6	5	5
Percent interviews rated excellent or satisfactory				
Reading questions as worded	30%	83%	72%	84%
Probing closed questions	48	67	72	80
Probing open questions	16	44	52	69
Recording answers to closed questions	88	88	89	93
Recording answers to open questions	55	80	67	83

Source: F.J. Fowler, Jr. and T.W. Mangione, "The Value of Interviewer Training and Supervision," Center for Survey Research, 1985.

were of these types. Viewing these acts within the context of a model of social interaction, Cannell gave weight to them as "feedback" to the respondent by the interviewer. In contrast to the findings of psychological experimentation on shaping of subjects' behavior through experimental feedback, however, it was observed that interviewers tended to provide only positive feedback and to do so with no particular schedule of reinforcement of good respondent behavior. Somewhat disturbing in this early work was the tendency for interviewers to give the most positive feedback when the respondent refused to answer a question or terminated efforts to recall the information required by a question (e.g., "Oh, that's ok, don't worry about it; we'll go on...").

These early results led to several years of experimentation with the programming of feedback that interviewers provide to respondents. In a

Table 8.8 Reported Media Use by Interview Procedure

	Interview Procedure	
Reports of Media Contact	Instructions Feedback Commitment	Control
Percentage watched TV previous day	86%	66%
Percentage listened to radio previous day	67%	65%
Percentage read newspaper previous day	83%	77%
Percentage had seen an X-rated movie	61%	51%
Mean number of books read in last 3 months	2.9	5.3

Source: C.F. Cannell et al., "Research on Interviewing Techniques," *Sociological Methodology, 1981*, Table 10.

set of split sample experiments, interviewers were guided by the questionnaire to provide programmed feedback to the respondents. This feedback was sometimes contingent on the initial responses given. If the respondent quickly provided a "don't know" answer to a question demanding factual recall, the interviewer was instructed to say "you answered that quickly; perhaps if you thought about it more, you could...." Often this manipulation of feedback was accompanied with more elaborate instructions to the respondent about the purposes of questions and with a initial commitment on the part of the respondent to try hard to answer completely and accurately. Table 8.8 presents results that typify those of the general line of research.

8.10 SOCIAL PSYCHOLOGICAL AND SOCIOLOGICAL EXPLANATIONS FOR RESPONSE ERROR ASSOCIATED WITH INTERVIEWERS

The previous section described efforts to reduce interviewer effects on measurement errors. These efforts are based on hypotheses regarding the source of interviewer effects. Training efforts assume that behavioral guidelines communicated in the sessions, if followed, reduce or eliminate interviewer effects on measurement error. The placement of programmed feedback into the questionnaire similarly asserts that standardization of that behavior will act to reduce error.

Another set of research on interviewer effects attempts no intervention into the measurement process to reduce errors. Instead, it examines the correlates (or hypothesized causes) of interviewer effects. This work resembles the efforts discussed in Section 8.6 to discover correlates of interviewer variance statistics. We review studies of the effects of interviewer expectations of respondent answers, of interviewer race, and of interviewer gender.

8.10.1 Effects of Interviewer Expectations

There is a small body of literature that posits a specific cause of measurement errors associated with interviewers: the fact that the interviewer expectations about likely answers of respondents influence the survey data. Some hypothesize this to be a cause of the alternative manners of presenting the questions to the respondent, wording changes, tones of voice, and other attributes of the delivery that influence the respondent in various ways. Others hypothesize that interviewer expectations influence recorded responses mainly through failure at the probing or recording stage of a survey questioning. Hyman (1954) was the first to investigate the role that interviewer expectations might play in affecting survey data. He argued that, before the survey begins, interviewers have a prior distribution of expected answers on questions and that their behavior during the survey affects such a distribution.

There were two similar attempts at empirical investigations of interviewer expectations. In the Sudman et al. (1977) work, the interviewers completed a questionnaire prior to the survey, measuring their perceived difficulty of asking the survey questions, how uneasy respondents might be about answering the questions, what proportion of the respondents would refuse to answer certain sets of questions, and how large the underreporting of sensitive events was likely to be. Unfortunately, interviewer assignments to sample persons were not randomly made, and the interpretation of interviewer effects is therefore complicated. Fifty-six interviewers completed 1077 face to face interviews in the study. A later work by Singer and Kohnke-Aguirre (1979) replicated large parts of the Sudman et al. research in the context of a larger study on informed consent. Both studies found that interviewer expectations with regard to reporting errors and difficulty of administration were, by and large, not strongly predictive of respondent behavior. Table 8.9 presents estimates of the proportion of persons reporting various kinds of sensitive behaviors for groups of respondents of interviewers with different assessments of the difficulty of asking the items. The table shows a small tendency toward lower reporting of these behaviors to interviewers who expected difficulties with the questions.

Another study of interviewer expectations about respondent reluctance to answer questions was conducted around one March supplement to the U.S. Current Population Survey on income measurement. *After the survey had taken place*, interviewers were sent a questionnaire that asked them to assess their own and their respondents' ease of using certain of the income questions. In reporting the survey, Stevens and Bailar (1976) show that levels of item missing data were somewhat higher for interviewers who thought it inappropriate to ask respondents about the amounts of their income. Unfortunately, there was no randomization of assignment of interviewer to sample person in this study. An additional weakness was the decision to measure interviewer attitudes after the survey was completed, which complicates the interpretation of relationships between attitudes and performance. It is possible in this study that some interviewers concluded that the asking of some information was inappropriate because of the large number of refusals they obtained during the survey.

From one perspective, the interviewer expectation hypothesis is a proposed cause of interviewer variance. The expectation attribute lies farthest back in a causal chain which leads to different styles of delivering the question, probing, and recording. The fact that the empirical studies yield so little support to this hypothesis should not be viewed as strong refutation of the theory. Instead, all studies seem to suffer from weak operationalization of the expectation concepts. The Sudman et al. and the Singer and Kohnke-Aguirre work measured interviewer expectations regarding the difficulty of sets of questions, not individual questions. To the extent that their expectations regarding difficulty of administration, for example, varied across questions in a set, the indicator of interviewer expectations is weakened. Furthermore, expectations were measured for all respondents, yet expectations might vary for different subgroups of the population. The ideal measurement would be of interviewer expectations of the individual respondent being interviewed. This perspective is admittedly different than Hyman's view of a prior distribution of expected answers influencing the interviewers' behaviors, but it brings the expectation effects closer to the survey measurement process itself. This viewpoint would argue that the past empirical efforts are weakened tests of the expectations hypothesis, which could usefully be improved.

Table 8.9 Sudman and Singer Results on Selected Items by Perceived Difficulty of Interviewer Administration

		Interviewer Expectations of Difficulty			
Statistic	Study	Very Easy	Moderately Easy	Neither Easy or Difficult	Difficult
Proportion petting and kissing last month	Sudman et al.,	0.81	0.75	0.80	0.71
	Singer and Kohnke–Aguirre	0.76	0.65	0.70	0.69
Proportion having intercourse past month	Sudman et al.,	0.74	0.67	0.70	0.66
	Singer and Kohnke–Aguirre	0.69	0.56	0.65	0.60
Mean earned income	Sudman et al.,	$12,216	$13,531	$14,041	$11,747
	Singer and Kohnke–Aguirre	$14,685	$11,462	$10,827	$11,945
Proportion ever smoked marijuana	Sudman et al.,	0.21	0.20	0.21	0.22
	Singer and Kohnke–Aguirre	0.28	0.20	0.24	0.29

Source: E. Singer and L. Kohnke–Aguirre, "Interviewer Expectation Effects: A Replication and Extension," *Public Opinion Quarterly,* 1979, pp. 245–260, Table 4. S. Sudman et al., "Modest Expectations: The Effects of Interviewers' Prior Expectations on Responses," *Sociological Methods and Research,* Vol. 6, No. 2, 1977, pp. 171–182, Table 2.

8.10.2 Demographic Characteristics of the Interviewer

The survey interview is an interaction that is conducted within the context of a complex set of social norms guiding interaction among individuals. Although there are efforts through interviewer training to enforce specific behavioral guidelines for the interviewers that may be contrary to their natural inclinations in such interactions, there are a variety of cues that respondents use to make judgments concerning their own behavior. Some of these cues may correspond to demographic characteristics of the interviewers. Following this reasoning, there have been some investigations of the effects of interviewers' gender and race (as well as some other attributes) on survey responses.

The largest literature concerns effect of interviewers' race. The first studies by Hyman and his associates (1954) in Memphis and New York found that black respondents gave more "patriotic" responses on some foreign policy questions when interviewed by a white than by a black. In contrast, blacks tended to characterize themselves as of a higher social status when responding to black than to white interviewers. Effects were larger in Memphis than New York. In Hyman's studies as well as others (Pettigrew, 1964; Williams, 1964) there were some racial attitude questions that showed strong differences by race of interviewer and others that did not. There appeared to be no distinguishing characteristics of the two groups.

Schuman and Converse (1971) used randomized assignments of 17 white student interviewers and 25 older black interviewers to a cluster sample of Detroit black households, yielding about 500 personal interviews. As shown in Table 8.10 the study found that race of interviewer as a predictor of responses to other questions explained more than 1 percent of the variance in 15 to 40 percent of the questions. Racial attitude questions were found to be most susceptible to these effects (for 32 percent of them race of interviewer explained more than 2 percent of the variance). The authors were not able to identify subsets of black respondents who showed unusually large race of interviewer effects, except for the tendency for lower status blacks to show more effects.

A later national study of race of interviewer effects reported by Schaeffer (1980) is limited by the lack of randomized assignment of interviewers but has the appeal of utilizing a national area probability sample of housing units as part of the General Social Survey conducted by NORC. Prior to estimating race of interviewer effects, the author analytically controlled for differences in respondent's education, income, size of place of residence, region of residence at age 16 (South versus non-South), status of residence at age 16 (rural versus nonrural), religion (Baptist versus other), and gender. Given that attempt to overcome the

Table 8.10 Distribution of Percentage Variance Explained by Race of Interviewer by Type of Question

Category	Number of Questions	Percentage of Questions			Total
		Less than 1% Variance Explained	1 to 2% Variance Explained	2% or More Variance Explained	
Racial attitudes	40	63%	5%	32%	100%
Racial facts	14	57	29	14	100
Nonracial attitudes	29	86	11	3	100
Nonracial facts	47	81	11	8	100
All questions	130	74	11	15	100

Source: H. Schuman and J. Converse, "The Effects of Black and White Interviewers on Black Responses in 1968," *Public Opinion Quarterly*, Vol. 35, 1971, pp. 44-68, Table 2. The authors note that explained variance estimates of 1% are statistically significant at the 0.05 level and those of 2% at the 0.01 level, under assumptions of a simple random sample assigned at random to individual interviewers.

absence of randomization, the implicit assumption required for interpretation of race of interviewer effects is that there are no other variables that affect the survey questions examined or that, if some do exist, they are uniformly distributed across the interviewer race groups. Schaeffer analyzes the data at the interviewer level to reflect the clustering of cases into interviewer workloads and finds that for black respondents a wide variety of questions, both with and without racially sensitive content, are affected by race of interviewer.

Hatchett and Schuman (1975) and Schaeffer examine the race of interviewer effects on white respondents. The former work again used a randomized assignment of interviewers in a Detroit sample and found that whites tended to give more "liberal" answers on racial attitude questions to black interviewers than to white (Table 8.11). Schaeffer found that items tapping racial liberalism showed effects but others with racial content (e.g., questions about anticipated problems in interracial marriages) showed no effects.

The literature on race of interviewer effects suffers from two weaknesses. First, confounding of race with other attributes of the interviewers complicates inference from the studies. Without randomization black interviewers typically tend to interview in higher

Table 8.11 Percentages of White Respondents in Various Response Categories by Race of Interviewer

Category	Black Interviewers	White Interviewers
Would not mind if relative married a Negro	71.7%	25.5%
Believes Negro and white students should go to same school	90.7	56.0
Would not be disturbed if Negro of same class moved into block	100.0	68.6
Believes Negro and white children should play together freely	92.5	84.0

Source: S. Hatchett and H. Schuman, "White Respondents and Race-of-Interviewer Effects," *Public Opinion Quarterly*, Vol. 39, No. 4, 1975-76, pp. 523-528, Table 1.

density black areas. In the Schuman and Converse work (that did randomize assignments) all the white interviewers were students, doing interviews for the first time as part of a graduate seminar. Hence, empirical estimates of race of interviewer effects might alternatively be interpreted as effects of interviewing experience or professional status of the interviewer. Second, usually the estimates of standard errors of interviewer effects ignore the clustering of observations into interviewer workloads. That is, since the measurement error component being investigated has its source in interviewer behavior, the stability of estimates is a function of the number of interviewers used in the study and the variability within and between interviewers' workloads. The work of Schuman and associates treats the cases as a simple random sample, as if the number of interviewers equaled the number of interviews. Dijkstra (1983) clearly indicates this strategy can lead to incorrect inference that race of interviewer effects exist when they merely reflect natural variability among interviewers. Schaeffer's study avoids this by creating an interviewer level data file and using the interviewer as the unit of analysis. The results probably overstate standard errors of estimated race of interviewer effects.

An interesting reaction to the race of interviewer results has been the decision by many researchers to match race of respondent and race of interviewer whenever possible. Indeed, results that show differences between responses of black respondents to white interviewers and those of black respondents to black interviewers are often cited to demonstrate better reporting among pairs of respondents and interviewers of the same race. For example, Schaeffer (1980, p. 417) says, "It appears that the practice of matching race of interviewer with race of respondent when racial topics are included in the survey is well founded." There is no attention given to the possibility that interviews among same-race actors may yield overreports of extreme positions on the racial attitude measures. Given the perspective of this book, such a prescription effects a movement of variable errors into a bias, by design. That is, any errors attached to same-race pairs of interviewers and respondents will consistently be present in all replications of the survey. The variable errors resulting from uncontrolled assignments have been eliminated.

If these questions were being used to predict support for policy proposals, for example, it is not at all clear whether the response of a black to a white interviewer or a black interviewer would be a more accurate predictor. That is, the correspondence between the environment of the survey interview and the target environment is an important issue. This perspective brings the race of the interviewer into the model of the survey measurement, viewing it as an important causal factor in response formation, but one for which it is unclear which design is preferable. By forcing race matching of interviewer and respondent, the implicit assumption is made that same-race pairs yield better data. This view, in short, is less compatible with the Shaeffer perspective than with that of Schuman and Converse (1971): "Such effects can usefully serve as an indicator of evolving areas of interpersonal tension between blacks and whites, and deserve to be treated as a fact of social life and not merely as an artifact of the survey interview."

Another characteristic of interviewers that has been hypothesized as affecting responses is gender. Nealon (1983) found in a survey of farm women that male interviewers obtained lower average reports of farm value, reports of more involvement in the work of the farm and in farm organizations, and reports of greater satisfaction with farm programs than were obtained by female interviewers. Although the initial sample was randomly assigned to male and female interviewers, refusal conversions and callback cases were sometimes reassigned. In other studies, female students gave more feminist responses to male than to female interviewers on questions of women's rights and roles (Landis et al., 1973); whereas male respondents offered more conservative opinions on that subject to male interviewers (Ballou and Del Boca, 1980). Colombotos et al. (1968) found that male interviewers obtained higher scores on two psychiatric measures, from male as well as from female respondents, although these differences were not significantly large. In another study, the percentage of respondents giving "sex habits" as a possible cause of mental disturbance was greater for female interviewers (Benney et al., 1956). Rumenik et al. (1977) in a review of the experimental literature find that male experimenters appeared to elicit better performance from adult subjects given verbal learning tasks. However, experimenter gender effects were not clearly established in the data on clinical and psychological testing. In a decision-making simulation enacted with married couples, the wives were found to offer more problem-solving statements when the observer was a woman (Kenkel, 1961).

Groves and Fultz (1985) cumulated 2 years of telephone survey data from the monthly Survey of Consumer Attitudes in which randomized assignments of cases were made to a total of 40 male and 80 female interviewers. On a variety of factual items there were no effects of interviewer gender observed. Table 8.12 shows that respondents, however, expressed greater optimism toward a variety of economic indicators when the interviewer was male. The differences between the percentages of positive responses recorded by male and female interviewers are significantly large for 7 of 19 items examined (using standard errors reflecting the clustering of respondents into interviewer workloads). These 7 questions cover diverse economic issues rather than representing a single domain. The authors investigated three alternative hypotheses to interviewer gender effects: increase in the proportion of male interviewers during a time of greater consumer optimism, other attributes of the interviewer related to gender (e.g., age, education, tenure in the job), and larger nonresponse errors among male interviewers. None of these hypotheses, to the extent they were addressable with the data, explained the observed effects of interviewer gender. In contrast to earlier studies, there appeared to be no interaction effect involving interviewer gender and respondent gender. That is, both female and male *respondents* gave more optimistic reports to male interviewers.

A norm of agreement may be activated when the respondent does not have strong prior guidance to his/her answer. The norm is implemented by offering that answer which would cause the smallest amount of disapproval by the interviewer. This is most obvious in questions for which the characteristics of the interviewers are strong clues to the value they would give in response to the question. This might be viewed as a corollary to social desirability effects to the extent that the interview situation defines a social system of limited duration, the norms of which are partly defined by the external system.

In addition to race and gender, other attributes of the interviewer are sometimes evoked as causes of response error. For example, some studies have found effects of the level of experience of the interviewer. Stevens and Bailar (1976) found that missing data rates on the March Income Supplement for the Current Population Survey were lower for

Table 8.12 Percentage of Respondents Giving Positive or Negative Response on Various Consumer Attitudes by Gender of Interviewer

| | Percentage Giving Positive Response | | | | Percentage Giving Negative Response | | | |
| | Sex of Interviewer | | | | Sex of Interviewer | | | |
Question	Male	Female	Difference	Standard Error	Male	Female	Difference	Standard Error
Better/worse off now than 1 yr ago	36.0%	33.7%	2.3%	2.0%	36.6%	35.9%	0.7%	2.0%
Better/worse off now than 5 yrs ago	54.3	51.2	3.1	2.4	33.9	35.7	-1.8	2.2
Better/worse off in 1 yr than now	36.5	34.6	1.9	2.0	14.2	15.5	-1.3	1.5
Better/worse off in 5 yrs than now	56.8	51.1	5.7a	2.4	13.0	14.1	-1.1	1.6
Good/bad times financially next 12 months	37.5	34.6	2.9	2.0	50.7	55.4	-4.7a	2.1
Good/bad times compared to 1 yr ago	28.2	26.7	1.5	1.9	63.2	64.4	-1.2	2.0
Better/worse business conditions in 1 yr	42.8	38.4	4.4a	2.0	17.7	18.9	-1.2	1.6
Government economic policy good/poor	21.8	22.2	-0.4	1.7	28.2	29.3	-1.1	1.9
Less/more unemployment next 12 months	26.3	21.4	4.9a	1.8	34.6	38.2	-3.6	2.0
Interest rates in next year down/up	39.3	38.2	1.1	2.0	25.2	26.2	-1.0	1.8
Prices last year down/up	6.0	4.5	1.5	0.9	72.9	76.8	-3.9a	1.8
Prices down/up next 5–10 yrs	7.2	8.0	-0.8	1.2	78.3	74.1	4.2a	2.0
Income up more/less than prices last year	19.0	16.7	2.3	1.6	49.5	50.9	-1.4	2.1
Income up more/less than prices next year	22.2	18.7	3.5a	1.7	34.7	37.0	-2.3	2.0
Income up/down next year	63.6	59.3	4.3a	2.0	15.5	14.7	0.8	1.5
Good/bad time to buy house	40.5	34.7	5.8a	2.0	55.2	60.7	-5.5a	2.1
Good/bad time to buy durables	55.4	53.1	2.3	2.1	34.2	36.7	-2.5	2.0
Good/bad time to buy car	53.1	48.9	4.2a	2.1	39.6	43.1	-3.5	2.0
Probably will/won't buy a car	18.3	16.7	1.6	1.6	71.6	74.3	-2.7	1.9

aEstimate is more than twice the standard error. Standard errors are estimated using a jackknife replication method, treating interviewers as independently selected clusters of observations.

Source: R.M. Groves and N.H. Fultz, "Gender Effects Among Telephone Interviewers in a Survey of Economic Attitudes," Sociological Methods and Research, Vol. 14, No. 1, August 1985, pp. 31–52.

interviewers using the supplement for the first time. Booker and David (1952) found few differences in results by experience of the interviewer.

8.11 SUMMARY

Interviewers are the medium through which measurements are taken in surveys. Traditional approaches to survey design have attempted to induce complete uniformity of behavior of interviewers in all possible interviewing situations. Through training, interviewers are given guidelines for reading the question, clarifying misunderstanding on the part of the respondent, probing incomplete answers, and dealing with questions of the respondent. The goal of this strategy is to remove any influences that variation in interviewer behavior might have on respondent answers.

Central to this notion is the reading of the questions exactly as they are written and strict adherence to other questionnaire scripts. This moves the burden for elimination of measurement error away from the interviewer to the author of the questionnaire. To be strictly consistent with this philosophy, the questionnaire and related training guidelines should provide instructions for interviewer behavior covering all possible interview situations. The two should be totally comprehensive so that interviewers are only the unvarying instruments of the research design. If this is attained, then the measurement error properties of the survey are totally a function of the questionnaire and interviewer behavior design chosen by the researcher.

There are two problems with this strategy, however, one practical, the other more theoretical. The practical side comes from cost constraints. To train interviewers to ask the questions as written, as well as to behave in many other interview situations in exactly the same way, would demand large increases in interviewer training time. The theoretical problem comes from the inability to identify all interview situations which might affect measurement error features of a survey. Because of this, most training guidelines attempt to give interviewers principles of behavior (e.g., nondirective probing) and let them apply the principles to the myriad of situations which arise in real interviews.

Other fields have different philosophies about the role of the interviewer. One, flowing more from clinical or anthropological interviewing, is that the interview content should be adapted to the two specific individuals interacting. Questions, even though worded precisely the same, may have different meaning to different respondents. The interviewer should be trained in the concepts inherent in the questions and be allowed to probe, rephrase, and adapt the questionnaire to individual respondent needs. The questionnaire is a basic launching point in this philosophy, a script which ad-libbing may improve in specific situations. Interviewers should be encouraged to do so.

Survey interviewers tend to behave somewhere between these two extreme positions. They are not given full instructions on appropriate behavior in all situations, hence they must make decisions on how to obtain the desired data in unusual circumstances unanticipated by the researcher. On the other hand, they are also not given the freedom to change questions substantially (in most surveys).

The survey literature has identified demographic characteristics which appear to produce measurement biases (existing over replications of the survey with interviewer corps of similar demographic mixes). These demographic effects appear to apply when the measurements are related to these characteristics and disappear when other topics are discussed. These issues appear to be outside controllable survey design features other than deliberate exclusion of certain demographic mixes in certain surveys. Race, age, or sex matching of interviewer and respondent assume that an interviewer who resembles the respondent on those characteristic can obtain better data than one who does not.

The survey literature also documents variation in interviewer results using the same instrument. This interviewer variance component is often small, in surveys with comprehensive training of interviewers, but appears related to how rigorous is the specification of their behavior in reaction to certain responses. Hence, the absence of prespecified probes in open questions may be related to higher interviewer variance on those measures. There is also evidence of larger interviewer variation among elderly respondents.

Interaction coding techniques attempt to record whether interviewers follow training guidelines in administering the questionnaire. It has been shown that much interviewer feedback to respondents is unrelated to motivating good respondent behavior and may reinforce low attention to the response task. Interaction coding can identify questions which many interviewers find difficult to administer. It can also permit comparison of interviewer performance.

Measured effects of interviewer training on survey data quality are rare in the literature. Results do suggest that supervision after training is an important part of interviewer performance. Some measurement errors have been reduced through increasing instructions to the respondent about the intent of a question or giving feedback about the quality of the respondent effort to respond.

The chief failure of the existing literature on the interviewer is its ignoring the relationship between interviewer training guidelines and the measured variation in interviewer results. This is another example of the separation of those interested in reducing errors (here through training) and those interested in measuring error (here through interviewer variance estimation). New research efforts should be mounted to investigate whether researchers instructions to interviewers do indeed reduce measurement error.

CHAPTER 9

THE RESPONDENT AS A SOURCE OF SURVEY MEASUREMENT ERROR

Zen is like a man hanging in a tree by his teeth over a precipice. His hands grasp no branch, his feet rest on no limb, and under the tree another person asks him: "Why did Bodhidharma come to China from India?" If the man in the tree does not answer, he fails; and if he does answer, he falls and loses his life. Now what shall he do?

"Achilles" quoting Kyōgen in D.R. Hofstadter, Gödel, Escher, Bach: An Eternal Golden Braid, 1979, p. 245

Chapter 8 dealt with the interviewer as a major vehicle through which the survey measurement process is conducted. This chapter examines the other major actor in the interview, the respondent, in order to understand how the respondent's attributes and actions may affect the quality of survey data. Attention to the respondent as a source of error has recently been given revitalized emphasis in the application of theories of cognitive psychology to survey measurement. This perspective is attractive because it focuses on how people encode information in their memories and how they retrieve it later. Since many surveys ask about prior events or about one's attitude on current events, cognitive processing appears to be central to survey measurement error.

9.1 COMPONENTS OF RESPONSE FORMATION PROCESS

Both traditional models of the interview process (Kahn and Cannell, 1957) and newer cognitive science perspectives on survey response (Hastie and Carlston, 1980) have identified five stages of action relevant to survey measurement error:

1. Encoding of Information. Sometime in the experience of the respondent, certain knowledge relevant to the question may have been

obtained, processed, and stored in memory. Without this knowledge none of the following steps in the response formation process can overcome this deficit.

2. Comprehension. This stage involves the assignment of meaning by the respondent to the question of the interviewer. This meaning is influenced by perceptions of the intent of the questioner (as we have seen in Chapter 8), the context and form of the questions (as we shall see in Chapter 10), and other characteristics of the environment.

3. Retrieval. This stage involves the search for memories of events or sets of knowledge relevant to the question.

4. Judgment of Appropriate Answer. This stage involves the choice of alternative responses to the question. In a closed question the respondent determines which fixed category is most appropriate for his/ her answer. In an open question the respondent chooses which evoked memories to summarize in his/her report.

5. Communication. This stage involves consideration of all other personal characteristics and social norms that might be relevant to expressing the chosen answer. In that sense, this stage acts as a filter to the reporting of the respondent's value on a questioned attribute. It forms another stage of judgment prior to communication. Finally, it involves the ability to articulate the response chosen. This may be a simple matter for many questions, but for others it may entail a combination of several separate ideas.

This chapter is structured about these sequential actions of the survey respondent. In using this structure we appeal to three different literatures: (1) research in cognitive psychology on memory and judgment, (2) research in social psychology on influences against accurate reporting, and (3) evidence from survey methodology research regarding response errors in surveys.

These three literatures are needed because none of them singly offers full understanding of the measurement error properties of survey data. While the cognitive psychological literature offers rich theoretical perspectives on the role of memory in survey response, there are four weaknesses in the literature that plague its application to surveys. First, the empirical studies used to test the alternative cognitive theories have used laboratory subjects who were for the most part university students. The generalizability of results to other population groups cannot be justified from the data themselves but can only be asserted by the theory of the individual investigators. Second, the tasks given to the subjects in some of the experiments are far removed from the question-answering tasks given to survey respondents. Many of the experiments measure the ability of subjects to recall words or visual images, not real-life events. Most retrieval tasks use relatively short time periods (i.e., minutes or hours) between exposure to the material and measurement, in contrast to the weeks or months of recall often asked of survey respondents. The data therefore may not speak to the problems faced by survey researchers. Third, the literature emphasizes properties of the object being remembered or the storage properties, not the nature of the communication process. The work often implicitly assumes homogeneity of cognitive and response behavior across persons given the same task. It deemphasizes individual differences in response behavior arising because of other psychological influences (e.g., social desirability effects or more generally the role of affect in judgment). Fourth, at this writing there is no *single* well-accepted cognitive psychological theory that is applicable to observed survey response errors. The problem is not the absence of theories but the multitude of theoretical perspectives, each obtaining some support from one or more experiments. Hence, the choice of theory to apply to particular survey measurement problems often requires careful examination of alternatives, speculation about the possible impact of exogenous variables not manipulated in experimental settings, and assessment of which experimental tasks are most similar to those of survey respondents. This is not easy.

In contrast to cognitive psychology's focus on input and output from memory, the social psychological literature concentrates on influences on the communication of answers to survey questions. These apply well to those circumstances in which respondents might be embarrassed to reveal their true answers to a survey question or feel influenced toward one answer or another because of the interviewing environment (e.g., the interviewer, the purpose of the survey, the presence of others). They also apply to understanding why some respondents appear to work more diligently at obtaining the correct answer (in cognitive terms, why they engage in deeper processing in performing the task). On the other hand, social psychological theory does not directly explain failures to recall facts about oneself under full motivation to do so.

The literature in survey methodology has documented many examples of response errors and has identified respondent groups and response tasks most susceptible to those errors. Furthermore, it has identified some survey design features that reduce the errors found in other designs. Finally, the survey methodology literature provides examples of measurement error properties associated with respondents that are not well explained by either cognitive or social psychological theory and thus represent challenges to theoretical developments in those fields. In contrast to the other two approaches, however, it relies more on models of measurement of response errors, not explanations for their presence.

9.2 THE ENCODING PROCESS AND THE ABSENCE OF KNOWLEDGE RELEVANT TO THE SURVEY QUESTION

The failure of respondents to give accurate survey reports may arise both because they fail to engage in the appropriate cognitive activities at the time of response and because they do not possess the relevant knowledge. The process of forming memories or retaining knowledge is called by some the "encoding process." Some researchers posit different encoding and storage properties for different kinds of information. Munsat (1966) and Tulving (1972), for example, make distinctions between "episodic" and "semantic" memory. Markus and Zajonc (1985) refer to these as "event" memory and "conceptual" memory, respectively. Basically, episodic (event) memory is concerned with "the storage and retrieval of temporally dated, spatially located, and personally experienced events or episodes, and temporal-spatial relations among such events" (Tulving and Thomson, 1973, p. 354). Semantic (conceptual) memory is "the system concerned with storage and utilization of knowledge about words and concepts, their properties, and interrelations" (p. 354). In Linton's (1982) use of the terms "episodic" and "semantic" some memories of specific events (presumably first stored in an episodic fashion), if repeated often over a long period of time, may transform themselves into more semantic memories. That is, the distinguishing features of each event are lost; instead there is a memory of a class of events, often repeated, with many shared features. The memory becomes the definition of the event type for the person involved.

This viewpoint of the organization of memory is shared by schema theory. "Schemata" (plural of schema) are sets of interrelated memories, organized so that relationships are represented among attributes of events or pieces of knowledge. Schema theory contains an economy of storage notion. If characteristics of an event are shared by many other previously encoded entities, they will not be stored separately as part of the memory of the individual event. Instead, the distinctive features will be encoded, and other shared characteristics will be remembered as features of the larger set of events of which this is a part (e.g., the dental visit for a checkup in which a substitute dental hygienist was used may share many of the features of all other dental checkups). There is some evidence that memories of events have some hierarchical structure of characteristics. That is, some characteristics are encoded as more central or important to the event; others are less so. This view receives experimental support in studies of recall of textual material with logical connections among ideas (Thorndyke, 1977). Kintsch and van Dijk (1978) theorize that important ideas are those more often referenced by recall of other ideas or for other reasons spend more time in working memory than others. Hence, these develop more connections with other ideas over time and are more easily evoked (perhaps because they are evoked by a greater number of different cues). Furthermore, there is evidence that the richness of the respondent's existing schema affects how fully remembered an event might be. For example, those who have some previous experience in a doctor's office tend to remember the nature of the tests run; those experiencing it for the first time tend to attend to other details, those fitting schemata more typical of their past experiences.

The "script" is one method of economical storage of frequently experienced events (Schank and Abelson, 1977). A script is a temporally ordered set of detailed memories that describes an episode in one's life (e.g., eating in a restaurant, going to a doctor). These scripts contain the series of actions performed during the event under most circumstances; in that sense, they are a type of schema of relevance to survey researchers (who ask respondents to report past events like searching for a job, visiting a doctor, or experiencing a victimization). Unusual orders of the events (e.g., a waiter clearing the table before you finished your meal) are stored as properties of individual memorable episodes. Scripts therefore are viewed to act as organizing structures for individual experiences and to free the person from storing every detail of each event experienced. Schemata become relevant to survey research by affecting the retrieval of information.

There is evidence from the experimental literature that the quality of recall is a function of whether the schema evoked at the time of recall is the same as that used at the encoding phase (Anderson and Pichert, 1978). Consistent with this, the quality of recall can be enhanced with alternative methods of recall. Hasher and Griffin (1978) induced respondents to alter their initial schemata used to retrieve information and demonstrated improved recall. Schemata may, however, lead to poorer reporting about past events. Bower et al. (1979) found in a series of experiments with students concerning stories about common activities (e.g., eating in a restaurant, visiting a doctor) that the subjects tended to recall attributes of the stories that were never communicated. That is, the schema previously held by the subject concerning a visit to the doctor provided details for the memory trace of the particular story. Subjects believed that they had read details that did not exist in the story itself. The more elaborated was the set of characteristics of the event, the more effective was that framework for stimulating recall of the event. Anderson and Pichert (1978) provide evidence of a corroborating sort. They investigated the independent effects of the nature of cues on recall by using two different cuing strategies for the same material. They were asked to read a story about activities of two boys in a house, either taking the perspective of a burglar or that of a potential home buyer. After performing some intervening tasks, they were asked to recall as much of the story as possible. Finally, after performing another intervening task the subjects were asked again to recall the story, at this point half of them using the other perspective. These subjects were told, "This study is being done to determine whether or not people can remember things about a story they thought they had forgotten if they are given a new perspective on that story.... Please try to think of the story you read from this new perspective" (p. 5). It was found that the perspective taken by the readers influenced what kinds of material they recalled. Features of the house relevant to the burglar (e.g., the presence of a television) tended to be recalled better when the burglar perspective was taken than when the house buyer perspective was taken. However, subjects who changed perspectives on the second trial also reported material relevant to the new perspective that was not mentioned on the first recall attempt. This is taken as evidence that there are independent effects of retrieval cues on recall. Material apparently forgotten on the first trial is remembered when a different perspective is taken. These results may be useful to survey researchers in that they demonstrate that the nature of questioning *may* have an effect on the quality of reporting. It might be difficult in a survey, however, to implement the methodology of these experiments, asking respondents to try again to recall something, the second time using a different perspective.

Research on learning has shown that recall of learned material is enhanced by measurement in the same environment as was experienced during the learning. S. M. Smith (1979) defines the environment broadly as consisting of "the location, size of room, objects and persons present, odors, sounds, temperature, lighting, and so on" (p. 461). Indeed, one of his experiments found that encouraging the subjects to recall (or mentally reconstruct) the learning environment improved recall.

The theories of schemata and scripts would prompt efforts of survey researchers to learn the schema that organizes the information they seek. In contrast to experiments that involve both encoding and retrieval, surveys are retrieval mechanisms only. The task of learning a relevant schema is also complicated by the respondents' ignorance of their own schemata.

Indeed, a fundamental difficulty with discussing the encoding process as a separate step from the retrieval process is confounding encoding and retrieval failures. Most tests of theories of encoding use some kind of retrieval experiment. Hence, retrieval procedures are the vehicle through which the encoding of memory is studied. Inextricably intertwined in the debate on whether (1) information was never encoded, (2) "forgetting" has taken place, or (3) memories are permanent is the fact that no observation of memory can be made without some retrieval effort, by definition. Hence, the failure to recall can always be attributed to the failure to encode, inadequate cues, or forgetting. Because of this, one of the unresolvable debates in cognitive psychology concerns the permanence of encoded information.

Do memories, once created, last forever? Those theorists who assert that they do often cite evidence from research observing the recall of trivial past events—through the electrical stimulation of the brain, the use of hypnosis, psychotherapy, and the everyday experience of sudden recall. They assert that apparent "forgetting" is merely the failure to utilize the appropriate cue to evoke the memory. Loftus and Loftus (1980), however, argue that much of the evidence from the above research is based on small numbers of unusual and ambiguous cases. Instead, they note that important features of memories of past events can be altered in a permanent way by later actions, citing evidence from eyewitness testimony in court cases. This finding appears to be most important for the less salient details of an event (Loftus, 1979). Fischhoff (1977) cites experimental evidence that demonstrates that new and demonstrably true information about a previous event replaces the old, incorrect encoded information in a manner such that the subject cannot recognize the old version of the memory. However, if the subject is later told the evidence is unreliable, the former characteristic can be retrieved (Hasher et al., 1981). This suggests a priority of encoding.

Obviously, from the survey researcher's viewpoint, it makes no difference whether a piece of information relevant to a survey question was never retained by the respondent or has subsequently been forgotten. In either case, no survey design can overcome this loss. (In this sense, we are concerned only with limited application of cognitive theories.) If, however, the respondent has retained the information, but it can be recalled only with some difficulty (i.e., with most measurement techniques it appears to be absent from memory), then the attention of the survey researcher should be focused on aiding the retrieval process through the interviewer, the questionnaire, and the mode of data collection. With that perspective, the more understanding we have about how memories and knowledge are stored, the easier the task of constructing survey procedures to access them.

There are two small literatures in survey research that are relevant to the encoding of information relevant to the survey question. The first is concerned with measurement errors associated with a particular respondent rule, the use of one person to report characteristics of another person. These are sometimes called "proxy reporting rules" or in general "respondent rules." Most of this literature comes from surveys measuring observable characteristics of family members (e.g., health service usage, unemployment status). The second literature stems from the measurement of attitudes. It examines measurement errors on opinion items among persons with poor knowledge of the topics being studied. This literature has come to be known as the study of "nonattitudes."

9.2.1 Respondent Rule Effects

By "respondent rule" is meant the eligibility criteria for persons to answer the questionnaire (as distinct from being measured by the questionnaire). For most attitudinal surveys all respondents report for themselves only. The implicit assumption is that no other person is fully privy to the sample person's thoughts and feelings that are elicited by the survey questions. Surveys asking about behaviors or observable attributes of persons (e.g., health, unemployment, economic surveys), however, sometimes use one household member to report for all persons in the household. For example, the Current Population Survey uses "any responsible" adult member of the household to report on the employment-related activities of all persons 14 years or older in the household. The choice of respondent rule is relevant to our current concern with the encoding of memories. One attribute of a desirable respondent is the encoding and retention in memory of information relevant to the survey questions. In addition, there is evidence that social psychological forces act on the reporting of survey answers independently from the existence of information relevant to the survey question.

Several experimental investigations have observed that the encoding of information about the self may be affected by differences in the perceptions of behavior about oneself versus others. Jones and Nisbett (1972) assert that self-schemata differ from those for others because the visual images possible for the self are limited. We see *others* moving, performing actions, and doing tasks. Yet the visual observations of *ourselves* are more limited, except on those occasions when we find ourselves in front of a mirror. This fundamental perceptual difference, they argue, affects the organization of memories about ourselves versus those of others. These observations directly address the organization of memory but also relate to the type of information that is encoded.

Although many of the experiments in this area may be of only marginal relevance to survey measurement, some offer intriguing hypotheses for surveys. Lord (1980), for example, finds that self-schemata may be better at aiding recall for self-attributes because of their greater elaboration and that images of the self most often contain notions of perceiving the actions of others (e.g., "standing, hearing, looking, ...," p. 265) but schemata about others included their actions. Lord also noted that subjects varied in the amount of self-monitoring they performed, however, and that degree of self-monitoring was related to the accuracy of recall of self-descriptors. One interpretation of these findings is that the memories about oneself are organized about central emotional states or other internalized attributes, but that memories about others are organized about observable traits.

These findings might be generalizable to naturally occurring events, most often the target of survey measurement. If so, characteristics of a person which involve physical action (i.e., movement, speech) might be better reported by others than would characteristics that involve more internalized states (e.g., chronic versus acute health ailments, "looking for work" versus losing one's job). Conversely, self-respondents may be less accurate than proxy respondents when an event was inconsistent with the internalized states dominating the self-schema (e.g., memories of having the flu among persons who believe themselves to be very healthy).

Social psychology offers very different explanations of respondent rule effects in surveys. One is based on differential influences of social desirability on responses about oneself and about others. This reasoning would assert that when the levels of information held by two persons are equal, if the trait being reported is judged socially undesirable, then it will be less often reported about oneself than by another. In addition to effects of social desirability, social psychologists note that different "roles" within families provide to their incumbents different information about family events. For example, health researchers have argued that knowledge about health-related events is more compatible with some self-schemata (or roles) than others. The notion of a family "health monitor" is one used frequently in survey research on health. In the United States it is often the case that one person (typically the female spouse) assumes the responsibility of making appointments for health care, obtaining medicine when necessary, and providing home care for sick household members.

Mathiowetz and Groves (1985) reviewed the literature on respondent rule effects for health reporting. They find that, contrary to prevailing beliefs, self-respondents are not consistently found to provide more accurate health data than proxy respondents. In some studies "more accurate" means better agreement with medical records. In other studies the criterion of evaluation is the extent of difference between self-reports and proxy reports. The literature on respondent rules in health surveys (and presumably in other fields) is plagued by serious design flaws:

1. The self-response treatment is often used only for those persons at home at the time the interviewer calls and those in single-person households. Proxy response is used for those absent in multiple-person households. Respondent rule effects are thus confounded with true health differences in the two treatment groups.

2. Medical records are assumed to contain the true values on the health measurements, but they are subject to underreports when the respondent has not revealed a condition to a physician.

When a fully randomized experimental study of respondent rule effects was mounted, they found larger *proxy* reporting of events such as "days spent in bed in the last 2 weeks because of illness" and the number of acute health conditions they experienced in the last 2 weeks. To eliminate the problems that plagued some of the earlier literature, the self-proxy comparison was made for persons who lived in families with two or more persons *and* for whom the self-response condition was randomly assigned.

Although Mathiowetz and Groves did not entertain the hypotheses noted above for self- and proxy reporting, their data do bear on one of the hypotheses. Jones and Nisbett (1972) argue that self-schemata are dominated by memories of internalized states and other schemata, by memories of actions. Chronic health conditions (e.g., arthritis, rheumatism) are less likely to lead to professional medical care or atypical home medical care than are acute conditions (e.g., influenza, some sort of physical trauma). The Mathiowetz and Groves data show higher proxy reporting for all health activities (e.g., medical care visits, days spent in bed) and for acute conditions. The only significant difference showing more self-reporting is the higher reports of persons with at least one chronic health condition over the past 12 months. This finding is consistent with the perspective offered by cognitive psychology.

The Mathiowetz and Groves data also permit a weak test of the social desirability hypothesis of measurement error: that self-respondents will be more reluctant to report embarrassing health conditions about themselves than would proxy respondents about others. The authors classified both chronic and acute conditions by their level of potential embarrassment using judges. There was no difference in levels of reporting threatening condition between self- and proxy respondents.

The two-category model of respondent rule effects (self versus proxy) is clearly a naive one. From the perspective of a potential proxy reporter, both the likelihood of exposure to knowledge about an event *and* perceived importance of the event should be a function of social relationships with the sample person. For example, spouses may be better informants about each other than about their adult children. Some support for this comes from a hospitalization validity check study (Cannell et al., 1965). Table 9.1 shows that more sample hospitalizations are recalled for spouses and younger children (about 90 percent) than for older children and other relatives in the household (about 75 percent). This result may be related to the concept of "role." It may be more fully part of one's accepted role to monitor the health of spouses and young dependents than of others. This affects both the likelihood that a reporter would have learned about the hospitalization and their encoding of the information in memory. Hence, from a cognitive perspective, one's stored memories about those persons are more fully elaborated for health-related attributes. The result is less reporting error.

There are also self-proxy comparisons from other topic areas. In a reverse record check study of survey reports of motor vehicle accidents, about 20 percent of the proxy reports omitted reference to the accident but only 9 percent of the self-reports did (Cash and Moss, 1972).

In summary, the literature from survey methodology on self- and proxy reporting, at least in the health field, has tended to use research designs that do not permit an unencumbered measurement of the marginal effects of the respondent rule. Cognitive and social psychology offer theoretical insights into the process of memory storage and communication of response that may help understanding of respondent rules.

9.2.2 Nonattitudes

Converse (1970) observed a group of survey respondents who appeared to provide inconsistent reports of their opinions on various political issues. Converse judged that their answers were unreliable in the sense that they did not appear to accurately reflect any relatively permanent attitudinal state, as represented by repeated questions in a panel survey. Some of these persons gave substantive responses to the questions, while others answered that they had no opinion on the issue in question. Converse labeled these groups as holders of "nonattitudes." It was argued that for many of these people the measures concerned topics for which little prior consideration had been given. Either the persons did not comprehend the question in a consistent way (this is discussed in Section 9.3) or had not assembled associated pieces of knowledge that permitted a consistent stand on the issue. They chose a substantive response to a closed question in a process well described as random. Over conceptual trials, their answers would jump across different response categories. Converse

Table 9.1 Percentage of Sample Hospitalizations Reported by Proxy Respondent, by Relationship of Proxy Respondent to Sample Person

Relationship of Proxy Respondent to Sample Person		
Respondent	Sample Person	Percentage Reported
Spouse	Spouse	90% (254)
Parent	Child <18 years	88% (330)
Parent	Child 18 or older	71% (55)
Other relative	Other relative	76% (73)

Source: C.F. Cannell et al., *Reporting of Hospitalization in the Health Interview Survey*, National Center for Health Statistics, Series 2, No. 6, 1965, Table 9.

finds that holders of nonattitudes disproportionately report low levels of interest in political events and disproportionately have little education.

The Converse work spawned a large set of attempts to predict the existence of "nonattitudes" on one or more topics. There were mixed results on the relationship between general political involvement and interest on one hand and inconsistency on individual items on the other (e.g., Asher, 1974; Erikson, 1979). Many of the studies, as did Converse's, used correlations or differences between two waves of a panel study to assess consistency in response. Hence, the researchers had to assume how much of the inconsistency in results was a function of true change and how much was a function of unreliable judgments. Schuman and Presser (1981) showed that specific indicators of intensity of feeling about an issue are good predictors of the consistent reports of opinions, but that general interest measures do not perform well.

Converse (1964) used metaphors often employed in cognitive psychology to describe the information storage features relevant to measurement error. He described respondent knowledge as an interlocking network of nodes (pieces of semantic memories, concepts, arguments relevant to an issue). The survey question stimulates this network. The respondent either reports a judgment well rehearsed in the past or forms a judgment by reviewing arguments and counterarguments previously encoded. The problem the survey question poses to someone with little information stored about the issue is that a strong network of related concepts does not exist and only weak ties between concepts and arguments of minor relevance are activated by the survey question. Which of these ties are judged by the respondent to be most important in forming an opinion on the issue is not the same over time, because few of them can be distinguished on their strength. Because there are many weak ties, the response process appears random over replications of the same question.

Smith (1984), in a comprehensive review of the nonattitude literature, suggests two interesting remedies to the problems of nonattitudes. The first is the use of prior questions that inquire whether the respondent has given careful thought to the issue prior to the interview. These are so-called "don't know" filters, discussed more fully in Chapter 10. These would be used to encourage respondents explicitly to provide a "don't know" answer if they had not formed an opinion on the item. Second, Smith suggests using follow-up questions about the intensity of respondents' expressed opinions (how strongly they hold to their positions on the issue), any behaviors based on the opinion, and knowledge about the topic. Abelson (1986) in a similar vein proposes follow-up questions about respondents' experiences defending their positions and trying to convince others of their merits. Others suggest that the problem of nonattitudes lies with the ambiguity of the question.

Every idea mentioned in the literature involves an increase in the number of measures used as indicators of the attitude being investigated in the survey. This is another example of the interplay of costs and errors in surveys. To more fully understand the measurement error present in describing attitudes of the population, longer questionnaires, asking for more detail about the topics of interest, must be implemented. Longer questionnaires increase the cost of data collection (or limit the number of other variables measured). A rational decision about how many questions to add must be based on the level of information about the measurement errors obtained by each and the cost of adding one more question to the survey.

9.3 COMPREHENSION OF THE SURVEY QUESTION

The discussion of nonattitudes above has already suggested that the meanings of questions are not fixed properties, constant over all persons in a population. Respondents give meaning to a question through their impressions of the overall purposes of the interview, the intent of the individual question within that context, the behavior of the interviewer in delivering it, and semantic memory associations of words and phrases in the survey question. Survey research "works" because for many questions all respondents make these attributions in similar ways. When this is not true, however, respondent answers are not equally good indicators of the single attribute intended to be measured by the survey question.

The comprehension phase of the survey question is not often studied by traditional cognitive psychology but is a topic of great relevance in psycholinguistic approaches to understanding oral presentation (Clark et al., 1983). Indeed, the approach taken in that work often includes many influences on respondent behavior that are common to the viewpoint of traditional survey methodologists and social psychologists. It views the meaning and intent of survey questions as influenced by shared beliefs and assumptions about the world between the interviewer and the respondent. Obviously, in most survey settings, the respondent must infer the beliefs of the interviewer about relevant topics with very little information. Typically, there has been no prior meeting between interviewer and respondent, the interviewer has limited her statements to those relevant to the survey task, and the respondent therefore must use visual cues (e.g., interviewer age, gender, dress, facial expressions) and audio cues (e.g., tone of voice, pace) to gather information about the meaning of the question (as discussed in the last chapter). The respondent also uses descriptions of the purpose of the survey to identify the intent of the question. In a broader sense, consistent with perspectives of symbolic interactionism, the respondents are constantly reinterpreting their role in the interview situation (Stryker and Statham, 1985). Each action on the part of the interviewer and each succeeding question offer more information about the purpose and intent of the meeting and therefore aid in the assignment of meaning to questions. Because the question comprehension step is so fully linked to the words used in the survey measure, a full discussion of it appears in Chapter 10.

9.4 RETRIEVAL OF INFORMATION FROM MEMORY

The cognitive theories of schemata and scripts suggest that incomplete details of an event will be stored in memory; people will reconstruct the event using the probable details of the larger generic group of events of which the target event is a member. This reconstruction process thus is a source of error in recall of event characteristics (Alba and Hasher, 1983)—a tendency to report event characteristics that are typical of those of the general class. However, there is evidence that recall is more affected by the activation of the relevant schema than is recognition (Yekovich and Thorndyke, 1981).

Alba and Hasher review the four classical features of the recall step that have been found to be important indicators of whether accurate retrieval is experienced:

1. The nature of the connections established during encoding.

2. Differences in the number and timing of rehearsals.

3. The nature of cues used to aid retrieval.

4. The order in which elements or sets are recalled (Alba and Hasher, 1983, p. 220).

"Rehearsal" means recall of material, an active processing of the memory at a later point (e.g., a retelling of a victimization experience repeatedly after the event).

Under one theoretical perspective retrieval from memory has different characteristics when the task is relatively free form (e.g., "Tell me how many times you went to the doctor in the past year," versus when the task is to recognize whether one has a particular attribute in question (e.g., "Are you a high school graduate?"). The distinction between these tasks appears to be one of how many cues are supplied to the respondent. Tasks of "recall" are sometimes deficient in cues relative to those of "recognition." (They also do not structure the judgment task for the respondent into a fixed set of alternatives.) One view of cognitive researchers was that recall tasks involved the retrieval of a set of candidate memories and the evaluation of each of them regarding their eligibility for some set (e.g., whether the event was a criminal victimization). In contrast, recognition tasks merely involved the latter, with no retrieval involved. Following that view, the finding that recognition tasks were better performed than recall tasks was explained by the greater depth of cognitive processing required by the recall task. More recent evidence (e.g., Tulving and Thomson, 1973) demonstrates, however, that recognition tasks may under some conditions yield poorer retrieval of information than recall tasks.

Hasher and Griffin (1978) note that one of the problems in recall of past events is the difficulty of discriminating between characteristics of the event itself and results of reflections (subsequent thoughts and feelings) about the event. Their work is in the tradition of Bartlett's (1932), which noted that persons tend to abstract the material presented in prose passages that they read, retaining the perceived central meaning or theme of the presentation but discarding some details, especially those that were contrary to the major theme. This viewpoint is also consistent with schema theory's notion of memory economy, the tendency to retain in memory those attributes that are most likely to be recalled in the future.

Some claim that memories of episodes are not collections of many details of the event, but that the details of the event are "reconstructed" at the time of recall. The work by Hasher and Griffin might be cited to describe how some of this reconstruction might take place—through recall of subsequent reflections on the event. Dawes and Pearson (1988) review other evidence that persons do indeed reconstruct the details of a past event when reporting. They describe studies showing that currently depressed persons tend to repeat earlier reports of being depressed at some point in their lives more than do those who are currently not depressed. Markus (1986) finds that current political attitudes are better predictors of current memories of past political attitudes than are the recorded past attitudes themselves. Current psychological states appear to be a filter through which the past is interpreted. They are the medium through which the reconstruction of past events is accomplished.

Within survey research literature the recall of past events has been found to be influenced by (1) the length of time prior to the interview that the respondent is asked to recall, (2) the importance of the event in the respondent's life, termed the "saliency" of the event, (3) the amount of detail required to be reported, termed the "task difficulty," and (4) the level of attention, the depth of processing (in cognitive terms), or the motivation (in social psychological terms) on the part of the respondents searching their memories to answer the question.

9.4.1 Length of Recall Period

Memory studies often find that recent events are recalled better than earlier events. The most influential early experiments in memory research were those of Ebbinghaus (1913), which found a negatively accelerating "forgetting" curve. That is, failure to report events rapidly increased in the days following an event but did not increase dramatically over time. Figure 9.1 shows an idealized result.

Further work in cognitive psychology has sought replication of the early Ebbinghaus findings with mixed results. Linton (1982) recorded two events in her own life each day for a 6 year period. Once every month items were drawn (in a "semirandom" fashion) from the list as it existed at the time. Linton attempted to identify the chronological order of randomly paired sample events and to date them exactly. Although the study concentrated on the accuracy of dating events, Linton observed two kinds of recall failure (despite the fact that the short descriptions of the

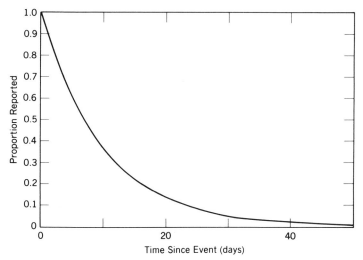

Figure 9.1 Proportion of events expected to be reported under exponential decay model.

events were used as cues). The first is the difficulty of distinguishing the event from other similar events; the second was the complete failure to recall the event, given the description. Many times the description evoked the memory of a whole class of events (e.g., a particular type of meeting). Instead of the negatively accelerating forgetting curve of Ebbinghaus, Linton found a linear trend (see Figure 9.2), about a 5 percentage point loss per year, over a 6 year period. Either these data contradict the Ebbinghaus findings or the nonlinearities in the forgetting function occur for this kind of material after 6 years have passed. It is also possible that rehearsal of some of the events of the past days (i.e., recall of the events for purposes unrelated to the study, prior to the experimental recall task) acted to interfere in the normal decay function.

Sudman and Bradburn, in an often cited paper (1973) on retrieval of memories of past events, posit an exponential model to describe the proportion of events that are reported (r_o) as a function of the length of time between the event and the interview:

$$r_o = ae^{-b_1 t},$$

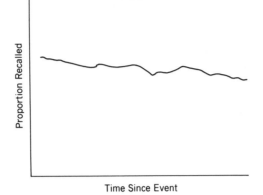

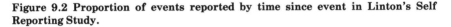

Time Since Event

Figure 9.2 Proportion of events reported by time since event in Linton's Self Reporting Study.

where a = proportion of events reported among those occurring
at the same time as the interview (this could be less
than 1.0 if other influences, like social desirability
effects, prevent recalled events from being
reported);

e = exponential number;

b_1 = factor reflecting the rate of reduction in recall;

t = length of time between the event and the interview.

Note that when b_1 is a large number, the proportion of events recalled
rapidly declines. Figure 9.3 plots these recall curves for different values of
b_1. When b_1 equals 0.0, the proportion recalled is constant over all time
periods and is equal to a, the proportion recalled instantaneously. To the
extent that b is a large positive number, shorter reference periods for the
survey measurement can have dramatic impact on underreporting.

There are several record check studies in the survey methodology
literature which measure the accuracy of reporting past events. Most of
these studies are reverse record check studies (sampling from the records
and determining whether record events are reported in the interview).
Hence, they are well suited to measuring underreporting but are
insensitive to overreports because for the most part the record systems
are not meant to be complete.

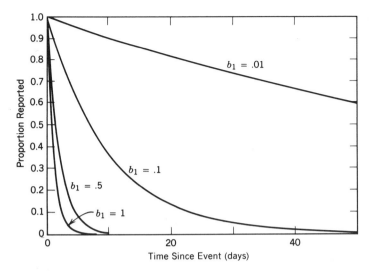

Figure 9.3 Proportion of events reported by time since event under different values of b_1 in the exponential decay model, $y = ae^{-b_1 t}$.

Cannell et al. (1965) describes a reverse record check study for the reporting of hospitalizations. Interviews were conducted with 1505 persons whose hospital records were sampled for visits that may have occurred as much as 1 year earlier. The sample also contained cases not sampled from hospital records, so that a blinding of interviewers to the existence of a documented hospital visit could be performed. A questionnaire resembling that in use for the National Health Interview Survey was used by experienced Census Bureau interviewers (ones who had interviewed for NHIS previously). Table 9.2 shows the results of the survey with regard to the most important measure, reports of hospitalizations. About 13 percent of all hospitalizations on the sampled records were not reported by respondents.

Figure 9.4 shows the percentage of hospitalizations that were not reported in the interviews by the length of time between the event and the interview.[1] The figure shows a gradual increase in nonreporting, starting with less than 3 percent for events within a month of the interview, but increasing to about 15 percent for those about 45 weeks prior to the interview. For the 131 hospitalizations sampled from those 51 to 53 weeks before the interview, nonreporting jumps to over 40 percent of the recorded hospitalizations. It is not known how much of this very high

---

[1] This figure ignores respondent errors in the dating of an event, except those leading to judgment that the event occurred outside the reference period.

Table 9.2 Reverse Record Check Study on Hospitalizations

Result Category	Number	Percentage of Record Hospitalizations
Total number of interviews	1505	—
Hospitalizations reported but not on records	59	—
Hospitalizations reported and on records	1600	87.2
Hospitalizations not reported	233	12.8
Hospitalizations on records	1833	100.0

Source: C.F. Cannell et al., *Reporting of Hospitalization in the Health Interview Survey*, National Center for Health Statistics, Series 2, No. 6, 1965, p. 5.

nonreporting arises from uncertainty concerning the timing of events relative to the beginning of the reference period. In any case, however, the forgetting curve from these data do not resemble that of the exponential function described by Ebbinghaus and modeled by Sudman and Bradburn (1973). Instead, like Linton's results the curve is almost linear, monotonically increasing over time until the period quite near the starting date of the reference period.

The results of a reverse record check study in three counties of North Carolina regarding motor vehicle accident reporting is presented in Cash and Moss (1972). Interviews were conducted in 86 percent of the households containing sample persons identified as involved in motor vehicle accidents in the 12 month period prior to the interview.[2] Table 9.3 shows higher proportions of persons not reporting the accident among those experiencing it many months before the interview. Only 3.4 percent of the accidents occurring within 3 months of the interview were not reported, but over 27 percent of those occurring between 9 and 12 months before the interview were not reported. These results also fail to follow the gradual decline in the rate of forgetting over time. It appears, in contrast, that the rate of forgetting is increasing over time.

In preparation for the National Crime Survey, a reverse record check study of police reported crimes was conducted by the U.S. Census Bureau

---

[2] There is no explicit description about whether interviewers were aware that the sample persons had been involved in accidents. The respondent rule used in the study was similar to that of the NHIS, permitting proxy responses for those persons who were absent at the time of the interviewer's visit. Proxy respondents failed to report accidents at a rate twice the overall rate of self-respondents.

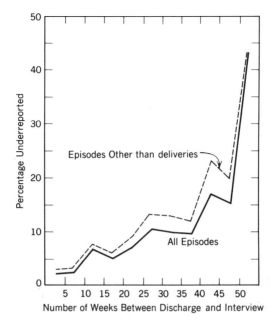

Figure 9.4 Percent of hospital episodes not reported by number of weeks since episode, for all episodes and those not involving delivery of babies.

Table 9.3 Percentage of Respondents Not Reporting the Motor Vehicle Accident by Number of Months Between Accident and the Interview

Number of Months	Percentage Not Reported	Number of Persons
Less than 3 months	3.4	119
3-6 months	10.5	209
6-9 months	14.3	119
9-12 months	27.3	143

Source: W.S. Cash and A.J. Moss, *Optimum Recall Period for Reporting Persons Injured in Motor Vehicle Accidents*, Series 2, No. 50, 1972, Table C.

in San Jose, California, in 1971. The validation study was part of a larger household sample survey using the same questionnaire. Interviewers were not told that the names for the record check study had been taken from crime files. Of the 620 victims sampled, only 394 (63.5 percent) were located and interviewed. There were large variabilities in response rates by type of crime committed against the person (e.g., assault victims had lower response rates than burglary victims, both because of change of residence and refusals). Figure 9.5 presents the proportion of victimizations sampled from records that were not reported by the victims interviewed. The expected form of the graph would show the highest rate of omissions for the early months (the left side of the graph) at a declining rate over the months. The rate of decline should increase moving to the right, with expected omission near zero at the point furthest to the right. If the graph were truncated at 2 months, this form would roughly apply. The anomaly in the data concerns the large amount of omission for crimes occurring in the past month, nearly one-quarter of them omitted from the survey reports.

An earlier test of the questionnaire used roughly the same design in Washington, DC (Dodge, 1970). Of 484 sample cases, interviews were taken with 326 respondents (67 percent). Only with crimes of robbery was there a monotonic increase in nonreporting with the time since the event (from 4 percent of those occurring 3 months before not reported, to 20 percent of those occurring 11 months before). Other crime types exhibited no relationships between time interval and nonreporting (see Table 9.4).

All the studies cited above were designed to estimate measurement bias of means and proportions calculated on the sample. Measurement error variance was not addressed. Measurement error variance can arise in this area both if respondents vary their ability to recall *and* if over replications of the measurement individual respondents tend to make different errors. In a record check study of employment-related characteristics of male steel company workers, Duncan and Hill (1985) explored both bias properties of the measurement and the variance of errors over respondents. This work is particularly relevant to the present discussion because it compares the errors in reporting earnings and work hours for the *past* year with those for 2 years previous to the date of the interview, early 1983. That is, the respondents were asked about total annual earnings for 1981 and 1982, total unemployment hours for 1981 and 1982, total work hours for 1981 and 1982, and average hourly earnings for 1981 and 1982. Table 9.5 shows very consistent patterns. With only one exception, measurement errors (both biases and variances) are higher for the reports about events 2 years earlier (1981) than about those 1 year earlier (1982).

It is not at all clear from empirical estimates of nonreporting of events that the gradually declining percentage of events remembered over time applies to the myriad of recall measurements used in surveys. We have tried to cite as evidence only those studies that separated the identification of reported events from the accuracy of the dating of those

Figure 9.5 Percentage of police reported crimes not reported in the survey by number of months since the incident.

Table 9.4 Percentage of Crimes Not Reported in Interview, Washington, DC, Reverse Record Check Study, 1970

Number of Months Before Interview	Crime Type				
	Assaults	Burglaries	Larcenies	Robberies	Total
3 months ago	18%	7%	36%	4%	14%
(December 1969)	(22)	(27)	(14)	(23)	(86)
6 months ago	53	17	17	5	22
(September 1969)	(17)	(23)	(18)	(19)	(77)
11 months ago	40	11	20	20	22
(April 1969)	(15)	(18)	(15)	(15)	(63)

Source: R.W. Dodge, "Victim Recall Pretest," Unpublished memorandum, Washington, DC, U.S. Bureau of the Census, 1970, Table E.

events. Of course, the two cannot be separated successfully at the boundaries of the reference period. For example, if the reference period contains the last 12 months, and an eligible event occurred 11 months earlier but is misdated as occurring 13 months earlier, the misdating error creates an underreporting error. In addition, in practical survey work there are hosts of other conditions in addition to length of recall period that can affect the completeness of recall.

Table 9.5 Error Measures for Validity Study of Employment Data for Steel Mill Employees, 1983

| Measurement | Sample Means | | Bias Ratio | Ratio of Error Variance to Variance of Record Values |
	Interview Report	Record Value		
Earnings				
1981	$29,579	$29,873	−.0098	.301
1982	$29,917	$29,972	−.0018	.154
Unemployment hours				
1981	39	63	−.38	.518
1982	169	189	−.11	.129
Work hours[a]				
1981	1880	1771	+.062	.919
1982	1693	1603	+.056	.366
Hourly earnings[a]				
1981	$14.71	$15.39	−.044	1.835
1982	$16.31	$16.97	−.039	2.801

[a]"Work hours" and "hourly earnings" apply to hourly workers only.
Source: G.J. Duncan and D.H. Hill, "An Investigation of the Extent and Consequences of Measurement Error in Labor Economic Survey Data," *Journal of Labor Economics*, Vol. III, No. 4, October 1985, Tables 2 and 3.

There is other information from surveys relevant to the effects of the passage of time on recall ability. For example, Horvath (1982) compared two surveys (the Current Population Survey and the Work Experience Survey) from the same organization of roughly identical design. One measured unemployment in the week prior to the interview (and is conducted each month of the year); the other measured unemployment in the prior year (and is conducted in March each year). Combining data between 1967 and 1979, she found lower estimates of unemployment from the year-long reference period. The longer reference period reduces estimated person-years of unemployment by 19 percent on the average over these 13 years. These might be regarded as relative changes in biases associated with the longer reference period.

9.4.2 Saliency of Events to Be Recalled

The traditional definition of a salient event is one that marks some sort of turning point in one's life, an event after which some portion of one's life is different from what it was before. Such events often are associated with strong emotional states at the time of their occurrence. Sudman and Bradburn (1973) propose that salient events are less likely forgotten than nonsalient events. Hence, other things being equal, events that are of more importance in the life of the respondent tend to be reported in surveys. This line of reasoning is compatible with schema theory. Salient events are likely to have experienced a rich encoding of details in memory at the time of their occurrence. Furthermore, it is likely that the memories of such events would be recalled more often by the respondent, in thinking about the event or perhaps in telling others about the event.

One error commonly associated with salience is a poor report of a date for a past event. The error is based on the perception that the event occurred later in time than it really did, termed "forward telescoping."[3] Note that this error concerns a property (the date) of an event that has been successfully recalled. There is evidence that more salient events are more easily telescoped forward in their dating. Brown et al. (1985) propose the hypothesis that events for which the respondent recalls more details are perceived to have occurred more recently. They use the concept of "accessibility" to note that one criterion of judgment about the date of an event is the richness of the detail of the memory of the event. Other things being equal, most memories rich in detail pertain to recent events. This strategy produces problems when the detailed memory is the result of the saliency of the event, the frequency of rehearsals of the event, or other properties unconnected with its date of occurrence. Brown et al. (1985) present several experiments that support the notion that "more important" events (e.g., the shooting of President Reagan) are reported to have occurred earlier than "less important" events (e.g., the shooting of Pope John Paul II). This tendency, the authors note, is dampened by the fact that one of the details stored in the memories about very salient events may be the date of occurrence itself. The hypothesis is only partially tested by the experiments, however, because there is no independent measure of how much is recalled about events that are more radically forward telescoped .

"Interference theory" was an early explanation for failure to report past events (Underwood, 1957) and offers slightly different interpretations of the effect of saliency.[4] Interference theory holds that events occurring after some target event can alter the memory of the target event. When

---

[3] There are also examples of "backward telescoping," the placement of an event at an earlier time than it occurred.

[4] Interference theory is one of the arguments cited above by Loftus and Loftus (1980) concerning the cause of "forgetting."

events that are similar to a target event occur after the target event, the memory of that single event may be altered. Confusion between the unique characteristics of any one of the similar events and properties of the class of events occurs. Thus, the word "interference" is used to describe the retrieval problems. "Salient" events by definition are unique in their impact on the individual but may not be unique in many of their other characteristics. Interference theory does offer an explanation why frequently repeated events are not well recalled. It does not offer an explanation why unique, but less important, events are not recalled.

9.4.3 Task Difficulty

Some information required for a survey response is easily accessible for the respondent. For example, the question may concern an important part of one's self-schema (e.g., marital status) or a fact that the respondent often thinks about (in cognitive terms a "well-rehearsed" memory). The information required to form an answer to the question is easily accessible in the respondent's memory. Other questions ask the respondent to retrieve many separate details of past events. They seek information or judgments that require novel combinations of memories of events and knowledge based on the respondent's life experience.

One illustration of the variation in response burden is the different sources of income measured in the Current Population Survey, the labor force survey of the United States. Some sources of income appear to pose the respondent less difficulty in assembling the information than others. For example, reported levels of wages and salaries (for most employed persons a relatively stable number) in CPS form 99.0 percent of the totals from independent estimates (U.S. Department of Commerce, 1985), but estimates of interest, dividends, and rental income (totals that may require summing several numbers together) form only 45.4 percent of the independent estimate. Similar results are obtained in other subject areas. Morgenstern and Barrett (1974) in a study of reports of unemployment find evidence that those persons who move in and out of the labor force are most susceptible to failure to report spells of unemployment.

One interpretation to the effect of task difficulty on errors in reporting is that respondents seek to satisfy the requirements of the survey with as little effort as possible. As Krosnick and Alwin (1987) state, people are cognitive misers, who seek to provide minimally satisfactory answers and to avoid additional cognitive processing that would not be necessary for the task. When the response task requires recall of many separate entities or arithmetic or logical combinations of them, some respondents may terminate their cognitive processing prematurely. In the next chapter, we discuss changes in the questionnaire format to reduce the number of distinct recall steps to provide an answer for one question, reducing respondent burden in an attempt to improve reporting accuracy.

9.4.4 Respondent Attention or Motivation

Gauld and Stephenson (1967) found that the nature of the instructions to subjects regarding the recall task has separate effects on the quality of recall. Subjects asked to be very accurate in their recall of details of events did indeed perform better. Furthermore, in reviewing their reports about a prior event, subjects can distinguish parts of the reports that were guessed by them from those that were indeed part of the original event. The time intervals investigated in this work were quite short by survey standards. Thus, it remains unknown whether people can identify guessed attributes of events occurring far in the past as with recent events. If so, survey respondents themselves might provide indicators of measurement errors affecting their reports.

A laboratory-based method of inquiry is probably particularly poor at measuring effects of motivation on the part of those measured. It is based on volunteer subjects who have distinctive characteristics (Rosenthal and Rosnow, 1975) and for the most part are committed to performing the experimental tasks well.[5] There is little concern with traditional social psychological variables that measure the respondents self-initiating behavior to provide accurate responses. In contrast, these variables are of central concern in some survey methodological approaches to respondent-produced measurement error.

For example, the procedures of Cannell et al. (1981) reviewed in Chapter 8, which alter the behavior of the interviewer, have as the goal to increase the depth of cognitive processing of the respondent. In cases in which the respondent failed to recall an event *and* provided the answer very quickly, the interviewer is instructed to say, "You answered that quickly" and ask the respondent to give more time to the question. This technique treads a narrow line between encouraging the respondent to overreport and instructing and enforcing minimum standards of consideration to each response in the survey.

---

[5] In this regard, it is understandable why cognitive psychology offers little insight into influences about the depth of processing in completing experimental tasks.

9.5 JUDGMENT OF APPROPRIATE ANSWER

One of the most rapidly developing fields in cognitive psychology in the last decade studies the process by which humans form judgments about alternatives and draw inferences from individual experiences (Kahneman and Tversky, 1971; Tversky and Kahneman, 1974; Nisbett and Ross, 1980). In repeated surveys it is shown that persons have a tendency to avoid burdensome, intensive thought about alternatives when forced to choose among them. They use easily accessible information about the alternatives to determine if sufficient discrimination can be made among them. In a sense they accept a risk that they will be wrong in return for much less effort and time to make the decision. They are "cognitive misers" in the terms of Krosnick and Alwin (1986), always making minimal effort to fulfill the standards demanded by the particular task at hand.

Tversky and Kahneman (1974) use the notion of "heuristics" as general guidelines or rules that are used to reduce cognitive processing in decision-making. These are shortcuts that are often efficient because they yield the same judgment that would result from more intensive thought. On the other hand, in some circumstances they yield erroneous conclusions, and thus they may be of interest to us in understanding measurement errors in surveys.[6]

The most relevant heuristic that can affect survey responses is labeled the "availability heuristic." This refers to a tendency to choose as most important, recent, or relevant that alternative that is most accessible in memory. Sometimes this is because the memory is more *vivid* than others (e.g., a presidential candidate cries about press attacks on his wife and the memory of the television news report dominates the recall of potential voters). Other times the cognitive task at hand resembles another one just performed (and "available") in some respects, and hence the person attempts to form the judgment using the same logical steps as performed in the prior one.

In many situations the most available procedures for making the judgment are the best. That is, the heuristic works in most cases and for that reason, people trust it. The most accessible memory is often that rich in details, most recently accessed, and most connected to strong emotions. Often this easy accessibility *is* a good indicator of importance. For example, consider the question frequently appearing in political attitude surveys: "What are the most important problems facing the

---

[6] We saw earlier (Chapter 5) that these notions also may apply to nonresponse errors, in decisions by sample persons about cooperating with the survey request.

United States at this time?" It is likely that the problem which most quickly "comes to mind," that is, which is most available to the respondent, is also that which has been considered most often and is related to most other concerns that the respondent may have about current affairs.

On the other hand, the availability heuristic can cause poor judgments. If the most accessible relevant memory is atypical of the experience of the respondent, he/she might be overly influenced by it. For example, let us examine the case of reporting of acute health conditions in a survey. A recent painful health episode may act to encourage reports of other such events in the reference period. The easily available memory self-labels the respondent as one who has suffered from illness and offers a reason to search for memories of past experiences. To the extent that these past episodes are seen as related events (through their diagnosis, duration, short- and long-term effects), recall is improved for a question measuring the total number of such events in a time period. On the other hand, if the most easily accessible health attributes are unrelated to acute health episodes (if the last episode occurred early in the reference period), despite the fact that the respondent might have had the same number of episodes during the reference period, fewer reports might be given. The easily accessible memory is a poor indicator of experiences throughout the entire reference period. Measurement error arises because the most available memory is atypical.[7]

It is generally assumed in the research on availability heuristics that the respondent is attracted to the accessible memory as an effort-saving strategy. There appears to be little concern in most experiments with the level of effort that respondents are willing to expend in obtaining an answer. Indeed, sometimes subjects are limited in the time allowed to answer the question. Because of this, the applicability of the apparent judgmental errors to the survey situation might be limited. A valuable research strategy might investigate the effects of vivid memories on the recall of other memories.

Another useful concept from the work of Kahneman and Tversky is that of "anchoring and adjustment." This refers to the process of answering a question by choosing a preliminary estimate or approximation (the anchor) and then adjusting it to specific differences applicable to the question. For example, Ross et al. (1975) find lower self-ratings of performance among subjects originally so evaluated, even after they were told that the original rating was a hoax. It appears their

---

[7] It is important to note that these thoughts are unsupported speculations because the data needed to test them (a full design record check study) apparently do not exist.

ratings are adjustments about the anchor of the first rating. They tend to underadjust by discounting the information about the first rating.

Anchoring and adjustment may be relevant to survey responses which must be based on some estimation process. Bradburn et al. (1987) cite the example of questions about the frequency of dental visits for a respondent. They note that some respondents appear to take as the anchor the well-publicized norm of twice yearly dental visits and then adjust that by an impression about whether they go more or less often than they should. Anchoring and adjustment may produce overreports or underreports, under this perspective. It is likely that this heuristic is used most often when the work involved on the part of the respondent in enumerating the entire set of answers is too burdensome (or judged by the respondent to be too error-prone). Thus, answering a question about the number of dental visits in the last month may prompt use of enumerative techniques but a question about the last 5 years might tend to be answered with an estimation process using anchors.

Finally, Kahneman and Tversky (1971) described the "representativeness" heuristic as a tendency to overgeneralize from partial information or small samples. For example, as Dawes (1988) notes, people misinterpret evidence that the conditional probability of teenagers using marihuana, given that they use any drugs, is high. They infer from this that the conditional probability of using other drugs, given that they use marihuana, is also high. Such problems of judgment may be relevant to attitudinal questions about the effect of changes of public policy (e.g., reduction of traffic deaths through enforcement of drunk driving laws) and to expectations about future events (e.g., likelihood of victimization).

9.6 COMMUNICATION OF RESPONSE

Within "communication of the response" we include any influences on accurate reporting occurring after the respondent has retrieved from memory relevant to answering the survey question. Social psychology provides many concepts that provide insights into this process. They include (1) the effects of social desirability or the degree of "threat" posed by the question and (2) acquiescent response behavior. These concepts were also discussed in the chapter on the interviewer as a source of measurement error. Indeed, they are seen to influence respondent behavior both through the interviewer and through the pressure of societal norms of behavior.

9.6.1 Social Desirability Effects

"Social desirability" is a judgment about how highly valued a particular attribute is in the culture. Some attributes of individuals are negatively valued by societal norms (e.g., lewd behavior, poverty, criminal activity, abuse of alcohol or drugs). Other attributes are positively valued (e.g., honesty, voting in elections, church attendance). The concept requires an evaluation by a person concerning what the society judges as a desirable trait. These judgments need not be consistent over persons in the same group.

"Social desirability" relates to survey measurement error by the reluctance of persons to reveal that they possess socially undesirable traits. It is argued that, other things being equal, people will prefer to present a positive, highly valued description of themselves to others. As DeMaio (1984) notes in a comprehensive review of research on social desirability effects, there is some debate in the literature about whether social desirability effects are highly variable over persons (i.e., that susceptibility to social desirability effects is a personality trait) or whether they are characteristic of traits. Some researchers (e.g., Gove and Geerken, 1977) conceptually separate these two interpretations and label the personality trait as "need for approval" and the property of the attribute as "trait desirability." They demonstrate relationships between the two. Some persons are more susceptible to social desirability effects than are others, but there is also consistency across persons in which traits are most and least susceptible to these effects.

The most frequently encountered operationalization of social desirability is the Marlowe-Crowne scale, a set of 33 statements for which the respondent is asked to express agreement or disagreement. The items have the property that they are either judged socially desirable but never true of anyone (e.g., "I never hesitate to go out of my way to help someone in trouble") or socially undesirable but true of everyone (e.g., "I sometimes try to get even, rather than forgive and forget"). The scale was originally developed for laboratory settings to measure the subject's need for experimenter approval. The literature is somewhat split on whether the scale is useful in separating people with different types of behavior or separating people who might be subject to differential measurement error. The first notion suggests that persons with high need for approval act in ways to maximize that approval. Hence, correlations between a social desirability scale and reports of socially desirable characteristics may be accurate. The other position suggests measurement error results from higher need for social approval. At the operational level it notes that people who distort their responses on the Marlowe-Crowne scale (where true values are well asserted) are also likely to distort their responses to questions about other socially desirable characteristics.

Bradburn et al. (1979) find little support for the response error argument in an examination of national survey data and a reduced Marlowe-Crowne scale. Instead, they repeatedly find evidence that those scoring high on the need for approval measure actually behave differently on a host of measures (e.g., marital status, frequency of social activities). The data show lower reports of the socially desirable attribute among those with high measured need for approval. This research does not claim to refute the social desirability influence on measurement error. Instead, it questions the Marlowe-Crowne scale as an indicator of that measurement error.

In the survey methodology literature, social desirability effects are often inferred to be present when traits the researcher judges socially undesirable are underreported (or traits positively valued are overreported). For example, there are many studies which match individual voter registration and vote records with interviews of survey respondents. The survey questionnaires asked about voter registration and voting in a prior election, the public records are assumed to be without error, and measurement errors are assessed by measuring differences in estimates based on the records and estimates based on the interview data. Table 9.6 presents data from Katosh and Traugott (1981) on voting in the 1976 presidential election and the 1978 off-year election. Although 72 percent of the survey respondents claim to have voted, only 61 percent of the respondents are recorded in the records as having voted.[8] The measured correlates of the apparent overreporting of voting were a somewhat controversial aspect of the results. The percentage of misreporters among all respondents show very few differences by major demographic groups, except for race (nonwhites showing more misreporting than whites). If the analysis is confined to those persons who were verified *not to have voted*, there are strong relationships with education (higher education nonvoters tending to report voting), announced expectation to vote (those expecting to vote overreporting the act), and other measures indicative of felt norms of voter participation (Silver et al., 1986).

Strong support for the conclusion that social desirability effects are heavily influenced by characteristics of the item is obtained from a study

---

[8]These data exclude those respondents who did not supply their names (and hence could not be matched to voting records), those respondents who did not answer the vote questions, and those respondents whose registration records contained no voting information. These exclusions form 6 percent of all interviewed cases. Those persons for whom no record of registration was found are assumed not to have voted.

Table 9.6 Comparison of Survey Reported and Record–Based Estimates of the Percentage of Respondents Who Voted

Source of Estimate	1976[a]	1978[a]
Survey report	72%	55%
	2415	2292
Record data	61	43
	2329	2230

[a]Entries are percentages of respondents who voted. Record data entries based on percentage of respondents for whom matched record data were located.
Source: J.P. Katosh and M.W. Traugott, "The Consequences of Validated and Self-Reported Voting Measures," *Public Opinion Quarterly*, Vol. 45, 1981, pp. 519-535, Table 1.

by Bradburn et al. (1979). For two presumed socially undesirable traits, having filed for bankruptcy and having been charged with drunken driving, a sample of court records was taken (reverse record check study). For three presumed socially desirable traits, having voted in a primary, having a public library card, and being registered to vote, an area probability sample was drawn and library and voting records accessed (a forward record check study). Table 9.7 shows that 39 percent of the sample reported voting in the most recent primary election but no record could be found of the vote. Lower levels of overreporting occur for having a library card and being registered to vote in the precinct in which the respondent lived. The bankruptcy and drunken driving experiences are not reported by one-third to one-half of the people experiencing them.[9] This pattern of findings is used to support the hypothesis that undesirable acts tend to be underreported in surveys and desirable acts tend to be overreported.

9.6.2 Acquiescence, Yea–Saying, Nay–Saying, Response Tendencies

In addition to the respondent actions discussed above which can produce errors in survey reports, there are others that have been identified and

---

[9]The report of Bradburn et al. (1979) gives little detail about respondent selection procedures and matching problems with the record systems. We thus have no information about these problems as possible causes of inconsistencies between survey reports and the records.

Table 9.7 Net Underreport (−) or Overreport (+) for Five Attributes in Face to Face Survey

Attribute	Percentage Underreport (−) or Overreport (+)[a]
Voted in a primary	+39 (80)
Having a library card	+19 (93)
Registered to vote	+15 (92)
Filing for bankruptcy	−32 (38)
Charged with drunken driving	−47 (30)

[a]Case counts in parentheses.

Source: N.M. Bradburn and S. Sudman, *Improving Interview Method and Questionnaire Design*, Jossey-Bass, San Francisco, 1979, Table 2, p. 8.

studied. Several of these come from psychological measurement with closed questions (with "yes" or "no" answers), with statements to which the respondent is to agree or disagree, and with scale questions asking the respondent to choose a category from an ordered set (e.g., strongly agree, agree, neutral, disagree, strongly disagree). Some of these will be discussed in more detail in Chapter 10, but a few comments about acquiescent respondent behavior are relevant here.

In the early 1960s social psychological studies of personality identified a tendency for some persons to agree with all statements of another person, with apparent disregard to the content of those statements (Couch and Keniston, 1960). One theoretical explanation was offered for the tendency for the agreeing response set to be more frequent among lower educational groups—that deference to the interviewer would lead to greater agreement among the poorly educated group. Later work by several critics suggested that the magnitude of the agreement response set was not large (e.g., Rorer, 1965). Schuman and Presser (1981) find that agreement is generally reduced when instead of a statement, the respondent is offered a forced choice between the statement and its opposite. Their findings support the speculation that the agreeing set might be more a characteristic of the question than of the respondents.

9.7 SOCIOLOGICAL AND DEMOGRAPHIC CORRELATES OF RESPONDENT ERROR

Prior to the interview, the survey researcher is not privy to the cognitive characteristics of the respondent. Furthermore, few of the survey tasks give direct measurements of encoding or retrieval capabilities of the respondent on the specific survey topic. For both those reasons most of the attention of survey analysts regarding respondent-level correlates of measurement error concern observable traits of the respondent. Most popular are age and educational differences. Sometimes education is formally recognized as a proxy variable for the success of comprehension of the question and sometimes of the depth of cognitive processing of survey questions (e.g., Schuman and Presser, 1981; Krosnick and Alwin, 1987).

Age is one correlate that is popular both among survey researchers and cognitive psychologists. It will be discussed first. In that discussion, it will be seen that the effect of aging on response performance is increasing failure to retrieve information from memory. In the literature, however, the term "elderly" is used rather loosely. "Elderly" is sometimes used to designate persons who are 65 years of age or older, but sometimes, persons 70 or older. In some work one will see the term used for those 55 or older. Finally, still other researchers conceptualize the effects of aging to be continuous over the life span and place less emphasis on specific age thresholds.

One motivation for survey researchers to choose respondent age as a cause of measurement error is the common difficulty of keeping the elderly respondent's attention on the interview task. Interviewers often complain of elderly respondents not addressing individual survey questions but answering questions with statements of only tangential relevance. "Keeping the elderly respondent on the topic" is perceived as a real challenge. Interviewers complain that the elderly respondent does not follow the protocol of the survey interview, the rules by which the interviewer is designated as the asker of questions and the respondent as a reactive member of the team answering those questions. Instead, elderly respondents appear to wander off the track of the discussion more often.

There is an active research program studying changes in the processes of memory as humans age. Craik in a 1977 review of this literature cites over 150 separate research efforts studying age effects on memory. In summarizing that vast literature, several converging results appear. First, there appear to be larger deficits in the elderly relative to younger persons in recall from secondary or long-term memory than in recall from primary or short-term memory. Most of these experiments are based on recall from semantic memory—reports of words learned earlier, names, or faces.

Second, the reduced performance of the elderly appears to be more obvious on recall tasks than on recognition tasks. The reasons for these deficits vary depending on the theoretical perspective of the researcher. Some claim that the problem stems from poor organization of memories at the encoding stage (Smith, 1980); others explain this in terms of less deep processing of the material both at acquisitions and retrieval. Still others imply that the richness of the links between any particular retrieval cue and material in secondary memory among the elderly might produce diminished recall because of interference between the target memory and other memory traces. This follows the notion that the set of experiences grows richer as one ages, and interrelationships among memories are plentiful.

Most research does not explicitly discuss formal causal models of the effect of aging on encoding and retrieval activity. Other research more clearly suggests that chronological age is a proxy measure for attributes like loss of brain tissue, poor vascular circulation affecting brain activities, or, more generally, poor health (Hulicka, 1967).

We have already seen some evidence of different behavior among the elderly in surveys. We have documented higher nonresponse rates among the elderly (Chapter 5) and have observed higher susceptibility to interviewer effects in their responses to survey questions (Chapter 8). There is also survey evidence of poorer recall among the elderly. In a personal interview survey in a metropolitan sample, disproportionately drawn from among the elderly, Herzog and Rodgers (1989) asked respondents at the completion of the interview to name six physical functions that were the topic of questions earlier in the interview. In addition, they presented 20 survey questions to the respondent, only 10 of which had been administered to the respondent, and then asked the respondent to identify which, if any, had been asked of them. The performance on these recall and recognition tasks decline monotonically over the age groups, as shown in Table 9.8. These results remain but are somewhat weakened when controls on educational differences across age groups are introduced. Herzog and Rodgers also found monotonically decreasing favorable ratings of memory assessment across the age groups, by the respondents themselves, thus illustrating some introspective abilities to assess certainty of recall and recognition.

Gergen and Back (1966) examine survey data to show that the elderly tend to give more "no opinion" answers to survey questions. They argue that this is a symptom of the more general process of disengagement from societal activities that accompanies aging oftentimes in this society. They introduce some (weak) educational controls and the results remain the same. Glenn (1969), using survey data from several sources, refutes the conclusions of Gergen and Back by showing that more rigorous

Table 9.8 Performance on Survey Recall and Recognition Tests

		Age Groups			
Measure	Total	20–39	40–59	60–69	70+
Number of physical functions recalled among 6 discussed	1.31	1.60	1.44	0.77	0.46
Recognition of 10 questions among 20 presented[a]	4.44	5.37	4.39	3.27	2.16

[a]This is scored on a range of −10 (reflecting erroneous choice of all 10 nonadministered questions and none of the 10 administered) to +10 (reflecting completely correct identification).

Source: A.R. Herzog and W.L. Rodgers, "Age Differences in Memory Performance and Memory Ratings in a Sample Survey," *Psychology and Aging*, 1989, Table 2.

controls for educational differences eliminate the age differences. Indeed, Glenn finds that in many cases the elderly group appears to be more informed and expressive of attitudes and knowledge on current events than younger cohorts.

Another respondent variable used by survey researchers as a correlate of measurement error is education. The argument supporting education as an influence on measurement error asserts that the sophistication of the cognitive processing and the knowledge base of the respondent is a function of education. In that sense education is used as a proxy indicator of cognitive sophistication. Typically, years of formal education are used as the measure of education, despite acknowledgment that this measure itself can be a poor indicator of intelligence.

One measurement characteristic associated with education is answering "don't know." Sudman and Bradburn in their 1974 review of response errors in surveys present a summary of missing data rates from "don't know" and "no opinion" responses, based on a review of survey data presented between 1965 and 1971. Table 9.9 shows that the "don't know" rates tend to be higher for those with a grade school education. This is consistently true over ever topic category presented in the table, from attitudinal questions on the Middle East (for which many respondents give a "don't know" response) to those on morality (where the vast majority of people offer an opinion). Converse, in a multivariate analysis of "no opinion" behavior in Gallup surveys, included education as a predictor. As control variables she included an indicator of whether the question contained more than 30 words (long questions hypothesized as producing more "don't know's"), whether the question forced a choice of two response categories (such questions producing more "don't know's"), and whether the question topic concerned foreign political affairs (such questions producing more "don't know's"). Using opinion questions from both Gallup and Harris surveys, Converse found that education, among all four predictors, was the strongest predictor of the "don't know" response. It is important to note that neither Sudman and Bradburn nor Converse examined the marginal effects of education controlling for age differences. Hence, some of the differences over education groups may simply reflect age differences.

These data do not allow us to identify whether the origin of the problem is the comprehension of the survey question or the absence of cognitive associations with the topic *or* some social psychological influence from perceived differences in status between interviewer and respondent. Converse, in a separate analysis of the same attitudinal questions, concludes that the complexity of the language as measured by the Flesch scale (see Chapter 10) does not seem to explain differences across questions in "don't know" rates. However, the Flesch scale may not fully measure the complexity of words in survey questions. Ferber's study (1956), mentioned earlier in the section on comprehension, is also relevant here. Over the four topic areas explored by Ferber, there *was* evidence of smaller percentages of respondents with little education knowing the meaning of the terms in the question. However, there was also evidence of greater willingness to admit this ignorance with a "don't know" answer.[10] In short, the tendency for persons with little education to answer "don't know" may be an effect *both* of less knowledge about an issue and less perceived pressure to appear informed, relative to persons with more education.

The relationship between education and other indicators of measurement error, however, receives only mixed support over many investigations. The arguments *for* education effects use notions that education is "associated with somewhat more self-developed and stable concepts" (Schuman and Presser, 1981, p. 91) and that less-educated respondents would "be slower to comprehend the content of a question and therefore more apt to have their choice influenced by an irrelevant cue" (p. 71). In a large set of experiments on question form and order, however, they found very mixed support for the hypothesis that low education groups are differentially sensitive to question effects. There was a tendency, however, for those with little education to give different

---

[10] These results are not uniformly obtained, however. In an attitudinal question about a very obscure piece of legislation, the Agricultural Trade Act of 1978, Schuman and Presser (1981) find that higher percentages of those with college education give "don't know" answers than those with less education.

**Table 9.9 Mean Percentage of "Don't Know" and "No Opinion" Answers on
Attitudinal Questions by Education of Respondent**

	Education			
Subject of Question	Grade School	High School	College	n
Free speech	8.9%	3.9%	3.4%	14
Capital punishment	11.6	10.8	7.8	5
Vietnam war	15.8	9.4	6.4	32
Middle East	40.9	31.1	19.5	15
Racial problems	17.1	11.9	10.7	15
Civil rights demonstrations	12.3	8.3	5.3	6
Birth control	21.7	13.2	8.8	15
Morality	8.5	5.0	2.0	2
Religion	10.8	11.2	10.2	4
Cancer	11.0	6.0	3.0	1
Anti-Semitism	16.4	10.0	7.9	7

Source: S. Sudman and N.M. Bradburn, *Response Effects in Surveys*, Aldine, Chicago, 1974, Table 4.3.

responses on open questions than equivalent closed questions. That is, there is some evidence that either they are more influenced in their choice of answers by the offering of the closed responses *or* they have less ability to communicate their thoughts in open questions.

9.8 SUMMARY

As with the other sources of measurement error reviewed in this book (the interviewer, the questionnaire, and the mode of data collection), identifying unique properties of the respondent which produce response errors is not an easy task. After all, measurement errors are generally viewed as specific to a particular measure, a question posed to the respondent. Only by identifying response tendencies of respondents over many questions can inference about respondent influences on measurement error be made. Then only by comparing different respondents on the same task can characteristics of the respondents which produce measurement error be identified. The survey methodological literature is richer in identifying joint properties of questions and respondents which produce measurement errors (e.g., low education and questions on political knowledge), but weak on what characteristics of respondents are uniformly harmful to data quality.

The researcher has a similar difficulty in separating respondent and interviewer effects. The interview is a social interaction; the meanings of questions are partially determined by the social context of the interaction. The interviewer, as we saw in Chapter 8, plays a role in that context, to define the question intent, to clarify the minimal requirements for a satisfactory answer, and to encourage the respondent's attention to the interview task. Empirical investigations into respondent error are made with particular interviewers administering the questionnaire. Separating interviewer effects from respondent effects requires research designs that unconfound these two influences on measurement error.

This chapter reviewed two bodies of theory which speak to those influences on measurement error which are more clearly properties of the respondent than of the questionnaire or the interviewer. The cognitive psychological literature provides some conceptual structure to understanding the procedures that respondents use to answer survey questions. It offers experimental evidence that explains why some respondents use different bodies of information to interpret the meaning of questions. It can thereby alert a survey researcher to alternative phrases to use in a survey measure. The literature describes how the circumstances surrounding the encoding of memories affect retrieval of that information. This explains why some cues to recall past events are effective for some persons but not for others. It offers experimental evidence of how context can impact the meaning assigned to questions. Finally, the literature provides evidence of the process by which recalled memories of past events are reconstructions of those events, using both specific properties of the events encoded and respondent judgments about other properties based on the likelihood of joint appearance of those characteristics. It notes that respondents are poor judges of real and imputed properties of the recalled events.

The cognitive psychological literature is thus rich in potential hypotheses about survey measurement error produced by respondents. At this writing it offers a common theoretical orientation to these problems. However, the application of the theoretical principles is not a straightforward one. There is no one cognitive psychological theory; there are many. The theories are restricted in their conceptual breadth, applicable to small subsets of cognitive tasks. Empirical support for the theoretical assertions generally come from controlled experiments on college students. In general, the levels of motivation to follow the instructions of the researcher appear to exceed those encountered in survey interviews. The theories are often focused so completely on processes within a subject that the influence of social context is defined out of the concern of the investigator.

The impact of social context on respondent behavior is the conceptual domain of social psychology. It has offered the notion of social desirability to explain some errors of response in survey settings. It also (as we learned in Chapter 8) has studied the process of interpersonal influence. The effect of the interviewer and others on respondent cognitive processes are examples of relevant influences in the survey setting.

The social desirability effect, the tendency to overreport characteristics perceived to be valued by others, is a frequent post hoc hypothesis for survey measurement error. It is used to explain overreporting of voting, reading of books, and amount of charitable contributions. Too few empirical studies, however, have obtained measures from respondents regarding their own perceptions of the socially desirable attribute corresponding to a particular measure. Instead, the researchers, based on their own judgments, have labeled attributes as socially desirable or undesirable and interpreted responses to the questions using those labels. Such an approach is wise only when there is homogeneity in the population on the extent of social desirability of the attribute. Since many procedures to reduce social desirability effects (e.g., rephrasing the question to encourage reports of the socially undesirable trait) generally change the form of the measure for all respondents, this assumption of homogeneity needs to be critically examined.

Another influence on respondent errors arises from lack of attention to the question and failure to engage in deep processing in memory retrieval during the response formation process. Quick, superficial answers to questions partially comprehended often are poor responses. From a measurer's perspective there has been little work toward attempting to document how much effort the respondent is making in responding to the questionnaire or how various distractions affected the respondent's behavior. From a reducer's perspective, Cannell's techniques of obtaining commitment of the respondent to work diligently at the response task and feedback about that performance are relevant.

This chapter has mentioned costs in only one context: increasing the length of the questionnaire to enrich the cuing in memory retrieval. This is an example of the dominant method of decreasing respondent errors in surveys —an alteration of the questionnaire. The second most prevalent method is alteration of interviewer behavior through increased probing. Finally, efforts to instruct respondents in appropriate respondent behavior (e.g., thinking carefully, seeking clarification from the interviewer) also has cost implications. The most direct treatment of such cost implication is that reviewed in the next chapter on cost implications of altering the questionnaire (see Section 10.6). Very rarely have formal cost models of efforts to reduce the respondent source of measurement error been built, but they would no doubt include impacts on interviewer training, interviewing time, and questionnaire construction costs.

CHAPTER 10

MEASUREMENT ERRORS ASSOCIATED WITH THE QUESTIONNAIRE

Epimenides to Buddha: "I have come to ask you a question. What is the best question that can be asked and what is the best answer that can be given?"

Buddha to Epimenides: "The best question that can be asked is the question you have just asked, and the best answer that can be given is the answer I am giving."

R. Smullyan, 5000 B.C. and Other Philosophical Fantasies, *1983, p. 31*

This chapter examines a basic building block in survey data collection, the question, as a source of measurement error. In doing so, it focuses on three attributes: the words in the question, the structure of the question, and the order or context of questions. Inevitably, the discussion of this chapter overlaps with the preceding chapters on the interviewer and the respondent as sources of measurement error. This is because the questionnaire is used as a script for the interviewer during the data collection; thus, questionnaires are an integral component of interviewer behavior.

The question also identifies for the respondent the cognitive tasks to perform, which information to retrieve from memory, and what judgments are sought. Thus, questions are closely linked to the errors attributable to respondents. In the following sections we examine evidence of the magnitude and causes of measurement error connected with survey questions. Whenever possible we use studies that permit a separation of the effects of the question from the effects of interviewers or respondents.

A fundamental tenet of scientific measurement is that the measuring device is standardized over different objects being measured. One problem in survey measurement is that the object being measured (the respondent) must itself be actively engaged in the measurement process. In most of the physical sciences and some of the biomedical sciences,

449

measurements are taken by external comparison with a standard, sometimes without contact between the measurement instrument and the object. In surveys, in contrast, even though the measurement instrument might be consistently applied to all respondents, it may be understood differently by different respondents. By these differences standardized questions can be assigned different meanings. Language is the conduit of social measurement. Although the language of the survey questions can be standardized, there is no guarantee that the meaning assigned to the questions is constant over respondents. For that reason, much of this chapter concentrates on alternative meanings of the same sets of words.

By far the most attention to measurement error properties of survey questions has been given to attitudinal questions. Tradition in many survey questionnaires about behaviors and demographic attributes holds that measurement error is not as sensitive to question wording as to respondent capabilities in the particular substantive area. This chapter attempts to describe research of both attitudinal and factual character whenever possible.

10.1 PROPERTIES OF WORDS IN QUESTIONS

There are several different types of measurement error associated with words in questions:

1. No meaning can be given to a word (apart from cues from its context) by the respondent.

2. A word can be taken to mean different things by the same respondent.

3. A word is taken to mean different things by different respondents.

These categories are obviously not mutually exclusive. The first type of error stems from different vocabularies used by different groups in the population. Practical guidelines for interviewers acknowledge that meanings might vary across respondents. In some factual surveys interviewers are given definitions of terms (e.g., "by 'orthodontist' we mean a doctor who straightens teeth"). Most typically, interviewers are instructed to provide those definitions only when a respondent inquires about the meaning of a word. We have not been able to locate any studies of the frequency of such requests by respondents or of the characteristics of respondents who seek that information. In attitudinal surveys interviewers are sometimes instructed to avoid answering such questions.

For example, the SRC interviewer's manual states that, if asked about the meaning of a term, the interviewer is to reply "Whatever it means to you." This decision is based on the belief that (1) not all terms in every question can be simply described and (2) the stimulus presented to each respondent should be standardized, that providing such answers to those respondents who ask would by definition depart from uniform measurement procedures. This argument implicitly asserts that respondents are more capable of discerning the intent of a question on their own than they are with the assistance of an interviewer. This is potentially a large price to pay for standardization of interviewer behavior.

Writers for newspapers and magazines are often told to keep their sentences simple and to use words common to everyday use. One way to measure the complexity of the words used is the Flesch scale which characterizes both word length and sentence length effects on complexity. The Flesch scale weights the number of words per sentence and total number of syllables. Reading ease, RE, is estimated by the equation RE = $206.84 - 0.85$ (word length) $- 1.02$ (sentence length in words). Thus, low scores apply to material that is more difficult to comprehend, while high scores apply to material that is easy to comprehend. As a rule of thumb, the median years of school (12.1 years) is represented by a reading ease score of 50 to 60. Converse and Schuman (1984) examined questionnaires from several survey organizations and applied the Flesch scale to 100 to 200 questions from each organization (see Table 10.1). They found that about 200 questions from Harris surveys were rated most difficult to read. The easiest to read were about 150 from the National Election Surveys of the Center for Political Studies at The University of Michigan. After rating the different questions, they examined whether the complexity of the question was related to the proportion of "don't know" answers. There were no differences evident in the data.

It is rare for surveys to inquire directly from the respondent about the meaning they assign to questions. We saw one example of this in the Ferber study of attitudes toward a "guaranteed annual wage," where after an attitudinal question on the topic, the respondent was asked what it meant. There have been several studies, however, that have asked the meaning of key words in questions after the questionnaire is administered or as a special component of the questionnaire.

Bradburn and Miles (1979) asked people how often they experienced various "feeling states" (bored or excited) using closed categories: "never," "not too often," "pretty often," "very often." After 10 such questions, the interviewer asked about the meaning of the chosen response category for two questions. For example, if the respondent reported that he/she had been excited or interested in something during the last few weeks "very often," he/she was asked "how many times a day or week" they meant by

Table 10.1 Mean Level of Language Difficulty as Measured by Flesch "Ease of Reading" Scores (Lower Score = More Difficult Language)

	Gallup 1969–71	Harris 1970	ISR-NES 1972	NORC-GSS 1972–74
Mean	62.1	58.7	72.7	63.6
Standard deviation	18.1	18.6	20.5	22.6
Number of questions	206	196	157	200

Note: Short probes numbering 1 to 3 words and questions in series numbering under 5 words were excluded.
Source: J.M. Converse and H. Schuman, "The Manner of Inquiry: An Analysis of Survey Question Form Across Organizations and Over Time," Chapter 10 in C.F. Turner and E. Martin (Eds.), *Surveys of Subjective Phenomena*, Vol. 1, Russell Sage Foundation, 1984.

"very often." Thus, each respondent was asked to quantify their answer for the frequency of being "excited" and the reported frequency of being "bored." Interviewers reported that respondents were reluctant to give answers to these questions, but only 5 to 6 percent failed to provide some answer.[1] As shown in Table 10.2 the results of this effort are that people who answer "not too often" on the average report that this meant 6.65 times a month, once every 5 days for the "excited" measure, but 4.15 times a month, once every 8 days, for the "bored" measure. Those who answer "very often" for the two questions gave more similar meanings on the average: 17.73 times per month and 17.39 times per month.

The means are only one attribute of interest. Of greater interest in terms of measurement error is the variability in meanings over different respondents who chose the same answer (Table 10.2). This is evidence of measurement error that varies over respondents. We should not assert that all respondents who chose "pretty often" should mean exactly the same number of times per month. Since only three closed–response categories are offered, yet the number of times per month could range from zero to a very large number (near infinity), there must be a

---

[1] The result that respondents were reluctant to respond to this question is interesting in itself. One possible interpretation to this is that they view the question as a change in the rules of the interview interaction. That is, although the question was initially posed with less precise answers, now the interviewer wants to know more precisely how often the event occurred. The two possible intentions of the follow-up question might therefore be (1) to have the respondent define what is meant by the answer category or (2) to think harder about how many times he/she experienced the event. The second interpretation can lead to measurement error in the follow-up question.

Table 10.2 Mean Number of Times per Month Respondent Ascribes to Response Category by Answer Given to Feeling State Question

Response Category Chosen in Feeling State Question	Definition of Response Category (Times/Month) Feeling State Question[a]	
	Excited	Bored
Not too often	6.65 (.47)	4.15 (.24)
Pretty often	12.95 (.54)	13.72 (.94)
Very often	17.73 (.95)	17.39 (1.31)

[a]Standard errors in parentheses computed from information in tables, assuming simple random sampling.
Source: N. Bradburn and C. Miles, "Vague Quantifiers," *Public Opinion Quarterly*, Vol. 43, No. 1, Spring 1979, pp 92–101, Table 1.

distribution of "times per month" within each closed category. However, if there is evidence of great overlap in the three distributions, this could reflect measurement error. That is, two respondents who felt "excited" the same number of times per month may give different answers in terms of "not too often," "pretty often," or "very often." The only data presented in Bradburn and Miles (1979) relevant to this question are the standard deviations of answers. To illustrate one possibility, we assume normal distributions to the answers. The result is Figure 10.1, which shows large overlaps across the categories. That is, many people who answered "pretty often" reported the same number of times per month as those who answered "not too often."

The Bradburn and Miles results (Table 10.2) show that the context of the question may affect the attributed meaning of the response category. That is, "not too often" is associated with a larger average number of times per month for "excited" than "bored." There are hosts of other examples of this phenomenon. Pepper and Pryulak (1974) found that "very often" was given a meaning of smaller numbers for earthquakes than "missing breakfast." Other research has found that the word "few" meant more in the context of "reading a few poems before lunch" than in the context of "writing a few poems before lunch" or that the number of people seen was larger when "I saw a few people out the front window" than "I saw a few people through the peephole of the door."

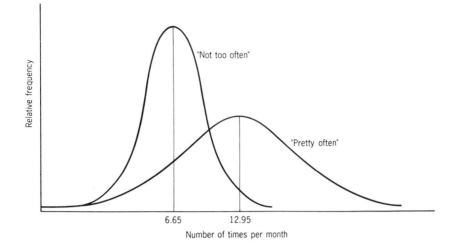

Figure 10.1 Hypothetical population distributions of number of times per month judged equivalent to "pretty often" and "not too often."

The reader should not infer that ambiguity in words of survey questions is limited to the "vague quantifiers" that Bradburn and Miles discuss for attitudinal questions. Questions that would often be labeled as "factual," that is, for which there are observable behaviors, are also subject to such problems. Another study gives direct evidence of the variability in meaning assigned to words in survey questions. Belson (1981) studied intensively 29 experimental questions by first administering them to respondents and then returning to question the respondents about how they understood various words in the questions. He inquired about the meaning of words like "you," "weekday," and "children," terms that have several meanings. He also probed for the understanding of concepts like a "proportion." Unfortunately, the research is based on a small number of cases. Belson presents tables similar to Table 10.3. This contains the distribution of reported meanings of the term "you" in several questions; for example, "How many times do you usually switch from one station to the other, when viewing TV on a weekday evening?" The term is taken to mean the respondent himself or herself 87 percent of the times. Belson does not reveal what the intended meaning of the term is in the study. It is noteworthy that the most frequent alternative interpretation would include other persons, a plural form of "you" that would include spouses or other family members. This variability in the perceived meaning of words would be productive of different answers to questions like that concerning the television viewing behavior. This is probably best conceptualized as an variable measurement error over persons

Table 10.3 Proportion of Instances in Which a Particular Meaning Was Assigned to the Term "You"

Meaning	Percentage of Instances
You, yourself, alone	87.4%
Evidence not conclusive, but probably as intended	0.9
You and your wife	1.1
You and your husband	0.9
You and your family	2.5
You and one or more of your family	1.0
You and others	0.3
You and your friend	0.5
You or your wife	0.2
You or your family	0.3
At least someone in your family	0.2
Your family (not necessarily you)	0.4
Your husband	0.2
Term omitted or overlooked	3.5
No evidence on interpretation	0.6
Number of instances	1225

Source: W.A. Belson, *The Design and Understanding of Survey Questions*, Gower, London, 1981, Table 8.3.

interviewed.[2] Belson found that over all questions less than 60 percent of the respondents interpreted them as intended by the researcher.

To summarize his research, Belson proposes 15 hypotheses that should, he argues, guide future research on question wording:

H1 When a respondent finds it difficult to answer a question, he is likely to modify it in such a way as to be able to answer it more easily.

H2 If a broad term or concept is used, there will be a strong tendency for respondent to interpret it less broadly.

H3 Under certain circumstances a term or concept may be widened.

[2] It would be interesting to obtain measures of the stability over replications of the same measure of the meaning of the terms.

H4 Part of a question may be overlooked under certain conditions.

H5 A respondent may distort a question to fit his own situation or position or experience.

H6 The general context or setting of a term in a question may wrongly influence the way that term is interpreted ('context' being the question and/or the questionnaire within which the term is set).

H7 A respondent may answer to what he regards as the "spirit" or "sense" of the question, rather than to its actual words, and so perhaps misinterpret the question

H8 A question may be wrongly interpreted if it has in it difficult words or words which mean different things to different people.

H9 A word or part of a word may not be heard properly and so lead to erroneous interpretation of the question in which it occurs.

H10 The word "you" is prone to be interpreted collectively (e.g., as "you and others") where it refers to behaviour in which the respondent is involved with others.

H11 A respondent may add to a question in order to be able to enlarge or to qualify his answer.

H12 Where a question is ambiguous, a respondent may select one of the two possible interpretations, without the interviewer necessarily being aware that this has happened.

H13 A qualifying clause may interfere with the consolidation of respondent's grasp of preceding elements of the question.

H14 When some concept presented to a respondent seems odd, he may well normalize it (e.g., "cartoon advertisements" may become "cartoons" or "cartoon programmes"; "advertising time" may become "the timing of advertisements").

H15 When a complex or thought demanding question is followed by a simple choice of answers (e.g., yes/no/?), the respondent is likely to give less care to his consideration of the detail of the question and so is less likely to interpret the question as intended.

These 15 hypotheses can be reduced without severe loss of content to two: (1) respondents interpret the interview rules to require them to answer each question, not merely to answer the questions they view as completely unambiguous. For the most part, they will follow these rules. (2) When faced with several possible meanings for a question, respondents will choose one for which an answer is more readily formed. This last point is consistent with the perspective which states that the respondent's activities are governed by goals of satisficing not optimizing, that the minimally sufficient cognitive activities to achieve an answer acceptable to the interviewer will be conducted.

The Belson work does not provide an empirical link between question meaning attributed by the respondent and the respondent's answer to the question. An early study by Ferber (1956) does provide such information. Ferber had interviewers ask attitudinal questions about specific political issues (e.g., "What is your attitude toward allowing labor to have a guaranteed annual wage? For, Against, Neutral, Don't Know."). These questions were then followed by a probe about the reason for the answer (i.e., "Why?"). The third question was one concerning the perceived meaning of the issue (e.g., "As you interpret it, what do the unions mean by a guaranteed annual wage?"), in the form of an open question. Judges examined the answers to the last question to code whether the respondent had a correct idea of the basic issue.[3] Table 10.4 shows that whether or not respondents know the meaning of terms in the question, they are willing to provide the requested opinion. Large portions of those who later admit ignorance about the meaning of the issue provided an attitudinal response (from 14 to 83 percent across the four issues). Those who are misinformed (provide an incorrect definition) respond at similar rates to those who are well informed.

Ferber's data also permit observations of whether the attitudes expressed by those who correctly interpreted the terms were different from others. Table 10.5 shows that those using the intended meaning offer distinctive opinions relative to the full sample. For example, only one-third of those who understand the meaning of "guaranteed annual wage" support it, but almost one-half (46.3 percent) of the total sample does. Ferber does not examine mulitivariate relationships involving these attitudinal variables, so we are not informed about the impacts of respondent ignorance on analytic statistics. The results offer strong

---

[3] On the question about minimum wage, the respondent was asked what the current figure for the minimum wage was; on another question about bond yields, how the bond yield compared to that of savings accounts. This causes some problems of interpretation across the issues in the importance of "incorrect definitions" of the terms in forming an attitude on the issue.

Table 10.4 Proportion Expressing an Attitude on an Issue by Their Ability to Define the Issue Correctly

Issue	Ability to Define the Issue		
	Admitted Ignorance	Incorrect Definition	Correct Definition
Minimum wage	62.1%	96.3%	91.5%
Guaranteed annual wage	47.0	83.6	90.6
Government bond yields	82.7	96.8	93.7
Fair trade laws	14.1	81.0	85.3

Source: R. Ferber, "The Effect of Respondent Ignorance on Survey Results," *Journal of the American Statistical Association*, Vol. 51, No. 276, 1956, pp. 576–586, Table 1.

Table 10.5 Percentage in Response Categories Most Affected by Inclusion or Exclusion of Respondents with Little Knowledge of the Issue

Issue	Response Category	Respondents with Correct Definition	All Respondents
Minimum wage	Raise	74.9%	63.7%
Guaranteed annual wage	For	33.3	46.3
Government bond yields	Raising would help sales	70.4	72.3
Fair trade laws	Abolish	48.0	31.8

Source: R. Ferber, "The Effect of Respondent Ignorance on Survey Results," *Journal of the American Statistical Association*, Vol. 51, No. 276, 1956, pp. 576–586, Table 2.

support for Belson's hypotheses that the respondent will answer survey questions despite little understanding of the question.

Another example of direct questioning of the respondent is that of Cantril and Fried (1944). On a relatively small sample, they used two different measurement methods to study meaning. The first followed a closed question and asked directly what the respondent took the question to mean:

> We are wondering whether the question I just read to you means the same thing to everybody who answers it. Would you mind telling me in your own words what you understood it to mean?

Only about 11 of the 40 respondents understood the question to mean what was intended by the researchers. This method of measuring understanding relies on the respondent's ability to articulate the perceived meaning of the question. The second method followed a closed question and asked respondents to define what they meant by their answers. For example, the question was asked,

> After the war is over, do you think people will have to work harder, about the same, or not so hard as before?

This question was then followed by

> When you said that people would have to work (harder, about the same, or not so hard), were you thinking of people everywhere and in all walks of life—labourers, white-collar workers, farmers and businessmen—or did you have in mind one class or group of people in particular?

This has the advantage of not asking directly what the respondent thought the interviewer meant by the question but what meaning they used in giving their answer. Slightly more than half of the respondents viewed people to mean everybody.

The more typical method of investigating the meaning of different words in questions is the split sample comparison. In this method two questions that use synonyms are administered to different half samples, and different response distributions are used to infer that different meanings are assigned to the words. One example was seen earlier, the comparison of the two questions,

> Do you think the United States should forbid public speeches against democracy?

> Do you think the United States should allow public speeches against democracy?

Radically different response distributions were obtained for these questions (see Table 7.2), but there was no direct attempt to measure what meanings were given to the question or the phrases "forbid" or "allow."

Another example of using the respondents to obtain information about the meaning of words in questions is that described by Martin (1986). The National Crime Survey asks respondents to report criminal victimizations that occurred in the 6 months prior to the interview. This is communicated to the respondents at the beginning of the interview by

> Now I'd like to ask some questions about crime. They refer only to the last 6 months—between ___ 1, 19_ and ___, 19_. During the last 6 months, did anyone break into ....

After the victimization questions, several minutes of other questions were asked of the respondent. At the end of the interview, a set of debriefing questions were asked of the respondents (these are listed in Figure 7.3). One asked the respondent what they understood the reference period to be. Approximately 5 percent gave an answer different from the 6 month period; about 15 percent replied that they did not know.

In one sense words in survey questions are shorthand, an utterance that is meant to evoke in the respondents' minds a consistent image. Words are imperfect mechanisms to perform this task because, as any dictionary exhibits, they have more than one meaning. One source of variability in response error over persons interviewed may be associated with different meanings given to words in the questions. One obvious way to eliminate this source of error, is to choose words with unambiguous meaning, but that goal may never be attainable.

Another approach is one common to other error sources that cannot be controlled perfectly. In the absence of reducing the errors, measuring them is the next best strategy. That measurement would involve the use of auxiliary measures, seeking from the respondent his/her perceptions of the meaning of the questions. We have seen that this approach has been tried by some researchers, but only in an experimental form. It appears to be a strategy worth studying more carefully.

10.2 PROPERTIES OF QUESTION STRUCTURE

Focusing exclusively on individual words in questions is not completely effective in understanding measurement errors associated with questions. Questions consist of phrases ("In general"), qualifying clauses ("Excluding the events you've already mentioned"), and separate sentences that give instructions to the respondent. Complex questions can be broken into multiple questions, and sets of questions can include or omit repeated phrases. This section reviews such properties of questions in order to learn whether they have been found to be associated with measurement error.

10.2.1 Length of the Question

There are two contrasting hypotheses about question length and measurement error. The first is that longer questions communicate more information about the nature of the cognitive task required of the respondent; they serve to explicate the meaning, providing more context to the individual words in the question; they serve to communicate to the respondent that the question is an important one, deserving of serious attention; and they give the respondent time to consider answers. The second hypothesis is that the amount of information so quickly communicated in a long question exceeds the capacity of the respondent to retain it, that multiple thoughts tend to contradict one another, and that confusion on the part of the respondent concerning the intent of the question results. In studies of employment interviews Matarazzo et al. (1963) found that the speech duration of the applicant patterned itself after the speech duration of the interviewer. When interviewer questions averaged 5 seconds, the respondents' replies averaged 30 seconds. When the interviewer questions averaged 15 seconds, the respondents' replies averaged over 1 minute. Findings such as these prompt examination of how survey question length might affect respondent behavior.

Bradburn and Sudman (1979) experimented with question length as part of an experiment on survey measures of sensitive behavior. They added an initial sentence to each question (e.g., "Occasionally, people drink on an empty stomach or drink a little too much and become intoxicated.") prior to asking the question (e.g., "In the past year, how often have you become intoxicated while drinking any kind of alcoholic beverage?"). There were no reliable differences found for questions with "yes" or "no" answers, which merely report on whether the respondent engaged in an activity at all. On responses about the frequency or magnitude of the experiences, however, longer questions led to more reporting of events like drinking or sexual activities. The authors claim this result suggests improved reporting because the resulting population estimates were closer to published reports of taxed alcoholic beverages.[4]

Laurent (1972) describes a reverse record check study of white female patients to a prepaid health clinic. Survey answers on measures of chronic conditions were compared to physician checklists on the same conditions completed on each visit to the clinic. Long forms of the questions had three sentences. They began with an introductory statement, describing the purpose of the question, with a different grammatical structure than the question itself. The second sentence either had some more information or merely communicated some obvious information. The third sentence contained the question itself. Laurent provides the example of the short form, "Have you ever had any trouble hearing?" and the long form, "Trouble hearing is the last item of this list. We are looking for some information about it. Have you ever had any trouble hearing?" Laurent estimated the probability of agreement between the physician record and the respondent for the short and long versions over 13 chronic conditions. Table 10.6 shows increases in agreement between the respondent reports and the physician reports when the long questions are used. The estimated probability of a respondent mentioning a condition recorded by the physician increases from .54 to .62 when long questions are used. Other experiments described by Laurent demonstrated that the amount of information given in response to long questions is more than to short questions; these seem to show that the long questions also yield lower measurement error.

Although the research above gave attention to making length uncorrelated with meaning in the questions, it is not clear whether that can be achieved completely. Additional words must contain instructions or supplementary descriptions of the question. These inevitably act to expand on the intent of the question as perceived by the respondent.

10.2.2 Open Versus Closed Questions

Open questions ask the respondents to phrase answers in their own words and the interviewer to record verbatim the answer. Closed questions ask the respondent to choose an answer from among the alternatives provided. In addition to these two extremes, there are two compromise structures. The "other" answer on a closed question allows some

---

[4] This is an example of an appeal to external data that permits assessment of net bias, but there is no evidence about whether the improvement in net bias is the result of overreports among some persons balancing underreports by others.

Table 10.6 Agreement Between Physician and Respondent on Chronic Conditions by Agreement Rates and Length of Questions

Probability of Agreement	Question Length		Difference	Percent Increase
	Short	Long		
Probability of respondent report, given physician report	.537	.622	.085[a]	+17.0
Probability of physician report, given respondent report	.477	.516	.039	+6.0
Overall probability of agreement	.338	.392	.054[b]	+16.0

[a] $p < .05$, one–tailed, based on Z.
[b] $p < .10$, one–tailed, based on Z.
Source: A. Laurent, "Effects of Question Length on Reporting Behavior in the Survey Interview," *Journal of the American Statistical Association*," June 1972, pp. 298–305.

respondents to volunteer an answer not provided. Finally, "field–coded questions" are presented to the respondent as open questions, in which the interviewer is instructed to classify the respondent's answers into one or more of the fixed response categories listed in the questionnaire. No verbatim recording is performed; the respondent is not alerted to the fixed response categories.

We discuss two experimental studies relevant to open and closed questions. The first explicitly was seeking measurement of response bias; the second sought evidence of response differences between open and closed questions, without formal specification of the nature of true values. Bradburn and Sudman (1979) found that open questions were more productive of mentions of sensitive behavior (e.g., drinking alcoholic beverages, sexual activities) than closed questions. They reasoned that the open question permitted the respondents to describe their behavior in their own words, sometimes providing spontaneous rationales or using qualifying words to label the behavior. For the alcoholic beverage questions, they obtained estimates of consumption closer to taxed quantities of alcoholic beverages.

Schuman and Presser (1981) conversely were interested in differences in response obtained from two questions identically worded except for closed and open structures. Despite several iterations that used an open version of a question to identify major answers for a closed version of the same question, they repeatedly failed to control differences between the two structures. Furthermore, there was evidence that relationships with other variables are affected by the change in question structure. There appeared to be some evidence of larger effects among respondents with less formal education than among others. From one perspective it appeared that the closed question acts to provide more context to the question, defines the possible answers to the question, and provides answers about which the respondent had not yet thought.

Collins and Courtenay (1983) describe a split sample experiment using three different forms of seven different questions: (1) an open question, with answers coded by coders after the survey was completed, (2) a field–coded version, and (3) a closed question. In addition to the split sample character of the study, interviewer assignments were made at random in order to measure the interviewer component to response and recording variability. When the office coding used the same categories as interviewers used in the field, no large differences were found between the open question and the field–coded question. The frequency of "other" answers was higher in the field–coded version, probably because of the difficulty of coding on the spot. The component of variance associated with interviewers was also higher in the field–coded version (2.9 versus 0.6 percent) than in the open–question version. There were no estimates provided of the component of variance associated with coders in the open question. The largest limitation of field–coded questions relative to open questions cited is that the code categories must be prespecified in the field–coded version, and this leads to higher aggregation of answer categories.

10.2.3 Number and Order of Response Categories in Fixed Alternative Questions

In fixed alternative questions the construction of the question involves both phrasing of the interrogatory part and specification of which responses will be explicitly given to the respondent. This is a problem that affects both factual and attitudinal surveys. In surveys about behaviors, the researcher must decide the level of aggregation of categories of answers (e.g., should telephoning for medical advice and annual medical physical examinations be combined as "doctor visits"?). In surveys of attitudes, the number of scale points must be determined, whether or not to provide a neutral position must be decided, and the degree to which the entire set of possible answers is made explicit must be determined. In addition to these properties, both for attitudinal and factual questions, there has been some study of measurement error affected with the order of the response categories and of the use of a set of subquestions, viewed to be simpler cognitive tasks, instead of one global question.

On the attitudinal front, the results are rather uniform and resemble those for the offering of an explicit "don't know" option. If explicit alternative response categories are offered (with whatever meaning) they will be chosen by respondents who would have opted for other responses without them. At this level the interpretation of the finding is close to a tautology, but several examples illustrate the point. Kalton et al. (1980) report on a variety of split sample experiments which varied the existence of an explicit neutral category. For example, Table 10.7 shows the results for the question, "How would you say you sleep these days?" In one half-sample the response categories read to the respondent were "very well, fairly well, rather badly, or very badly." In the other half-sample, the option "about average" was placed between "fairly well" and "rather badly." A large percentage of people chose the middle alternative when explicitly offered (18.9 percent), but no one volunteered the answer without an explicit mention. This is the typical finding of these experiments (see Stember and Hyman, 1949-1950; Schuman and Presser, 1977; Presser and Schuman, 1981).

A separate literature exists concerning the number of scale points to use in Likert-type measures. Much of this is based on self-administered ratings, and scales from 2 to 19 or more points have been used, sometimes having an explicitly labeled neutral point, sometimes not. These scales have been subjected to test-retest reliability studies, measures of internal consistency, measures of the proportion of scale points used by respondents, and measures of predictive and concurrent validity (see Chapter 1) (Bendig, 1954; Guest, 1962; Matell and Jacoby, 1971, 1972). Over all respondents more of the categories of two and three point scales tend to be used than those with larger numbers of points. Beyond three categories, however, few differences are observed in the proportion of categories actually used by respondents. There is some tendency for smaller proportions of respondents to choose the neutral category as the number of scale points increases. There are no differences in reliability or validity estimates across measures with different numbers of points.

Recent attempts to adapt questions that use response cards (i.e., visual aids to present the response categories) to telephone survey usage have explored an unfolding technique. For example, Groves and Kahn (1979) and Miller (1984) altered a seven point fully labeled scale question into two questions. For example, Miller used

Table 10.7 Comparison of Response Distributions for Split Sample Experiment on Offering a Middle Alternative to the Question, "How would you say you sleep these days?"

	Form of Question	
Response Category	With Middle Alternative	Without Middle Alternative
Very well	41.1%	46.7%
Fairly well	21.4	34.0
About average	18.9	0.0
Rather badly	13.2	13.1
Very badly	3.9	4.6
Other	1.4	1.5
Total	100.0	100.0
n	(793)	(800)

Source: G. Kalton, J. Roberts, and D. Holt, "The Effects of Offering a Middle Response Option with Opinion Questions," *The Statistician*, Vol. 29, No. 1, 1980, pp. 65–78.

> Now, thinking about your **health and physical condition in general**, would you say you are **satisfied, dissatisfied**, or **somewhere in the middle**?

If the respondent replied "satisfied" to the first question, the second question was,

> How satisfied are you with your health and physical condition—completely satisfied, mostly, or somewhat?

If the respondent replied "dissatisfied" to the first question, the second question was,

> How dissatisfied are you with your health and physical condition—completely dissatisfied, mostly, or somewhat?

If the respondent replied "somewhere in the middle" to the first question, the second question was,

If you had to choose, would you say you are closer to being **satisfied**, or **dissatisfied** with your health and physical condition, or are you **right in the middle**?

Note that this format of the question sorts the responses into the same seven categories that would exist with a single seven point scale. The question addressed by Miller is whether the measurement error properties of the two methods are similar. His evaluative criteria included mean scores on five questions that formed the experiment, standard deviations of scores over the seven points, patterns of intercorrelations, and empirical relationships with other variables. In the terminology of Chapter 1, the standard deviations would be affected if measurement error of a type uncorrelated with true values of the estimates would be different in the two methods. With such a result, intercorrelations and relationships with variables of theoretical relevance to the scales would be attenuated. The reader should note, as in the above, that if the measurement error were correlated with the values (i.e., if those with high values on the scales tended to make different kinds of response errors than those low on the scale), standard deviations, intercorrelations, and relationships could be either weakened or strengthened because of the measurement error. Nonetheless, such evaluation using multiple criteria is a useful exercise. On the whole, the two methods tend to yield similar results. The intercorrelations are slightly smaller in the two–step method, and the means revealed slightly more satisfaction with personal health, but the differences are smaller than those expected by sampling variability alone.

Similar unfolding procedures for measurement of income were studied by Locander and Burton (1976). In a split sample study they compared different ways of asking several "yes/no" questions about income. The first began with a low income level and asked the respondents whether their incomes exceeded that amount (i.e., "Was it more than $5,000?"). The questions that followed asked about successively higher levels of income. There were a maximum of six questions asked. This is labeled the "more than" form. The second set began in the middle of the income range and asked whether their incomes were under or over the given amount (i.e., "Would it be $15,000 or more or would it be less than that?"). There was a maximum of four questions asked. This is similar to the unfolding approach used by Miller (1984). The third form is the "less than" form, starting at a low income level (i.e., "Was it less than $5,000?") and continuing for a maximum of six questions. The fourth form was a "more than" form, starting at the highest category (i.e., "Was it more than $25,000?") and then continuing down the income levels for a maximum of six questions. Locander and Burton found that the median income level from the survey was closest to the U.S. Census provided figure for the unfolding scale (Form 2) and the "less than" scale. The first form ("more than" starting at the low income) underestimated the median income levels, and the fourth form ("more than" starting at the high end) overestimated the median income.

10.2.4 Explicit "Don't Know" Response Categories

The phenomenon of nonattitudes (reviewed in Chapter 9) has encouraged research to identify those persons who have not thought about issues measured in the survey to have formed an opinion. Similarly, the failure of a respondent to recall any answer for a question (e.g., "When was the last time you went to a doctor?") has prompted research into question structure that may assist respondents in obtaining such information. The methodology used to study the "don't know" problem has been the split sample experiment both in factual and attitudinal surveys, but the guidance to practical researchers from the two areas is somewhat different. This difference concerns the recommended amount of effort desirable in urging the respondent to provide a substantive answer. In attitudinal questions it appears very common for respondents uninformed about the real issue to provide an answer based on surface meaning given to a question (e.g., Converse, 1970), but in factual questions most researchers appear to believe that it is preferable to minimize "don't know" answers.

In a variety of split sample tests in attitudinal surveys, one half-sample was given an explicit opportunity to say "don't know" to a question and the other half-sample was forced to volunteer the "don't know" answer. For example (from Schuman and Presser, 1981, p. 117):

> The Arab nations are trying to work for a real peace with Israel. Do you agree or disagree?

or

> The Arab nations are trying to work for a real peace with Israel. Do you have an opinion on that?
>
> (If yes) Do you agree or disagree?

The rather typical result of these experiments is that larger proportions of the sample answer "don't know" with the explicit filter.

Another technique is to use "quasifilters," again from Schuman and Presser (p. 121):

Do you think that **quite a few** of the people running the government are crooked, **not very many** are, or do you think **hardly any** of them are crooked?

or

Do you think that **quite a few** of the people running the government are crooked, **not very many** are, **hardly any** of them are crooked, or do you not have an opinion on that?

The phrase at the end of the second version, "or do you not have an opinion on that," constitutes the "quasifilter," and the typical result is that "don't know" responses form a higher proportion for that version than for the first. However, usually the quasifilters are less powerful producers of the "don't know" response than are the full filters.

Following the tradition of psychological measurement, with the absence of external information with which to estimate measurement error, these experiments often evaluate the quality of questions by the strength of measured relationships between variables which substantive theory asserts should be related to one another. Unfortunately, the theories are not specific enough to specify the level of relationship that is expected under a system of measures without measurement error. One possible result is based on a model of random errors in the versions without the filters. The model states that those respondents without real attitudes to report choose the "agree" or "disagree" in a stochastic fashion with equal probability. If this model were true then the measured relationship (using almost any discrete data statistic) would be lower between the target variable and another variable, related to the target under a substantive theory. This follows a similar argument as that of attenuated correlations between measures in the presence of random error. The result of these empirical examinations, however, do not identify unambiguously the form of the question with lower measurement error. Some empirical relationships appear to grow stronger under the filtered form; others are weaker. The researcher is left with three possibilities: the substantive theory is incorrect or the nature of the errors are correlated with true values of the measures or correlated across the two variables being examined.

The fact that differences are routinely observed with the "don't know" filters is often interpreted to mean that those with less well formed or "crystallized" thoughts on the issue answer "don't know." Tests using education as a weak proxy for such attributes generally fail to identify greater susceptibility to filtering effects among the lower educational groups.

In surveys asking the respondents to report past behaviors or experiences, researchers have been more aggressive in reducing the "don't know" answer through change in questionnaire structure. Many of the Cannell et al. procedures alter the question format with this end in mind. For example, if a respondent answers the following question with a quick "don't know" answer (Cannell et al., 1979, p. 361) :

> Please give the *date* of your most recent visit to a dentist, dental office or clinic, as close as you can.

a follow-up question, with negative feedback content, is given:

> Maybe if you think about it a little longer, you will be able to remember a time.

The second probe generally induces more reporting; for example, Cannell et al. report that the precision of the date was improved for the split sample portion receiving the second feedback question. The highest precision was assigned to those answers with month, day, and year provided.

Another example concerns respondents' difficulty in providing the data of an event recalled. For example, Martin (1986) introduced a series of aids to prompt the memory of a reported criminal victimization (reported to have occurred in the 6 months prior to the interview). After the respondents reported that they could not remember the date of the incident, they were asked:

> Try to figure out when it happened, for instance whether it was before or after some important event, or by using some other way of remembering the month it happened.

Then, if no memory was stimulated,

> Do you think it was in the coldest winter months—December, January, February—or in the Spring, Summer, or Fall?

Then cues concerning holidays were attempted,

> Was it before or after Thanksgiving?

if the respondent remembered the incident as occurring in the Fall. This procedure follows the theory that respondents associate experiences in a temporal way, that stimulating the recall of seasons, holidays, or other events with known time spans may prompt the recall of the date of occurrence of a specific event.

10.2.5 Question Structure in Measurement of Sensitive Topics

In Chapter 9 we discussed in detail the influence that perceived social desirability of certain responses might have on measurement error. There we learned that net underreporting often occurred for events and attributes that are viewed as "socially undesirable." A variety of questioning strategies have been attempted in order to reduce these apparent effects.

The most explicit of these techniques is the "loading" of the question, that is, the explicit assertion in the question that the behavior is exhibited by many people or an explicit assumption that the respondent *did* behave in the manner in question.[5]

Another technique, used by Bradburn and Sudman (1979) is to use open questions that permit respondents to describe the sensitive behavior in their own words. This experiment found greater reporting of sensitive behavior with open questions.

An ingenious statistical device for sensitive questions was suggested in 1965 by Warner, but has encountered practical problems in surveys. This is called the randomized response technique. The technique introduces explicit randomization at the person level so that some respondents are asked the sensitive question and some are asked another question.[6] The choice of which question is asked is left to a box of colored balls or some other method. For example, the respondent might be presented with a box of blue and yellow balls. When the box is shaken a single ball will appear in a window built into the box. The respondent is told that if a blue ball appears that the question,

Have you ever had an abortion?

---

[5] A humorous and informative article on changes in question structure to improve reporting by males of whether they have killed their wives can be found in Barton (1958).

[6] This is the so-called unrelated question version of the randomized response technique, suggested by Horvitz et al. (1967). The technique, as originally suggested by Warner, randomly asked the respondent either to agree or disagree to having the sensitive attribute or its complement (e.g., "I have had an abortion" or "I have not had an abortion").

should be answered with a "yes" or "no." If a yellow ball appears in the window, the respondent is to answer,

Were you born in the month of September?

with a "yes" or "no" answer. The interviewer is instructed to illustrate to the respondent that when the box is shaken, sometimes a yellow ball appears and sometimes a blue ball appears. The interviewer asks the respondent to shake the box of balls, observe which color appears, and answer the corresponding question. The respondent is asked to perform these steps in a way that will not permit the interviewer to see which color ball appears in the window. The interviewer then instructs the respondent to shake the box and reply with a "yes" or "no" to the appropriate question.

The technique thus obtains an answer to an unknown question from each respondent. Although information at the individual level is limited, an estimate of the overall proportion of persons having an abortion can be made, because two features of the system have been controlled: (1) the number of blue and yellow balls is known, and (2) the proportion of persons born in the month of September is known. The estimated proportion who had an abortion is

$$\pi_a = [\lambda - \pi_b(1 - P_1)] / P_1$$

where π_a = true proportion having had an abortion;

P_1 = probability of asking the abortion question;

π_b = true proportion born in the month of September;

λ = proportion of "yes" answers.

Randomized response is an attempt to reduce response bias (by freeing the respondent from fears of disclosure of embarrassing information) but increases the variance of estimated proportion. Bradburn and Sudman estimated change in measurement bias for both questions about characteristics they viewed as socially desirable and those they viewed as socially undesirable. For both types they obtained "validating" information. They used a forward record check study to examine whether persons correctly reported their having a Chicago Public Library card, being registered to vote, and having voted in a recent primary election. For example, for each person interviewed in a household sample, they accessed records in the Chicago Public Library. They use a reverse record check study design for socially undesirable behaviors. Thus, a sample of persons was chosen from police records for drunken driving charges in the recent past and a sample of persons who have declared bankruptcy was obtained from court records. For each of these topics, various measurement methods were used. Of interest to us in this section is the fact that face to face direct interviewing was compared to the randomized response technique. Following typical findings of the past and the guidance from theories concerning social desirability, Bradburn and Sudman hypothesized that measurement biases would tend to produce underreports of drunk driving arrests and bankruptcies and overreports of library card ownership, voting registration, and voting behavior.

As shown in Table 10.8 the randomized response technique seems not to reduce overreporting of desirable characteristics but does seem to reduce underreporting of socially undesirable attributes. For example, there was a net 15 percent overreporting of being registered to vote with direct questioning and 11 percent with the randomized response, a difference that did not exceed that expected by sampling error, given typical criteria. The reader should note in Table 10.8 that the variances for the randomized response technique are higher than those of the direct question, both because of sample sizes and because of the added source of variability due to the randomization of question. Although the bankruptcy measure shows great improvement from underreporting problems with the randomized response, the drunken driving measure does not show statistically significant changes. Because of the small sample sizes used, the experiment has very little power.

Does the technique work in a practical sense; that is, do respondents accept and understand the approach? Bradburn and Sudman report that only 5 percent of respondents thought it was silly. At the end of the interview, interviewers were asked to rate the respondent's understanding of the box. Interviewers believed that 80 to 90 percent of the respondents understood it, and that 80 to 90 percent accepted the explanation of why the procedure was being used.

There are other examples of using the randomized response technique for admission of criminal activity (Fox and Tracy, 1986), drug usage, and abortion experience. Some of these are imbedded in methodological studies, with split sample comparisons; others merely compare results using the procedure with those of previous research using direct questioning. Many of the studies, like Bradburn and Sudman, demonstrate reduced measurement bias, but not complete elimination of bias. There has been too little research into the respondents' understanding of the procedure itself, whether they believe that the

Table 10.8 Proportion of Wrong Responses by Method for Five Characteristics (Standard errors)

Question	Question Method	
	Direct Questioning	Randomized Response
Voter registration	.15 (.037)	.11 (.058)
Library card	.19 (.04)	.26 (.08)
Bankruptcy	−.32 (.075)	.00 (.00)
Vote primary	.39 (.055)	.48 (.10)
Drunken driving	−.47 (.09)	−.35 (.14)

Note: Positive numbers reflect the proportion of cases who reported having the attribute who did not have it; negative numbers reflect the proportion who failed to report having it, when they did have it.
Source: N.M. Bradburn et al. *Improving Interview Method and Questionnaire Design,* Jossey-Bass, San Francisco, 1979, Chapter One, Table 2.

interviewer does not know which question was answered, and whether they play the game by the prescribed rules.

Finally, there is a variant of the randomized response technique which has an unobtrusive randomizing device. The "item count" technique uses a split sample design (Miller et al., 1986). For example, imagine interest in measuring whether respondents had ever used marihuana. Two half-samples are randomly identified. Respondents in each half-sample are asked to provide a count of the number of activities from a list which they have experienced. They are to report only the total number, without identifying which on the list they experienced. The list for one half-sample might include "took a trip outside the U.S.," "didn't eat food for a full day," or "went without sleep for more than 24 hours." The other half-sample receives the same list, with the addition of the item "smoked marihuana at least once." Data from each half-sample produces a mean number of activities experienced. The difference in means between the two half-samples is taken as the estimate of the proportion of persons who have smoked marihuana at least once. Many of the statistical properties of the randomized response technique are shared by the item count technique, and choice of the nonsensitive questions affects the variance properties of the final estimate.

10.2.6 Question Structure and Recall of Past Events

In Chapter 9 we reviewed the research on the relationship between the time interval since an event occurred and the reporting of such events in survey interviews. Some research into the role of the question format in this measurement error has been conducted. One line of research attempts to reduce the respondent burden by using more frequent measures. In measurement of television audiences, both recall of watching a program has been used and "coincidental" interviews have been conducted. The latter method has interviewers call respondents during or immediately after the program is over. Another method used to reduce the reference period over which events must be recalled retrospectively is the diary.

Sudman and Ferber (1974) present the results of a consumer expenditure survey of Chicago area. The research design included a split sample experiment which varied the method of interview (diary, telephone calls daily, use of tape recorder, or choice of method), compensation (5 dollars for 2 weeks), and stated auspices of the survey (U.S. Census Bureau or the University of Illinois Survey Research Laboratory). Of interest to us in this section is the use of the various ways of measuring products bought by the household. Table 10.9 presents the ratio of number of products bought, reported by telephone, to the number bought, reported in the diaries. Most of the categories of products show higher reports using the diary method. On the other hand, the average amount spent shows somewhat smaller differences. Expenditures may be more easily remembered than number of products.[7]

Another line of research concerning recall of past events compares the measurement error properties of summary questions on a topic and a set of detailed questions. This literature is quite compatible with the perspective that measurement errors and survey costs are related. It observes that different (generally more accurate) responses are obtained by dividing global questions into constituent parts. Thus, the results of many questions are compared to the results of just one question on the same topic.

---

[7] Although the sample sizes were very small, the response rates among the different groups appear to be similar. Using full cooperation with the 2 week survey as the criterion, the diary group had a 71.3 percent response rate, the telephone a 75.4 percent rate, and the choice of method group a 73.3 percent rate.

Table 10.9 Ratio of Reports by Telephone to Those by Diary by Type of Product

	Ratio of Telephone to Diary Reports	
Type of Product	Number of Purchases	Cost
Dairy and bakery	0.69**	1.43
Meat, fish, poultry	0.63**	0.73*
Fruits and vegetables	0.70*	0.69*
Beverages	0.71*	0.95
All other food	1.06	0.96
Total food	0.78**	0.95
Meals and snacks	0.62*	0.71
Clothing, linens	0.86	0.93
All other purchases	0.72**	0.78
Total purchases	0.75**	0.82*

*Difference is statistically significant at .05 level.
**Difference is statistically significant at .01 level.
Source: S. Sudman and R. Ferber, "A Comparison of Alternative Procedures for Collecting Consumer Expenditure Data for Frequently Purchased Products," *Journal of Marketing Research*, Vol. XI, May, 1974, pp. 128-35, Table 5.

The clearest examples of this work come from factual surveys. For example, Herriot (1977) compares four different forms of income questions. Although the research used self-administered questionnaires, the results suggest mechanisms that may operate in the interviewer-administered questionnaires also. The simplest form of the question asked respondents to choose one of 16 income categories for their total personal income in the prior year. The most elaborate form (1) asked for dollar amounts of earnings (wages, salary, commissions, bonuses, or tips) from three different sources, (2) asked whether or not they received any income from each of 12 other sources, and if so, asked the dollar amount, and finally, (3) asked the person's total income. The median income for the complex version was \$14,085 and for the one question version was \$13,543, a reduction of 3.8 percent (beyond that expected by sampling variance at the 0.05 level). Mean income was reduced by 4.5 percent with the simple question. This is compatible with the hypotheses that the component questions serve to define explicitly what the word "income" means (a more linguistic interpretation) or, alternatively, the separate questions force the respondent to recall separate components of income so that the final answer is the arithmetic sum of enumerated components not an estimate of the total (a perspective emphasizing the cognitive demands of the task). In contrast to these differences for summary statistics on the full population, the percentage of families below poverty was reduced by only 1.3 percentage points (9.3 to 8.0 percent) by the short version of the question.

Another study, which used a forward record check design, offers some explanation to the differences. Borus (1970) compared a single earnings question to the sum of separate earnings for episodes of a work history. He finds that differences between the two methods are greater for those persons who had a number of different jobs during the year of the earnings report and for persons with large earnings levels. The first of these results is consistent with those of Herriot. When the task for the respondent requires the recall of many components of income or earnings and the summing of them, helping the respondent to enumerate them will yield different results than a single global question. Thus, those subpopulations that have complex earning or job histories profit most from the set of component questions.

Failure to recall events is only one source of measurement error concern; there are also problems of misdating events that are recalled. This is most obviously a problem in surveys with long reference periods. For example, the National Crime Survey uses a 6 month reference period, asking the respondent to report victimizations that occurred during that period. There is evidence from record check studies that forward telescoping occurs when respondents recall events. To counteract this problem a bounding interview is taken and a seven wave panel design is used. The bounding interview is the first interview in the seven wave panel. It asks the respondent to report all criminal victimizations occurring in the 6 month period prior to the interview. *These data are never used in estimation.* Instead, the interviewer checks victimizations mentioned on the second visit to make sure they are not reports of incidents that actually occurred before the first interview. These are incident reports that would be telescoped forward in time. Through the check with the bounding interview these duplicate reports can be eliminated. In the seven wave panel, therefore, the prior interview acts as a bound for each interview. This technique will eliminate forward telescoping problems to the extent that the events are recalled in the first interview after they occur and that the descriptions obtained by the interviewers on the two waves are sufficiently similar to discern this.

10.3 PROPERTIES OF QUESTION ORDER

Methodological research on both factual and attitudinal surveys has observed differences in responses associated with different contexts of questions. In attitudinal research, despite assertions that context effects may be the cause of frequent differences among survey results (Turner and Krauss, 1978), there are both experimental demonstrations of such effects and some examples of failure to produce context effects despite prior arguments to the contrary. In surveys on behaviors and other observable attributes of the respondents, most context effects have been surprise results of changes in question flow introduced for other reasons.

10.3.1 Context Effects on Responses

There is an unfortunate diversity of language for the different effects of question context. In attitudinal surveys, "consistency" effects[8] are taken to mean the fact that when a question is preceded by certain others the respondent tends to give answers similar to those given to the preceding ones. "Contrast," sometimes "redundancy," effects exist when there is an influence to give an answer dissimilar to those of the preceding question. The phrase "saliency" effects might be applied to both factual and attitudinal survey situations. This implies that prior questions have led the respondent to interpret the question differently and hence respond differently. In one sense, both consistency and contrast effects might be considered saliency effects with different outcomes. The final effect noted is a "fatigue" effect or a "questionnaire length" effect, which is typically a hypothesis of larger measurement error for questions at the end of a long questionnaire.[9]

There are several posthoc hypotheses to explain context effects when they occur, but genuine puzzlement when they do not occur. Consistency effects are generally seen to be the result of a deductive process on the part of the respondent, moving from the specific to a general principle, to the answer of the target question. For example, an early result replicated several times involved the questions

> Do you think the United States should let Communist newspaper reporters from other countries come in here and send back to their papers the news as they see it?

---

[8]The paragraph above generally uses the term "effects" instead of "differences in measurement error" because most of the methodological literature is focused on attitude measurement, in which the researcher avoids identifying the preferred order of questions to minimize measurement error.

[9]A contrasting hypothesis to this is that lack of experience in the respondent role or lack of uniform context to the questions at the beginning of an interview leads to measurement error.

and

> Do you think a Communist country like Russia should let
> American newspaper reporters come in and send back to
> America the news as they see it?

Both the original work (Hyman and Sheatsley, 1950) and its replications
(Schuman and Presser, 1981; Schuman et al., 1983) found that support for
giving freedom to Communist newspaper reporters increased when it was
preceded by the question on American newspaper reporters in Russia.
The "norm of even-handedness" is given as the explanatory concept that
produces this consistency effect in this case (Schuman and Ludwig, 1983).
This essentially refers to principles that, *ceteris paribus*, persons in
similar circumstances should be treated similarly.

Contrast effects appear in the past literature for two related
questions, where one is a specific example of a larger issue and the other is
a global statement of the issue. For example, as shown in Table 10.10,
Kalton et al. (1978) found that ratings of current driving practices in
general were higher after a question about driving abilities of younger
drivers. That is, in the context or in contrast to attitudes about the skills
of younger drivers, drivers as a whole were rather careful. Similar findings
apply for a general attitudinal item about abortion rights followed by
attitudes to the specific case of abortion when there is a strong chance of a
defect in the baby (Schuman et al., 1981).

Despite these examples, there are others of a similar nature in which
no context effects are exhibited (Bradburn and Mason, 1964). At this
writing there seems to be no general theory that predicts when such
effects are to be expected and when they should not be expected.
Researchers who apply theories from the cognitive sciences are currently
examining such effects, and the next few years may yield some insights.
These approaches take the perspective that early questions can serve to
evoke "schemata" or sets of interrelated memories which serve as a filter
to interpret succeeding questions. In some sense they form a part of the
logical deliberations that occur internal to the respondent prior to the
response formation. Consistency effects might result when the questions
preceding the target questions are important components to the
respondent's feeling toward the target issue. Contrast effects might result
when the preceding questions refer to components that are exceptions to
the general feeling toward the target issue.

Other saliency effects manifest themselves in factual questionnaires.
In a random half-sample of a personal interview survey on criminal
victimization, a set of attitudinal questions asking the respondent to rate

Table 10.10 Percentage Reporting Driving Standards Are Lower: "Do you think that driving standards generally are lower than they used to be, or higher than they used to be, or about the same?" and "Do you think that driving standards amongst younger drivers are ...?"

	Position of Question		Difference (Standard Error)
Question	First	Second	
All Respondents			
Younger drivers	35%	35%	0
			(2.1)
Drivers in general	34	27	7
			(2.0)
Respondent aged 45 or more			
Drivers in general	38	26	12
			(2.7)

Source: G. Kalton et al., "Experiments in Wording Opinion Questions," *The Journal of the Royal Statistical Society, Series C*, Vol. 27, No. 2, 1978, pp. 149–161, Table 6.

the seriousness of various crimes was placed before a set of questions asking the respondents to recall whether they had been victims of various crimes in the prior 12 months (Cowan et al., 1978). It was found that the number of reported victimizations was greater when the recall questions were preceded by the attitudinal questions. There have been several *post hoc* hypotheses offered in explanation of this finding. One stems from laboratory research that suggests that if affective states (e.g., fear, anger) that accompanied a past experience can be induced, the respondent may be able to recall the past event more effectively (S.M. Smith, 1979). The second hypothesis is that the attitudinal questions serve to define for the respondent, in more detail than the direct victimization questions do, what the survey means by crime, that is, what types of events are within the scope of interest of the survey measurement. Hence, when the more global victimization questions are encountered (e.g., "In the past 12 months has anyone broken into your house/apartment?"), the respondent is armed with a set of examples of "breaking in" that would not have been available prior to that point.

Both of these hypotheses were investigated in experimental studies as part of the redesign of the National Crime Survey. One experiment introduced on a random half-sample a set of attitudinal items that asked the respondents to remember recent times when they were angry or afraid for a variety of reasons. There were no large differences in number of reported victimizations. Another experiment introduced a formal definition of the scope of the crimes that were to be covered by the survey prior to entering the direct victimization questions and then altered the questions to provide examples of different types of crime as cues to stimulate the recall of victimizations. This produced increased reporting versus a more traditional questionnaire with direct global victimization questions without formal definitions of the scope of crimes covered.

10.3.2 Position of the Question in the Questionnaire

With each study survey researchers inevitably face a decision about how many questions should be asked or how long the interview should last. Any researcher is capable of suggesting a new question on a topic of interest and thereby increasing the length of the interview.

Despite the ubiquitous nature of the problem, there is little research on the measurement error effects of questions being asked within a few minutes of the beginning of an interview and questions being asked after a substantial period of time has elapsed since the start of the interview. The two contrasting hypotheses on this issue are that the rapport between the interviewer and the respondent needs some time to develop and that the interview should begin with simple questions, in order to put the respondent at ease and to demonstrate that the interview will be nonthreatening and not burdensome. Similar arguments are made concerning the need for the initial questions of the survey to instruct the respondents in their role, to teach them the nature of the question and answer process, and to communicate the standards of appropriate response. Finally, some arguments are made that, for complicated topics, the early questions can be designed to provide a context for later questions, one that will serve to evoke memories of relevant experiences to later questions.

The counterarguments to these are that respondents lose motivation to attend to their tasks as the interview proceeds. The novelty of the interaction disappears with each succeeding question, and the level of detail required forces them to expend more effort at forming answers than they had anticipated. Following this argument, responses at the end of a long interview are likely to be subject to more measurement error than those at the beginning.

There are only a few studies that have investigated these hypotheses, often through split sample experiments. Most therefore look for differences in responses and utilize some model of measurement error in interpreting the differences. Clancy and Wachsler (1971) found no effect of placement of questions at the beginning or end of a 25 minute telephone interview. Their criteria of evaluation were evidence of agreement bias and the report of socially desirable attributes. Sudman and Bradburn (1974) find no effects of placement, in a review of the survey methodology literature.

It is not clear whether the past literature has investigated the type of questions likely to be most sensitive to the effects hypothesized. It would be expected that questions that are burdensome to the respondent, that require careful judgments or difficult recall, would be most susceptible to reduce motivation at the end of the interview. These might be open questions that do not channel respondent behavior as fully as closed questions. In support of this, we note a study in self-administered questionnaires that finds reduced reporting on an open question at the end of an 18 item survey (Johnson et al., 1974).

10.4 CONCLUSIONS ABOUT MEASUREMENT ERROR RELATED TO THE QUESTIONNAIRE

This chapter began with a statement about the role of standardization in scientific measurement. It noted that survey research faced problems of interpretation on the part of the subjects studied that induced lack of standardization in questions, despite their uniform administration to all subjects. Furthermore, the questionnaire was the tool most used by those interested in reducing measurement error.

Most of the examples of research in measurement error examine the dimension of the specificity of the stimulus of the survey question. "Specificity" is a rather nonspecific word in this context and is best defined by illustration. The measurement error properties of questions appear to vary by how much cognitive activity is required by the question. Sometimes this means (as in the cases of explicit balance in attitudinal questions) that the respondent is not asked to define for himself/herself the complement set of positions to the one in the simple agree/disagree form but rather is presented the explicit complement position. Other times, instead of adding all components of one's income, sets of questions ask about each component and the researcher does the adding. Instead of asking the respondents' viewpoint on each possible circumstance under which an abortion might be sought, a global question is made. From a cognitive viewpoint, the same set of circumstances may not have been reviewed by each respondent in the global version of the question and hence different answers would result.

This view is compatible with traditional psychometrics, which notes that the stability of a measurement is enhanced with many indicators of the underlying concept. It would argue that the reliability of answers to single, global questions is below that offered by a combination of many indicators. This assumes that the expected values of all the indicators are the same (after accounting for specific error variances due to scale or other measurement properties). An elaboration would state that different domains of a concept need to be measured explicitly in order to increase stability of response. One way to approach this in the attitudinal measurement enterprise is to enumerate components of the attitude measured. This would resemble the psychometric approach, as exemplified, by measures of intelligence attempting to cover the various domains of abilities (e.g., quantitative, verbal, analytic reasoning capabilities).

Much of the split sample comparisons of questions have the implicit null hypothesis that the expected responses of the different versions of the questions are equivalent. For example, "forbid" should be the opposite of "allow" in the questions concerning attitudes about speeches against democracy. However, very different empirical results are obtained in split sample experiments which varied the wording. Little of such methodological research characterizes the differences which result as reflecting measurement error. Instead, consistent with the psychometric view, the different versions are viewed to be indicators of slightly different concepts. Our point here is that these might be viewed as different domains of a single concept. On the "forbid" and "allow" wordings, for example, various shades of meaning might be explored. Phrases such as "requiring a permit from the police," "acquiring court permission," "requiring a vote of the community," "assuring that other views will be represented," "not permitting advertisements in the media," or "whenever desired, wherever, to any audience" might explore the "allow" version of the wording. Phrases such as "requiring a bond to be posted," "requiring an oath not to attempt to act to eliminate democracy," "specifying a fine," "imprisoning," "stopping speeches by force," "creating a law," or "using current laws to stop" might explore the "forbid" side of the question. The implicit hypothesis is that answers to these kinds of questions are less subject to variation in meaning across respondents and across trials within respondents.

The argument against this proposed strategy is not unlike that in the controversy about open and closed questions. Closed questions are most useful when the response categories fully enumerate the possible answers of the respondent population. Sets of component questions preceding a global question can improve the quality of measurement when they fully represent the various domains of the concept being measured. To the extent that one or more domains are absent, we would expect differences between the global measure and the combination of component measures. This observation has great intuitive appeal for factual items (e.g., review the literature on measuring earnings), but it is also applicable, one could argue, for measures of affective states.

10.5 ESTIMATES OF MEASUREMENT ERROR FROM MULTIPLE INDICATOR MODELS

We reviewed in Chapter 7 the use of multiple measures of the same concept taken on each respondent. The variability in results across the measures was used to construct estimates of measurement error variance. This approach uses confirmatory factor analytic techniques to provide estimates of coefficients in measurement models for individual questions in a survey. The approach has gained considerable acceptance in social science data analysis, and it is often used to combine estimates of substantive relationships between variables and estimates of measurement error. The explicit inclusion of measurement error parameters in the structural models (describing the effects of substantive variables on a dependent variable) is seen as a real advance in purifying the estimators of causal effects by eliminating the confounding effects of measurement error.

This section examines the results of several efforts to estimate measurement error through this methodology. It is placed as a separate section in this chapter because, for the most part, the measurement error models are not theoretically motivated by hypotheses about causes of measurement error. This is especially true relative to the experimental data cited above. Those experimental studies generally attempt to identify cognitive and social psychological influences on measurement error and obtain empirical evidence of their effects. Much of the multiple indicator research, in contrast, specifies measurement error models (see Chapter 7, Section 7.5.2) but does not identify or test hypothesized causes of the errors. The models can therefore be useful in understanding the imprecision, lack of reliability, or extent of error in answers given by respondents but not the cognitive or social psychological processes producing the errors.

The best example of this methodology is given by Andrews (1982, 1984), who used 14 different question structures on 106 substantive topics, asked of respondents in six different surveys, using face to face interviews, telephone interviews, and self-administered questionnaires. One goal of the work was to estimate measurement errors common to a particular question structure.

Because the measurement perspective taken by Andrews (and implicit in work that resembles it) is different from some of the studies reported earlier, we review it before describing the results of the study. First, the work is explicitly interested in analysis which estimates relationships between substantive variables, typically using some linear model. It observes that if the measurement errors in an independent variable are correlated with the measurement errors in the dependent variable, estimates of coefficients in the linear model can be biased. Furthermore, the coefficients can be overestimated or underestimated, depending on the nature of the correlation of the errors. This focus by Andrews on measures of relationships is the first distinction between his perspective and others reviewed above.

Second, one source of correlation of measurement errors across variables is that they are measured with the same question structure, or "method" in the Campbell and Fiske (1959) sense. Andrews examined a large set of structures which might produce correlated errors: yes/no questions, better/worse questions, five point goodness ratings, seven point satisfaction ratings, seven point delighted/terrible scales, seven point unfolding delighted/terrible scales, 10 step ladders, 0 to 100 point scales, five point agree/disagree scales, five point "extent" scales, four point frequency ratings, nine point frequency ratings, and days per month frequency ratings. He was interested in whether respondents reacted similarly to two measures on different topics if they shared the same question structure. This similar reaction could be characterized as a "method" effect.

Third, the technique used to estimate the method effects is factor analytic in nature. Figure 10.2 presents a simple version of the analysis. This model contains four questions asked of respondents (questions AX, AY, BX, and BY). The first two are posited to be indicators of the concept A; the second two of concept B. Thus, two different substantive topics are included in the analysis. Of interest to the analyst might be the correlation between A and B, represented by the curved arrow between them, labeled i. (Ignoring that measurement error properties of the four measures would bias the estimated correlation coefficient.) There are two different question structures represented by the four questions, structure X and structure Y. Each topic (A and B) is measured with both structures. In addition, each question is subject to a random error, uncorrelated with true values and with the common method effect.

Thus, the model specifies that a response to question j ($j = 1, 2, 3, 4$ in this example) is a function of the true value of the attribute (A or B) for the respondent, a common effect of the method (X or Y), and a random error. For example, the answer to the first question, AX, for the ith person might be described as

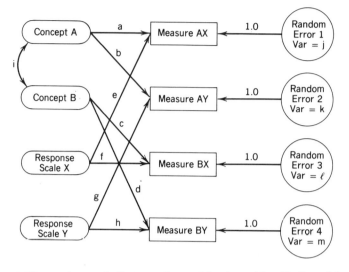

Figure 10.2 Illustrative path diagram for multitrait multimethod model used in Andrews Study of Measurement Error.

$$y_{iAX1} = \mu_A + \beta_{A1}X_{iA} + \alpha_{X1}M_{iX} + \varepsilon_{i1}.$$

It consists of an overall population mean on the attribute A (μ_A), a validity coefficient representing the effect of the true value of A for respondent (β_{A1}), a method effect coefficient (α_{X1}), and a random error (ε_{i1}). The values of μ_A, β_{A1}, and α_{X1} are viewed as shared by a population of respondents. X_{iA} and M_{iX} are common factors extracted using confirmatory factor analysis of the covariance matrix for the four survey questions.

The method coefficients can be interpreted as estimates of measurement error components when the only source of correlation between them is the shared effect of using the same question structure. In addition, the method effects are measured within the context of a specific measurement model, one that posits, for example, that AX and AY (in Figure 10.2) are indicators of the same underlying concept, and that BX and BY are indicators of another concept. The estimates of the effects therefore have the desired interpretation as components of variance only if the model is correctly specified. That is, if the component of variance shared by two indicators of the same trait (say, AX and AY) is not merely A, the target concept, but something else, the interpretation of the method effects might not be that desired. Hence, interpretation of validity coefficients and method effects must be made jointly with the evaluation of the measurement model.

As an example, Andrews presents three question structures for self–reports about eating too much:

1. Now turning to things you each and drink. Some people feel they eat too much. During the past month, how often do you feel you ate too much? Almost every day, every few days, once or twice, or not at all?

2. As you know we're trying to get the most accurate information we can so I'd like to ask you about a few things we have already talked about. These may sound like the same questions, but they give you different answers to choose from. Please tell me how often each has been true for you over the past *month*.

   .
   .
   .

   b. During the past month were there more than two days when you *ate too much*?

   .
   .
   .

3. Here are the last questions about things we asked earlier....

   b. On about how many days during the past month did you eat too much?

Under the model examined, these three questions are viewed as indicators of the same underlying trait—"eating too much"—and from their covariance matrix a common factor is extracted that represents that trait. Each of the questions is a representative of a different question structure and thus also brings with it the common effect of that structure.

Now that the error perspective has been reviewed, the results of the work can be discussed. Based on the combination of question types, substantive topics, and surveys, Andrews assembled over 2000 estimates of validity coefficients and method effect coefficients on models like that in Figure 10.2. He finds that about 66 percent of the variation in answers to the survey questions arises from variation in true score values on the traits, about 3 percent is due to method effects, and about 28 percent is due to error variance uncorrelated with method or traits. Thus, overall method effects appear to play a small role in that variation in measures not associated with the variation in true values of the traits across persons.

Using the 2000 validity and method effect estimates, an analysis was conducted to determine which characteristics of question structure were related to high or low method effect coefficients. As shown in Table 10.11, three multiple classification analysis models were constructed predicting the three coefficients. The coefficients presented in the table are β^2 coefficients, reflecting the strength of the relationship between the predictor and the dependent variable and controlling for other predictors in the model.

Interpreting the estimated effects in Table 10.11 requires some care. If a characteristic of a question is a strong predictor of the method effect coefficients, then there is evidence of influence on the respondent toward behavior that is similar across topics for the same question feature. In addition, however, there are examples for which the predictor is not strong for the method effect model but is for the residual error model. This might reflect the fact that questions of a particular type do not induce correlated errors but do induce more random error behavior, with respondents giving answers that reflect neither the underlying concept being measured nor the effects of the method. We must caution readers against uncritically accepting this interpretation, because residual error components can also rise if the indicators of the concept are themselves weak ones. That is, residual error can also rise if the measurement theory is poorly formulated (in our example, that AX and AY are indicators of the same concept, A).

As shown in the middle column, the number of response categories and the existence of an explicit "don't know" option were important predictors to the size of method effect coefficients. Unfortunately, the pattern of the effect of number of categories on the method effect coefficients is not monotonic and shows higher method effects associated with 2 and 9 to 19 categories (versus 3, 4 to 5, 7, 20 or more). This result does not fit any simple hypothesis of how method effects occur. If we examine the residual error model, however, there is clearer support that more categories are associated with less residual error.

An explicit "don't know" option leads to reduced method effects, as would be consistent with the work of Schuman and Presser (1981). On the other hand, the labeling of all categories on a response scale appears to increase method effects. Length of the introduction of a question and question length itself were combined into one predictor in the model. Questions with long introductions are associated with larger method effects. Questions in the middle of the questionnaire were found to have smaller method effects than those at the beginning or at the end.

Table 10.11 Multiple Classification Coefficients for Models Predicting Validity, Method, and Residual Error Components, Using Questionnaire Features as Predictors

Predictor	MCA Coefficients		
	Validity	Method Effect	Residual
Number of answer categories	0.56	0.68	0.74
Explicit "don't know" option	0.31	0.45	0.30
Category labeling	0.27	0.28	0.15
Explicit midpoint	0.01	0.06	0.00
Absolute versus comparative response	0.28	0.15	0.33
Length of introduction and question	0.13	0.35	0.10
Length of question set	0.17	0.19	0.44
Position of item in questionnaire	0.13	0.18	0.16
Mode of data collection	0.03	0.24	0.02
Sensitivity to social desirability	0.07	0.00	0.08
Content specificity	0.06	0.00	0.04
Experience versus prediction	0.01	0.00	0.01
Content salience	0.01	0.00	0.00
Adjusted R^2	0.66	0.72	0.67
n	2115	2115	2115

Source: F.M. Andrews, "The Construct Validity and Error Components of Survey Measures: Estimates from a Structural Modeling Approach," *Institute for Social Research Working Paper Series*, ISR Code Number 8031, May 1982, Exhibit 6.

These results give empirical support to hypotheses about the effects of question structure and order on measurement errors correlated across measures. Various properties of question structure were found to induce covariances across answers which were not properties of the concepts being measured but properties of the structures themselves. The analytic approach used is clearly in the "measurer" camp (Chapter 1). It provides estimates of correlated errors for use in adjusting estimates in structural models. It does not attempt to test hypotheses about how the various properties do affect respondent behavior. Andrews does offer advice, however, on avoiding those question forms which are associated with higher method effects or lower validity coefficients. Thus, it does yield itself to use by "reducers" also, in avoiding those features with harmful effects.

10.6 COST MODELS ASSOCIATED WITH MEASUREMENT ERROR THROUGH THE QUESTIONNAIRE

This chapter has identified several strategies for measuring or reducing measurement error associated with the questionnaire (e.g., questioning respondents concerning the meaning of questions, asking component questions as well as summary questions). Almost all of them require the construction, testing, and implementation of additional questions on each topic covered by a survey. To the extent that this approach has widespread applicability, the balancing of costs and errors becomes quite clear.

The costs attached to each question in a questionnaire flow from the conceiving and writing of the measure, the choice of position of the question in the questionnaire, the pretesting of both of those decisions, the training of interviewers on the question, and the interviewer time required to administer the question. These costs rise with each question added to the questionnaire. Thus, as we saw with other errors, the design issue could be phrased in terms of which set of questions minimizes measurement error, given fixed costs available for the survey.

There are only a few examples of past researchers taking a cost and error perspective in assessing the desirability of specific questions. For example, Matell and Jacoby (1971) investigated the measurement error properties of Likert scale items with different numbers of response categories (from 2 to 19). Although they found that measures of consistency and concurrent validity did not differ across items with different numbers of categories (Matell and Jacoby, 1971), they inquired whether testing time might vary by the number of categories. They were not directly concerned with costs of administration. They were most concerned with offsetting errors that might be associated with length of a test, the "warm-up, fatigue, and boredom" effects, but their approach can be adapted to the cost-error perspective. They find that the time required to respond to 40 such items was relatively constant over different numbers of categories. There was some evidence that items with 13 or more categories required slightly more time to answer.

It is surprising to most novices to survey research that the relationship between the number of questions asked in a survey and total survey costs is not a very strong one. Figure 10.3 is a scatterplot of the relationship between the average number of minutes that an interviews lasts (x axis) and the ratio of total minutes of interviewer time required to obtain the average interview (y axis). This figure results from over 90 centralized telephone surveys conducted by the Survey Research Center between 1978 and 1985. The figure shows that for each length of interview there were many different total number of minutes of interviewer time required. The large scatter of points for each length of interview is the result of variation in other factors that affect interviewer productivity. These include issues that were discussed earlier, like response rate targets for the survey (the higher the desired response rate, the larger the costs), the length of the survey period, the correspondence between number of interviewers hired for the project and the sample size, the population studied, the proportion of interviewers who are inexperienced, and a variety of other features.

When other features of a design are held constant, each additional question adds only a small amount to an interview. Each closed question might add about 15 seconds to an interview and open questions about 30 to 45 seconds. It is unlikely that the total amount of interviewer time required to complete each interview is a linear function of the length of the interview. When interviews become very long, two changes to the interviewing experience occur: (1) both in personal and telephone interviews, interviewers will make judgments about whether an attempt to obtain another interview at the end of an interviewing day will force them to work beyond their desired quitting time (thus, long interviews lead to shortened interviewing shifts); and (2) as the interview gets longer the proportion of cases that require second visits after initial contact increases (either because the respondent has difficulty identifying a free block of time sufficient to complete the interview or the interview is interrupted and taken in more than one session).

Linking survey cost models to error parameters associated with questions has not been formally attempted, to our knowledge. Within the perspective of a model of parallel measures (see Chapter 1), the perspective would be an extension of the work of Wilks (1962), sometimes called "matrix sampling." This approach concentrates on the relationship between the number of parallel measures chosen for administration and the reliability of an index combining the various individual measures. This flows from mental ability testing, where multiple measures are used as indicators of abilities of interest.

The simplest case is when the statistic of interest is a sum of variables obtained from an individual subject. This is the situation when tests are

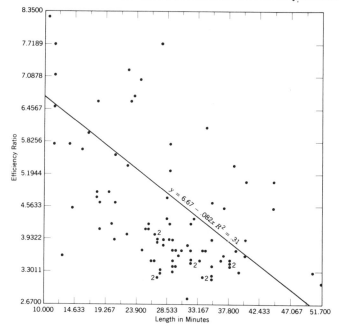

Figure 10.3 Ratio of total interviewer hours required for project to total interviewer hours spent questioning respondents by average length of interview in minutes.

administered to subjects and scored to evaluate their abilities relative to some norm. In a survey setting this is relevant to the situation in which a scale containing potentially many different items measuring the same concept is to be administered to a sample. The design decision for the researcher concerns how many questions to include in the interviews. The larger the number of questions included, the higher the reliability of the score. However, the larger the number of questions, the longer the interview and the higher the cost of the measurement.[10]

To illustrate this with a simple case, imagine that one possesses a set of M items measuring the same concept. That is, the M items are **parallel measures** (see Chapter 1, Section 1.5), with same expected value and error variance. Lord and Novick (1968) show that the reliability of a mean score of those M items is a function of the number of items used, $\overline{m}$. That is, if the mean score for the ith person over $\overline{m}$ items is defined by

---

[10] The problem could be made more complex by building an error model in which the reliability of the score increases with more questions, but other measurement errors (biases) increase with the number because of respondent fatigue.

$$\bar{y}_i = \frac{\sum_{m=1}^{\bar{m}} y_{im}}{\bar{m}},$$

the so-called Spearman–Brown reliability is

$$\frac{\bar{m}\rho_{yy'}}{1 + (\bar{m} - 1)\rho_{yy'}},$$

where $\rho_{yy'}$ is the correlation between pairs of items in the set for the population of which the ith person is a member. This expression notes that the reliability of a mean over items increases as the number of items increases. The increase in reliability, however, is dampened when the items are highly correlated (i.e., when the reliability across individual items itself is very high). This can be seen in Figure 10.4, which plots the reliability of the mean score for different number of items in the scale and different $\rho_{yy'}$.[11]

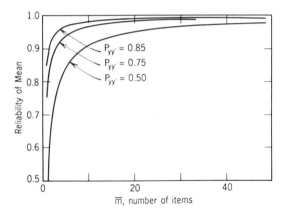

Figure 10.4 Values of Spearman–Brown reliability measure by number of items used in the measurement ($\bar{m}$) and pairwise correlation among items ($\rho_{yy'}$).

---

[11] Given the true score model, the reader will note that reliability of a measure or mean of a set of measures is related to its variance over persons. When the variance of the mean score is expressed in similar terms it is $[\sigma_x^2/\bar{m}][1 + (\bar{m} - 1)\rho_{yy'}]$. The reader should note how similar this is to the expression for the inflation of response variance associated with interviewers (see Chapter 7).

A cost model for the number of questions asked could be constructed as

$$C_i = C_{oi} + C_{mi}\overline{m} \,,$$

where C_i = total cost of interviewing the ith person;

$\quad C_{oi}$ = all costs required to interview the ith person not relevant to the scale questions;

$\quad C_{mi}$ = cost of asking one question in the scale of the ith respondent;

$\quad \overline{m}$ = number of scale questions asked.

The uniting of the cost and error model here allows us to pose the question, "What number of items will maximize the reliability of the mean score, subject to fixed costs?" The question is easily answered by the maximum number of items that can be afforded within the fixed budget.

A more complex question involving costs and measurement error properties can be posed when the problem involves a balancing of sample size and number of items. Here the conflict concerning resources and errors is that increasing the number of persons sampled will decrease the sampling variance of the sample mean and increasing the number of items asked of each person will decrease the measurement variance (i.e., increase the reliability) of the sample mean. These two features thus compete for the resources of the project. Should the designer use long questionnaires on a small sample or short item series on a large sample? This question can be addressed statistically when some portion of the questionnaire is viewed as measuring a single construct.

The most interesting perspective on this problem involves what has been termed "matrix sampling" (Wilks, 1962). That is, both a population of items and a population of persons are viewed to exist. The survey design involves decisions regarding how many people *and* how many items to sample. Thus, the image is of a matrix cross-classifying items and people. The design decision is what rows and columns to select from the matrix.[12] Note that this model is somewhat different from the model

---

[12] Obviously, a selection of rows and columns is only one selection scheme that is possible. Lord and Novick also review designs that sample different sets of items for different persons in the sample.

above because the variability across items in the questionnaire will arise because of selection from the population of items, a selection that is viewed to vary over replications of the survey.

The statistic of interest in this work is the mean over sample persons of their means over the scale items, a mean of scale means. When the scale items are dichotomous (with 0 and 1 scoring), Lord and Novick (1968) show that the variable of the proportion of items answered "1" in a sample of n persons asked $\bar{m}$ items each is

$$\text{Var}\,(\bar{y}) = \left(\frac{N(n-1)}{(N-1)n}\right) \frac{(1-\bar{m}/M)S_m^2}{\bar{m}}$$

$$+ \left(\frac{M(\bar{m}-1)}{(M-1)\bar{m}}\, \frac{(1-n/N)S_y^2}{n}\right)$$

$$+ \left(\frac{MN}{(M-1)(N-1)}\right)\frac{(1-\bar{m}/M)(1-n/N)\,Z(1-Z)}{\bar{m}n}\,,$$

where S_m^2 = variance over items in their expected values for the population of N persons;

S_y^2 = variance over person in their expected values for the population of M items;

Z = expected value on population of M items for sample of n persons.

The first term in this expression is the variance associated with sampling $\bar{m}$ items from the population of M items. The second term is the variance due to the sampling of n persons from the population of N. The third term is variance of the sample mean if sampling of items had been independently done for each person in the sample and vice versa.

A cost model that would apply to this case is

$$C = C_o + (C_{oi} + C_{mi}\bar{m})n\,,$$

where C_o = all costs that are not a function of the number of
            scale questions or number of respondents used;

C_{oi} = cost of administering all the other questions in the
           questionnaire to ith respondent;

C_{mi} = cost of administering each scale question for the ith
           respondent.

With such a combination of cost and error models one can address the
question of minimization of the sampling and response variance of the
overall mean, given fixed total resources.[13]   Figure 10.5 gives an example
of the change in variance of the mean over different combinations of scale
length and sample size. The figure plots values of the combined sampling
and measurement error variance of the overall proportion of items
answered "1" in the sample for different numbers of items asked of each
sample person. With more items used, fewer persons can be sampled
given fixed costs. The effect on the variance of the mean of using more
items itself is a function of the size of the variation across persons relative
to that across items. With larger variation across items than across
persons, longer questionnaires are desirable (e.g., $S_m^2 = 2$, $S_y^2 = 1$). Indeed,
because the cost of contacting another person (C_{oi}) is large relative to the
cost of adding one question (C_{mi}), only very large relative variance across
persons makes longer questionnaire less desirable (e.g., $S_m^2 = 1$, $S_y^2 = 100$).
    There are two features of the error and cost models that will seem
inappropriate to many survey researchers. First, on the error model side,
there is an implicit assumption that the reliability of any single measure is
not affected by the number of different measures used in a single
interview. This assumption is threatened by concerns about "warm-up"
or "fatigue" effects, that is, measurement errors that vary by the length of
the questionnaire. Second, the error model focuses totally on
measurement error variance, not measurement biases.
    On the cost model side, the models do not reflect the discontinuities in
the increases in per unit costs related to the need to take very long
interviews in two sessions or the increased difficulty to persuade
respondents. The appropriate cost model is therefore probably one with
an increasing marginal cost for each added question beyond a certain
point.

---

[13] It is useful to note that Lord and Novick label the variance expression as sampling
variance, because of sampling from both item and person populations. However, the
variation over items arises because they vary in their expected values (or **difficulties**, as
labeled by this literature).

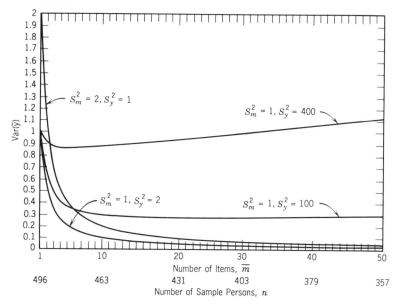

Figure 10.5 Combined sampling and measurement error variance of sample mean for different numbers of scale items, for different values of variance across items S_m^2 and variance across persons S_y^2. (Assumes $C = \$55,000$, $C_0 = \$50,000$, $C_{oi} = \$10.00$, $C_{mi} = \$0.83$, and $Z = 0.5$)

10.7 SUMMARY

Survey research stands on a foundation of language. Words and phrases are its measurement instruments. There is ample evidence that words are fragile devices, subject to multiple meanings and lack of comprehension on the part of respondents. Even common words yield problems, but many survey questionnaires are filled with vague phrases (e.g., "pretty often") that are interpreted differently by different respondents. There is observational and experimental evidence that respondents will answer survey questions despite gross lack of knowledge about words in the question. Thus, measurement errors invisible to the data analyst arise because of question wording. Survey methodologists do not yet possess the tools to predict which words or phrases will produce measurement errors, and split sample experiments in question wording sometimes show large differences because of apparently small wording changes.

The length of questions has been studied as a separate property affecting measurement error. Most results suggest that longer questions may improve response quality, by forcing the respondents to engage in deeper cognitive processing and allowing the respondents to deliver their answers in their own terms. Both reduced effects of social desirability and evidence of improved recall have been found with longer questions.

Open questions have properties distinct from closed questions. Attempts to construct closed questions to duplicate the response distribution of open questions have not been generally successful. Open questions also appear to produce higher reports of sensitive behavior.

On closed questions, when response alternatives are offered to respondents, some will choose them. This near tautology underlies the increased reporting of "don't know" when offered explicitly. It also underscores the effects of offering a middle alternative in attitudinal questions. The measurement error effects of this phenomenon are not well established in the literature.

A variety of techniques have been studied to improve the candor of respondents' answers to sensitive questions. Deliberate loading of questions to communicate how common the undesirable trait is, use of randomized response, and use of item count techniques generally show higher reports of sensitive behavior but not complete elimination of social desirability effects.

Answers to some questions appear to be affected by what questions preceded them. Cognitive schema theories are often used to describe how respondent comprehension of questions can be affected by context. Only rarely, however, do well-supported theories exist to explain context effects and offer predictions about their future occurrence.

The past two decades have seen an explosion in research on the impact of the survey question on measurement error. This research has examined individual words, question structure (e.g., balance of response alternatives, open versus closed form), and question order. The majority of this research has used split sample methods to compare results of alternatives. The work has been principally motivated by social psychological concerns with comprehension by respondents of the questions. This work appears to fall in the camp of the reducers more comfortably than that of measurers because the implicit goal is the discovery of the best question. In doing so it has tried to derive principles of question writing that transcend individual topics, although they appear to be limited in number at this point.

Contemporary with this work was the development of a set of statistical models and estimation procedures for measurement error. The multitrait multimethod model permits the investigator to measure the component of variance in an indicator of a trait associated with a particular method of measuring it. These estimates require formal measurement error models and more restrictive assumptions than the split sample methods, but in turn they provide comparative numbers of data quality for different question types. The approach requires the formulation of several questions, viewed to measure the same concept but using different question structures. Because of this burden and the lack of acceptance of some of the assumptions of the models, it has not been used as frequently as split sample methods to investigate measurement error properties of questions.

CHAPTER 11

RESPONSE EFFECTS OF THE MODE OF DATA COLLECTION

In May 1877, Bell and his partners distributed a circular assuring the reader that using the telephone was not difficult: "conversation can be easily carried on after slight practice and with occasional repetition of a word or sentence."

S.H. Aronson, "Bell's Electrical Toy..." in J. Pool, (Ed.),
The Social Impact of the Telephone, 1977, p. 25

Surveys involving interviewers are acts of communication between two people. However, such communication is unusually highly structured by the questionnaire and by guidelines for interviewer behavior prescribed by training. Some norms of behavior in personal communication (e.g., reacting to others' comments) are therefore not active in these conversations in the same way as in "normal" conversation. Other behaviors (e.g., avoidance of interrupting a speaker who has not yet completed his/her turn), however, are not specified either by interviewer training or the questionnaire to be any different from those of everyday conversations between strangers. Hence, one might look to more universal guidelines of behavior to understand parts of the communication process.

In the middle of this century (1940 through 1970) the dominant mode of survey data collection in the United States was the face to face or personal visit mode, in which an interviewer typically visited the housing unit of the sample household and, sitting nearby the respondent (across a kitchen table, on a living room sofa), asked questions of the respondent. In recent years in the United States, however, an increasing number of household surveys are conducted by telephone, based on samples of telephone numbers screened to identify the residential numbers. Often these interviews are conducted from a centralized location, where interviewers work under continuous supervision.

The telephone is a medium of communication obviously limited only to audio-based information. Because all the other sense data (touch, smell, vision) are removed from the medium, a distinctive set of norms for behavior might apply. These might well affect the error properties of the survey data. Despite this possibility, however, the procedures for respondent selection, question design, and interviewer behavior often used in telephone surveys mimic those in face to face surveys. Thus, there are important questions to ask about the impact of mode of data collection on the quality of survey data.

As with other design features that have implications for error, cost differences exist between the two modes of data collection. Telephone interviewers often are given larger workloads (because less of their time is spent in noninterviewing tasks like traveling). There is no need to coordinate large groups of interviewers dispersed geographically. Finally, the costs of telephone communication are generally smaller than the costs of travel between the sample households. Thus, the comparison of measurement errors by mode should acknowledge cost differences between the modes.

This chapter first reviews some experimental studies of differences between telephone and face to face communication. We then examine some formal studies of effects of mode of data collection on survey estimates. In that discussion we will address several hypotheses about joint effects of mode and other survey design features on measurement error. Finally, the chapter ends with examples of cost and error models for telephone and personal interviews.

11.1 TWO VERY DIFFERENT QUESTIONS ABOUT MODE OF DATA COLLECTION

The practical difficulties of studying the effect of mode of data collection on survey statistics can be highlighted by contrasting two questions:

1. What is the marginal effect of mode of data collection on the survey statistics (assuming all other attributes of the design are the same as a face to face survey)?

2. Would a telephone survey get the same results as a face to face survey (given all the differences in the way we conduct surveys in the two modes)?

The first question is one of primary interest to students of communication. It is motivated by interests in the process of verbal interaction of persons, focusing on distinctive features of the audio-only channel of communication. The question is answered by attempting to identify inherent properties of the audio channel which might produce differences between face to face interaction and communication through an audio channel only. The survey design decision of choosing a mode of data collection is just one manifestation of this phenomenon.

The second question is a practical survey problem. In contrast to the first, it describes a problem faced by a researcher choosing a mode of data collection for a particular topic on a given population. In addition, however, along with mode the researcher must choose a specific set of measures, callback rules, refusal conversion rules, respondent rules, and interviewer selection and training procedures. Not all these rules and procedures are equally well suited to the two modes of data collection. Different procedures tend to be chosen for one mode than for another. A telephone survey is one bundle of methodologies; a face to face survey is another bundle. The bundles *may* produce different results because the medium of communication is different. It is also possible, however, that differences will arise because of differences in other design factors, many of which we have shown in other chapters can affect coverage error, nonresponse error, sampling error, or measurement error.[1]

This difficulty of measuring "mode effects" is most clearly illustrated by describing the alternative methodologies used by survey researchers to study mode effects. One method is a panel study design in which the first wave of interviews of an area probability sample is conducted face to face and the second wave is conducted by the same interviewers using their home telephones. The sample for the second wave consists of respondents to the first wave who have telephones in their homes. Nonrespondents in the first wave and respondents without telephones are eliminated. Consider comparisons of analytic results between the first wave and the second wave. In addition to the medium of communication causing differences in results, there are four other possible causes:

1. The coverage error properties of the second wave sample are different because the nontelephone group differs from the telephone group on the survey statistics.

---

[1] The careful reader might observe that the problem of measuring the marginal effect of mode of data collection should not be any more difficult than measuring the marginal effect of other design attributes (e.g., interviewer behavior effects, incentive effects on respondent cooperation). The difference in the case of mode of data collection is that a telephone survey design that was identical in all respects with a face to face survey design except in medium of communication would have a very unusual combination of features (e.g., area sampling but telephone interviewing).

2. The nonresponse error properties of the two waves' estimates are different (n.b., the second wave's response rate is the product of the first wave's and the second wave's).

3. The experience of the first interview may have altered the response behavior in the second wave (i.e., a measurement error related to repeated interviews).

4. The first interview may have led respondents to change their actual behaviors related to the survey measures.

Sometimes this design is also used to provide a mode comparison only from the second wave data. That is, among respondents to the first wave who have volunteered their telephone numbers, one half-sample is assigned to personal interviewing, while the other half-sample is assigned to telephone interviewing. Mode effects are estimated by comparing the two half-samples from the second wave only. This is the design used by Rogers (1976) in measuring attitudes about city services. Here the two statistics being compared are identical in terms of coverage and wave-one nonresponse error, but the target population (all those who would consent to the first personal interview and volunteer their telephone numbers) has no simple relationship with the total household population (or even the total telephone household population).

Another design is sometimes labeled a "maximum telephone/ maximum personal" comparison. It too can use a panel design or it can be based on an area frame sample, supplemented with tracking of telephone numbers. We discuss the panel version of the design. In one wave of the panel based on an address sample (or area probability sample) efforts are made to maximize the proportion of interviews taken face to face. In the following wave, efforts are made to maximize the proportion of interviews taken by telephone. The typical result in the United States is that the maximum personal wave will have about 80 percent of the interviews taken face to face, and the maximum telephone wave will have about 80 percent taken by telephone. This is the design used by Woltman et al. (1980) in estimating method effects on reporting of criminal victimization and by Hochstim (1967) concerning health behavior data. In addition to the medium of communication causing differences in results, there are other factors that may be the causes:

1. The statistics from the two waves are merely different mixes of the two modes.

2. The nonresponse error properties of the two waves' estimates are different (n.b., the second wave's response rate is the product of the first wave's and the second wave's).

3. The experience of the first wave's interview may have altered the response behavior in the second wave (a measurement error related to repeated interviews).

4. The two wave's results could differ because real behaviors had changed between waves.

The three final designs are split sample approaches. They differ on which sampling frames are used for the two modes:

1. Both modes use an address frame or area frame, and the researcher seeks telephone numbers for the subsample assigned to the telephone mode.

2. The face to face interviews are taken on a sample from an area frame and the telephone interviews are taken from an independent sample from telephone directories.

3. The face to face interviews are taken on a sample from an area frame and the telephone interviews are taken from an independent sample from the full frame of telephone numbers (using random digit dialing methods). [This is the design used by the Groves and Kahn (1979) and Cannell et al. (1987) studies.]

These three split sample designs differ on what proportion of the telephone households are eligible for assignment to the telephone treatment. From the survey researcher's perspective they compare a face to face survey to three possible types of telephone survey (ones using frames with different coverage properties for the telephone household population). For one interested in the marginal effect of mode, they require analytic controls on differences between the population assigned the face to face mode and that assigned the telephone mode. Different conclusions about the effect of mode could result because of uncontrolled real differences between the populations interviewed in the two modes.

These last examples come closer to a simpler statement of the practical problem of measuring differences between telephone and face to face surveys. The differences can be measured only if specific definitions are supplied for "telephone survey" and "face to face survey." Figure 11.1

Design Feature	Important Question
Sampling frame	Same frame or frames with equivalent coverage of the telephone population used for the two modes?
Interviewer corps	Same interviewers used in both modes? If not, same hiring and training procedures? Same experience?
Supervision	Is telephone interviewing centralized? Is supervisory contact equivalent in the two modes?
Respondent rule	Are the same respondent selection procedures used in the two modes?
Questionnaire	Are identical questionnaires used? How are questions using visual aids in the personal interview handled on the telephone?
Callback rules	Are the same callback rules used? Are they equivalently enforced?
Refusal conversion	Are equivalent efforts made in both modes?
Computer assistance	Is CATI used in the telephone mode?

Figure 11.1 Design Features and Important Questions in Mode Comparison Studies.

summarizes some key attributes of the definitions that should be provided. These are survey design attributes already discussed as having impacts on survey errors. To the extent that the telephone interviews were collected with different procedures than the personal interviews, the differences among them will be reflections of the combined effects of all those differences. Mode may not be the only factor.

From the perspective of the survey researcher, the practical question of whether a telephone survey would yield different conclusions than a face to face survey is probably more important than the question about the marginal effect of the medium of communication on survey statistics. The burden on investigators comparing the two modes of data collection, however, is to compare a personal interview and a telephone interview survey which have "typical" features of surveys in the two modes. The experimental treatments should be realistic.

11.2 PROPERTIES OF MEDIA OF COMMUNICATION[2]

Considerable experimental research has been conducted on the differences between mediated and face to face interaction (Williams, 1977). While this research has sometimes identified media differences, no comprehensive theoretical framework has been introduced to explain them. There are, however, two main explanations that have been advanced to account for the differences between audio and face to face modes of communication, which are relevant to the comparison of telephone and personal interviews. First, audio and face to face communication differ in their "channel capacity," with the latter being capable of carrying more types of messages (e.g., nonverbal cues). Second, differences exist in the "intimacy" of the social interaction, which create an impersonal, distant quality for conversations that are mediated.

Investigations of nonverbal communication suggest that the visual channel is important in conveying affect and permitting evaluation of others (Ekman, 1965; Mehrabian, 1968). The nonverbal messages that are transmitted in face to face settings are those efficiently communicated by facial expressions or gestures while not intruding on the speech of others (e.g., an expression of puzzlement), those that social norms suggest should not be verbalized (e.g., boredom), and those that are unintentional (e.g., head nods). Other visual cues are socially defined messages (e.g., social class of interviewer inferred from style of dress). The complexity of the information communicated by nonverbal cues is clearly limited. Language offers a much richer tool. Furthermore, nonverbal cues vary across subcultures. Hence, it is typical for interviewers to be trained to control their facial expressions, head nodding, and laughter during the interview. Some of the interaction coding research (reviewed in Chapter 8) has shown that such behavior still exists, despite prohibitions.

Although the face to face mode is clearly richer in the nature of communication handled, it is not obvious that (1) the richness enhances respondent attention to the survey task or (2) any content handled nonverbally in the face to face situation is necessarily lost in audio–only communication (this is because many of the nonverbal cues in face to face meetings are combined with verbal messages). Affective impacts of limiting communication to the audio channel may be more important.

In many laboratory studies of group communication there appeared to be few effects on the transmission of factual material from eliminating the video channel. However, for activities concerned with interpersonal morale, there were differences (Champness, 1972). A series of studies

---

[2] This section draws on Groves et al., *Telephone Survey Methodology: A Review* (1980).

conducted by the British Post Office (Davies, 1971; Short, 1971, 1973; Williams, 1974; Young, 1974) compared communication using picture telephones (with audio and video communication) with traditional telephones. The transmission of factual information and cooperative problem-solving were relatively unaffected by the modality in which they occurred. There was some evidence, however, that the availability of visual cues affected impression formation and person perception, although the effect was not very strong and there were no consistent patterns of differences.

There *was* a consistent tendency for subjects to be less confident of their judgments in the no-vision condition of the telephone and to express a preference for the face to face condition. This included judgments by a speaker that listeners understood less of their comments in the audio-only condition than in the face to face condition (Young, 1975). Perhaps the face to face condition was preferred because of the greater social contact it afforded and the relative ease of judging intents and trustworthiness of another. We shall see later that similar comments are elicited from survey respondents in the face to face and telephone modes.

There also seems to be a tendency for visibility to heighten the effects of social status, with solutions offered by high-status members of problem-solving groups more likely to be adopted in face to face than in telephone conditions. This might apply to the survey situation through increased interviewer effects in face to face situations (a finding discussed in Chapter 8). It also suggests the need to identify those attributes of voice which communicate social group membership (e.g., pronunciation, intonation).

There is a tendency for those tasks which require rapid feedback about another person's reactions to exhibit differences across modes (Short et al., 1976). This may be related to the visual cues which speakers provide when they are preparing to yield the floor for the next speaker. The reliance totally on associated audio cues for this "turn-taking" warning may produce inefficiencies in communication.

Most experimental researchers also agree that the telephone is more "socially distant" in some sense, compared to face to face interactions, although they use different terms to describe this quality. Wiener and Mehrabian (1968) considered the media as differing in "immediacy." Short et al. (1976) used "social presence," and Morley and Stephenson (1969) used "formality." The fact that the telephone is less "personal" than the face to face condition may not necessarily be a drawback for the interviewing situation. Mehrabian (1968) suggests that the choice of media in regard to intimacy should be determined by the nature of the task to be carried out; if it is a relatively unpleasant task, too much intimacy should be avoided.

From studies of tape recorded speech there do appear to be differences in natural conversations occurring on the telephone versus those occurring face to face. One of the differences is the frequency of repeating phrases in telephone conversations versus face to face meetings (Wilson and Williams, 1975). This may be due to the absence of visual cues reporting understanding of utterances delivered. Alternatively, it may arise because of a felt need to avoid moments of silence on the telephone, to fill them with redundant information until the other party orally acknowledges receipt of the information. If the medium needs more redundant transmissions to be effective, then telephone surveys should have different question and probing structures. If the speakers are avoiding silence to assure their partner of their attention to the interaction, then relatively fewer implications arise for telephone survey. In any case, the social meaning attributed to silence appears to be somewhat different in the two media.

11.3 APPLYING THEORIES OF MEDIATED COMMUNICATION TO SURVEY RESEARCH

The discussion above yields some hypotheses about differences in telephone and face to face survey results. Appealing to the experimental literature as the sole source of theoretical motivation for survey mode effects, however, may be dangerous. Most of its empirical tests of the theoretical statements are based on heavily controlled laboratory experiments with tasks that are presented as shared problems for the different subjects. For example, it is clear that the number of different types of messages which can be transmitted to respondents is curtailed in telephone interviews. The variety of measurement procedures used in telephone surveys is thus limited. It is not clear, however, what compensating design features might exist for telephone surveys. Visual measurement methods are incorporated into survey practice only in two ways: some questions employ visual aids (e.g., response cards) and some surveys ask the interviewer to make visual observations (e.g., race of respondent, characteristics of the housing unit). These cannot be incorporated into a telephone survey. Furthermore, some of the other sources of measurement error are believed to act partially through visual stimuli (e.g., race of interviewer effects). These might be diminished in a telephone survey. Each of these, however, might be transferred to some type of oral presentation. What effects this transfer might have are appropriate questions for this chapter.

Some of the experimental findings above suggest difficulties with telephone surveys. The heightened "intimacy" of face to face interactions may be of great assistance to the interviewer at the initial moments of contact with the sample person. These moments are used by the interviewer to judge levels of understanding of the survey request, to tailor his/her persuasive comments to the perceived interests and concerns of the sample person, and to develop rapport with the person. Personal contact may lend more legitimacy to the interview, as evidenced by the typically higher response rates for face to face surveys (e.g., see Bercini and Massey, 1979; Steeh, 1979). Since the telephone is used mostly for business and friendship contacts in the United States, interviews may seem inappropriate, or at least unusual, for that medium. (Interviews are not common occurrences in face to face interaction either, but the interviewer may be able to make a better case for the contact in person).

On the other hand, all the other nonverbal cues (e.g., head nodding, smiling, laughing) on the part of the interviewer are generally targets of control by training guidelines. For example, interviewers are instructed not to show reactions of approval, disapproval, or surprise to answers provided by respondents. Most methodologists acknowledge that not all interviewers follow those guidelines, and some interviewer variance in statistics results from those violations. There are some questions asked in surveys which might be especially susceptible to such nonverbal cues— so-called sensitive or threatening questions (e.g., about drug usage, sexual practices). In response to questions that present a degree of threat to respondents, telephone survey respondents may be less likely to give a socially desirable answer because of the lack of personal contact with the interviewer.

The norms of behavior used in the face to face interview may be those developed for interactions with strangers in one's home, one's appropriate for guests at social occasions. For example, face to face interviewers are sometimes offered refreshments prior to beginning the interview. In terms of cognitive script theory (Chapter 9), the respondent might be employing the "visit from a guest" script versus, say, the script appropriate to a visit from a repair person or delivery person. The role of host carries with it some obligation to attend to the needs of the guest, to listen to them, respond to their inquiries, and to maintain a pleasant interchange.

Several properties of the telephone communication can be contrasted with those of the in-home conversation. The receiver of a telephone call does not formally invite the caller into his home and hence does not formally take on the role of host. Telephone calls received at home from strangers are typically for business purposes. They tend to be short interactions. The conversation is not elaborated but concentrates on the exchange of information needed for the task in question. Telephone respondents may be employing their "call of solicitation" script. The rules governing behavior in this script probably require much more immediate personal gain on the part of the respondent to justify the continuation of the interaction.

Furthermore, the telephone is an exclusive link to the household. Opportunity costs of a telephone conversation may be high in certain circumstances. While one telephone interaction is ongoing, no others can be initiated. This means that when a telephone conversation with an interviewer is taking place, calls from friends, relatives, and others cannot be obtained.[3] (It is likely that this pressure is differentially felt by those with dense social networks.) This feature is not present in face to face interviews, where other visitors may call on the sample household at any time. Both the exclusion of other possible interactions and the traditional use of the telephone for business calls probably affect the respondent behavior on the telephone.

11.4 FINDINGS FROM THE TELEPHONE SURVEY LITERATURE

In this section we examine the empirical evidence for direct effects on the respondent behavior of the mode of data collection. By direct effects we mean properties of the mode of data collection that appear to affect respondent behavior regardless of question form, interviewer behavior, and personal attributes of the respondent. There are four aspects of data quality that have been investigated by several comparisons of face to face and telephone interviewing:

1. Expressions of enjoyment of the interview.

2. Pace of the interview.

These are not direct indicators of data quality, but other factors come closer:

3. Detail of answers on open questions.

4. Item missing data rates.

---

[3] This ignores the existence of the facility for "call waiting" now available in most parts of the United States. "Call waiting" facilities alert someone using the telephone that there is another incoming call and permits the person to disengage the current conversation temporarily to discover who is calling. At this writing, only a minority of U.S. households purchase this facility.

First, we examine evidence of reduced feelings of enjoyment reported in telephone communication. A National Academy of Sciences project (1979) on privacy and confidentiality in survey research asked respondents about which mode of data collection they preferred (see Table 11.1). Over half of the respondents in this national face to face survey favored the face to face procedure. This compares to 30 percent preferring mail and only 7 percent preferring the telephone interview. Groves and Kahn (1979) used the same question in a split sample study of mode effects. In that study respondents in the face to face mode expressed even more preference for that mode (78 percent preferred that mode) relative to mail or telephone contact. When the same question was asked of telephone survey respondents, however, very different results apply. Only 23 percent favored the face to face mode, while 39 percent favored the telephone mode.

This comparison demonstrates that the mode used to ask the question affects which mode is reported as preferred. The nature of the effect is to increase reported preference for the mode being experienced. It is also possible that such effects are magnified in the face to face interview because of the physical presence of the interviewer. That is, norms of politeness encourage reports of preference for face to face contact when the questioner is physically present.

Reports about feelings of uneasiness while discussing certain topics, however, support the notion of greater enjoyment of the face to face mode. In itself the finding of greater enjoyment or preference for the face to face interview may not imply lower measurement error in that mode. The reasons given by respondents for preferring the face to face mode were that it is a "more personal" experience and the respondent can "give better answers," using the assistance of the interviewer (not clearly a feature that is helpful to survey measurement error). On the other hand, the reported reasons for the preference of the telephone mode are that (1) it is quicker and easier to do and (2) it does not require a visit to the respondent's home. These suggest that persons with low motivation to attend to the respondent role may prefer the telephone interaction. These too are reasons that may not suggest favorable measurement error properties.

Another difference between modes of interviews is the pace of telephone interviews. Three experimental studies compare the number of minutes required to complete a series of questions in face to face and telephone interviews. Groves (1978) reports that an identical section of questions required 13.7 minutes of interviewing time on the average in person and 11.8 minutes on the telephone. Kormendi (1988) in an experimental study in Denmark, reported that a 36.5 minute section in person required only 32.1 minutes on the telephone. Sykes and Collins

Table 11.1 Reported Preferences for Mode of Interviewing by Mode of Interview in Which Question Was Asked

	Survey and Mode of Questioning		
Preferred	NAS (face to face)	SRC (face to face)	SRC (telephone)
Face to face interview	51%	78%	23%
Telephone	7	2	39
Mail	30	17	28
Other	12	3	10
Total	100%	100%	100%
Not ascertained	14	14	69
Number of interviews	1187	1437	1696

Source: National Academy of Sciences data from National Research Council, Panel on Privacy and Confidentiality as Factors in Survey Response, *Privacy and Confidentiality as Factors in Survey Response*, National Academy of Sciences, Washington, 1979, Table 32. SRC data from R. Groves and R. Kahn, *Surveys By Telephone: A National Comparison with Personal Interviews*, Orlando, Academic Press, 1979, Table 4.2. Face to face interview data from telephone households only.

(1988) found that 41 percent of face to face interviews lasted more than 30 minutes, but less than one-third of the identical telephone interviews required that much time. It appears that without any counteracting training guidelines, telephone interviews will tend to proceed more rapidly. This may be related either to norms of telephone communication that prohibit long periods of silence in the medium or to the reluctance of the respondent to engage in a long interaction with a stranger on the telephone.

Are there measurement error effects of this quickened pace? Telephone respondents, particularly well-educated ones, appeared to give less detailed answers to open questions than did the face to face respondents in the Groves and Kahn and the Kormendi work. This tendency to "truncate" the responses may also be attributable to the greater sense of speed in the telephone interaction, or a desire to get the interview over with, or both.

Jordan et al. (1980) compared responses to health attitude and behavior questions in telephone and face to face interviews and found that the telephone respondents showed more acquiescence and extremeness response biases on the attitude items. The acquiescence and extremeness responses support the interpretation of the Groves and Kahn research that the telephone interview produces faster and less considered responses. To the extent that a question requires deep cognitive processing, the quickened pace of telephone interviewing may induce superficial responses. These studies do not address, however, whether such effects can be avoided by altering the pace of the interview.

Other experimental studies have compared the rate of missing data in the telephone and face to face modes. Jordan et al. (1980) found larger numbers of "don't know" and "no answer" responses to attitude questions among telephone respondents. Kormendi (1988) reports no differences in item missing data rates (except that on income, reported in Section 11.5.2). Groves and Kahn reported higher missing data rates in early telephone surveys, but gradually decreasing rates as the organization gained experience with the mode. Jordan et al. (1980) had the same experience. Sykes and Hoinville (1985) report no differences except on items viewed to be susceptible to social desirability effects (see Section 11.5.1).

Despite the above evidence, the typical finding from most studies comparing face to face and telephone interviewing is that there are few differences in statistics estimated in one mode versus another. This means that the net effects of all influences tend to be small on the average. That is, on estimates of population means and proportions, similar conclusions will be drawn from telephone and face to face interview surveys.

For example, Groves and Kahn (1979) in the United States estimated over 200 means on the total sample and found only a small number exhibiting differences between the modes. Sykes and Hoinville (1985) in Great Britain estimated 95 means on the total sample and found that only 9 of them exhibited mode differences larger than those expected by chance (at the .05 level). Kormendi (1988) in a study of Danish Postal and Telegraph employees examined 291 means and found only 11 exhibiting "statistically significant deviations." (p. 130). There seems to be little evidence that major differences in responses are obtained for most survey questions.

On the other hand, mode differences are often confounded with other differences in the survey administration. Cannell et al. (1987), in a study of mode effects in health surveys, compare data from U.S. Census Bureau face to face interviewing on the National Health Interview Survey with that from Survey Research Center centralized telephone interviewing on a reduced National Health Interview Survey questionnaire. On several indicators of health behavior, experiences of health conditions, and usage of health care services, there was a tendency for the telephone respondents to report more events than the face to face interview respondents. Studies by Madow (1967) and others, using reverse record check designs, have demonstrated a tendency for underreporting of these events in face to face interview surveys. Thus, one interpretation of the Cannell et al. results is that the telephone respondents were reporting more accurately.

An alternative interpretation is not favorable to the telephone mode—that higher reports of health events within the reference period arose from forward telescoping of events by the respondent. Table 11.2 from the Cannell study provides some indication of the magnitude of differences between modes for various estimates. Of the 13 statistics, chosen because of their importance to the National Health Interview Survey purposes, seven show higher reporting on the telephone to a degree beyond that expected by sampling error alone (a .05 percent level was chosen as the criterion). Twelve of the 13 statistics exhibit greater reporting in the telephone mode than in the face to face mode. Note that the larger differences appear on statistics which measure the percentage of the persons with one or more health events, not on the statistics measuring the mean number of such events. This results from most of the change in reporting between the modes arising from more people mentioning one health event (instead of none) in the telephone mode. The means are affected more by the persons who report many health events than just one.

This study is an exception to a large set of studies in mode comparisons. It seems likely that the cause of the differences between the two modes is not the medium of communication itself but rather the interviewer behaviors used in the two surveys. The SRC interviewers used questioning techniques like those described in Chapter 8, controlled instructions and feedback to respondents and procedures designed to communicate the importance of recalling events accurately. No such interviewing procedure was performed in the face to face survey by the Census Bureau interviewers. Hence, the interviewer, not the medium, may be the real cause of the apparent mode differences.

However, there is another reason to qualify the typical result of "no mode effects." Despite the fact that overall comparisons between modes of data collection do not reveal large differences, there may be subgroups of the population that behave differently in the two modes. Most large-scale experiments have compared major demographic subgroups in order to address this hypothesis. One popular hypothesis is based on differential cognitive demands of the two media of communication. It hypothesizes that the absence of visual cues increases the burden of the respondent to process the information in the question, to make correct

Table 11.2 Results from Cannell et al. Mode Comparison Study on Health Surveys

Statistic	Mode of Data Collection	
	Telephone (SRC)	Face to Face (Census Bureau)
Two week recall (% with at least one)		
Bed days	10.9%	7.7%
Cut down days	12.0[a]	7.1
Work loss	9.4[a]	4.5
Dental visits	7.4[a]	5.3
Doctor visits	17.8[a]	13.6
One year recall (% with at least one)		
Limitation of activity	26.5[a]	18.7
Doctor visits	73.4	73.5
Hospitalizations	12.6[a]	13.3
Rate per 100 Persons per quarter		
Bed days	223.0	208.0
Work loss days	228.8	111.5
Dental visits	59.2	40.9
Doctor visits	167.7	124.8
Acute conditions	132.0	68.3

[a]Statistically significant difference between telephone and face to face interviews at the 0.05 level.

Source: C. Cannell et al., *An Experimental Comparison of Telephone and Personal Health Surveys*, National Center for Health Statistics, Series 2, 1987, Table G.

judgments about the meaning of the question, and thus to provide answers that correctly reflect his/her own attributes. There are rarely good measures of these differential cognitive abilities, and researchers tend to use education as a convenient proxy variable. Groves and Kahn (1979) find no interpretable differences in mode effects across education groups. Cannell et al. (1987) with much larger samples find such variation across education groups, but no pattern of low education groups exhibiting larger mode effects.

A second frequently encountered hypothesis is that elderly persons will exhibit more mode effects than younger persons. Some researchers believe that this may stem from physiological impairments among the elderly, exacerbated by the telephone medium. Loss of hearing acuity may lead to comprehension failure and then to response error. Groves and Kahn, however, found no important differences among age groups; Cannell et al. found a set of uninterpretable age differences in mode effects.

Past research has examined other demographic variables to test for differential sensitivity to mode, but little consistent evidence has been assembled on such effects. Most of the analyses resemble those reported above. The researcher chooses a set of questions broadly representative of the questionnaire and searches for a general tendency across all questions for mode effects to be greater in one group than another. What is clearly absent in this approach is a guiding social psychological theory that would posit that some questions would be more subject to the effects than others. At this point in the literature it is safe to say that the general results over all measures is one of very small differences between modes. This similarity exists despite the response rate differences between mode, the different supervisory structure in the two modes, and some changes in questionnaire formats.

11.5 INTERACTION EFFECTS OF MODE AND OTHER DESIGN CHARACTERISTICS ON RESPONSES

In Chapters 8 through 10 we reviewed a variety of influences on survey measurement error. Some of these have been hypothesized to interact with mode of data collection in producing errors. That is, in one mode these factors appear to be more important than in another mode. These include social desirability effects, the tendency for respondents to describe themselves in favorable terms; the influence of the interviewer on respondent behavior; response error from large cognitive demands of the survey questions; and the impact of the question format on respondent behavior. This section discusses evidence concerning these interactions.

11.5.1 Social Desirability Effects on the Telephone

Many social psychological investigations of measurement error focus on the physical presence of the interviewer. Response bias is often seen to arise from societal norms of self-presentation, pressures to describe oneself in favorable terms to others. Following Hyman (1954), researchers initially treated responding in a telephone interview as analogous to responding to questions in a self-administered questionnaire, since both modalities eliminated the interviewer's physical presence. One should note, however, that the self-administered modality also eliminates the interviewer's delivery of the question and knowledge of the response, unlike the telephone mode, so the analogy is somewhat flawed. In any case, researchers such as Colombotos (1965) anticipated more social desirability bias in responses to questions asked in face to face interviews than in telephone interviews. This is similar to Argyle and Dean (1965), who argued that proximity and eye contact may be detrimental to discussions of intimate subjects.

Hochstim (1967) describes a "maximum telephone/maximum personal" comparison of face to face and telephone interviews, based on an area probability sample of households in Alameda County, California. Telephone numbers were obtained from sample households at the time of an initial visit by an interviewer to enumerate the household. Following that, subsamples were designated for the telephone or face to face treatment (and, in addition, a mail treatment). Hochstim found in three questions on drinking behavior that women were somewhat more likely to say that they never drank wine, beer, or whiskey in face to face interviews than in telephone interviews or in mail questionnaires. Other questions showed no difference. An interpretation of lower effects of social desirability on the telephone could be given to this result.

Other studies found no support for the general hypothesis. Colombotos (1965) in two studies of physician's ideologies and attitudes did not find consistent differences in the expected direction of social desirability or prestige enhancement in the responses given in mail, telephone, and face to face interview comparisons. Kormendi (1988) finds no mode differences in questions concerning marital relations or other sensitive topics. There was evidence, however, of higher refusals to respond to income questions (7 to 25 percent in person versus 13 to 33 percent on the telephone). There was some greater reporting of use of alcoholic beverages on the telephone but this was not beyond levels expected from sampling error. Sykes and Hoinville (1985) also failed to find large differences between mode in responses to sensitive items. However, the *direction* of the difference on items concerning self-reports of being stopped by the police, of racial prejudice, and of attitudes toward sexual practices all support the hypothesis of reduced effects of social desirability in telephone interviews. Finally, Bradburn and Sudman failed to find the hypothesized pattern of less social desirability bias in telephone interviews (Bradburn et al., 1979).

In the Groves and Kahn (1979) data there seemed to be a tendency for telephone respondents to express more optimism about the state of the economy and about the quality of their lives. At the same time, the responses from the telephone sample contained more missing data on income, and telephone respondents reported being more "uneasy" about discussing a variety of matters in the interview than did respondents in the face to face sample. Unlike the other research these findings tend to favor the face to face interview. If one argues that it is more socially desirable to express optimism about the state of the economy and one's life circumstances than to give negative opinions, the telephone survey may be seen as producing *more*, rather than fewer, socially acceptable answers.

One empirical result consistent with the hypothesis is that of Schuman et al. (1985), which is based on variation in social desirability effects across subgroups of the population. They examine data from white respondents on their self-reported attitudes toward blacks. They find that whites in the South exhibit greater mode differences than whites in other areas of the country. They interpret the finding by describing the likely field organization of national face to face interview surveys. In such designs interviewers are hired from among residents of the sample areas throughout the country. Their job is to interview sample people who live in the same counties as they do. In a loose sense, they interview their neighbors. Hence, the interviewer and respondent tend to share accents in speaking, knowledge of the area, and other traits that may or may not affect behavior in the interview. In contrast, the telephone interview data were collected from a centralized interviewing facility in Ann Arbor, Michigan. Telephone interviewers who live in the Ann Arbor area conduct the interviews.

Schuman et al. find that white Southerners express more conservative racial attitudes in face to face interviews than in the telephone interviews.[4] They infer that the respondents are providing more accurate answers in the face to face mode. This follows the reasoning that the white respondents tend to perceive that the attitudes held by the white Southern face to face interviewer are less liberal than those of the Northern telephone interviewer. Hence, the respondent feels "more free" to reveal his/her more conservative attitudes.

A more balanced interpretation of the finding might be that both modes are subject to bias, but the errors are in opposite directions. Under this interpretation, the two modes create two different social settings. The face to face interview is a set of questions asked by a local white resident. Conformity with the perceived attitudes of the interviewer may imply more conservative attitudes (regardless of the respondent's real beliefs). The telephone interview is a set of questions asked by a staff member of a Northern university. Conformity implies expression of *more* liberal attitudes. Both modes have their effects but in opposite directions. The resolution of these discrepancies lies in further analysis of the correlates of the expressed racial attitudes. That is, examination of the "construct validity" of the attitudinal reports might be helpful (see Chapters 1 and 7). This requires a theory of what personal attributes, not

---

[4] They find this result for three of four racial attitude items.

affected by the same response errors, might be related to the racial attitudes.

A meta-analysis of the mode effect literature by deLeeuw and van der Zouwen (1988) found a small tendency for higher social desirability effects in telephone versus face to face surveys. However, their analysis omits some of the more recent studies (Kormendi, 1988; Sykes and Collins, 1987, 1988), which found no differences in general between the modes on sensitive items. There may be counterbalancing influences on this kind of response error comparisons. Telephone interaction may be subject to greater social distance or less "intimacy." This may more easily permit frank discussion on sensitive topics. This favors the telephone. At the same time, telephone interviewers may have more difficulty establishing their legitimacy in seeking such sensitive information. Groves and Kahn (1979) found more admission of being uneasy discussing income and race relation questions in a telephone survey than in a comparable face to face survey. Some respondents acknowledge sufficient legitimacy to ask the survey questions, but not sufficient to eliminate social desirability effects. It is possible that different parts of the household population react differently to these two features of the mode. Some persons, suspecting the intentions of the caller, may not provide accurate information. Others, believing the stated purpose of the call, take advantage of the more anonymous character of the telephone interaction and respond frankly to sensitive questions.

We thus speculate that some of the confusion in the literature on social desirability effects and mode of data collection arises from the confounding of two variables, both of which differ between modes: (1) ease of establishing the legitimacy of the survey request, and (2) reduced influence of social desirability on responses. Future research should attempt to manipulate these two variables separately in order to determine whether they have counteracting effects on responses in the two modes.

11.5.2 Cognitive Demands of the Survey Question and Mode

Another popular hypothesis about mode differences concerns the complexity of the measurement task. It is sometimes asserted that questions in a telephone interview should be "simpler," that the questionnaire should be shorter and require less attention on the part of the respondent. This has been interpreted to mean that questions should be short, contain very simple concepts, and require retrieval of only easily accessed information. These are vague statements, ones whose operationalization is not well specified. Unfortunately, there have been no intensive studies of these assertions.

There are, however, several strains of research that relate to this question. One of them concerns the use of scale questions, those asking the respondent to give an answer on an ordinal scale. Sometimes each point on the scale is labeled (e.g., "Strongly Agree," "Agree," "Neutral," "Disagree," "Strongly Disagree"). Other times the respondent is asked to provide a number from the scale, after having been told the meaning of various points on the scale. For example,

> Now I'll ask you to give me a number between one and seven that describes how you feel about your health. "One" stands for completely *dissatisfied*, and "seven" stands for completely *satisfied*. If you are *right in the middle*, answer "four." So, the low numbers indicate that you are dissatisfied; the high numbers that you are satisfied.

> First, what number comes closest to how satisfied or dissatisfied you are with your *health and physical condition in general*?

> _____ Number

In the face to face interview, the respondent is most often supplied a response card, containing a listing of the response scale and its labels, and is asked to provide the interviewer with a number:

Completely	Right in	Completely
Dissatisfied	the Middle	Satisfied

1————2————3————4————5————6————7

One issue surrounding such questions is whether the telephone mode, stripped as it is of visual aids, produces different answers than those provided through the use of response cards in the face to face interview. The more important question for the survey researcher is not merely the measurement of differences but the identification of which method yields more accurate responses. It is important to keep in mind that the question being addressed here concerns two differences in measurement technique. First, the medium of communication is different. Second, the use and nonuse of a response card may affect responses. It would be helpful to measure the effects of use and nonuse of response cards within the face to face interview mode, in order to measure that effect singly, but we have not been able to locate studies that have manipulated this feature.

There is, however, some literature on the effects of adapting response card questions used in face to face interviews to full verbal presentation on the telephone. Groves and Kahn (1979) explored a seven point satisfied-dissatisfied scale on five life satisfaction items. Differences by mode were quite small. There was a small tendency for persons on the telephone to choose the same scale point many times (57.9 percent on the telephone chose the same answer three or more times versus 55.2 percent in person). Measures of relationships between pairs of the five items were similar in the two modes. Sykes and Hoinville (1985) also find few differences. In their experiment only 2 of 13 questions using response cards in the face to face interview show differences by mode when adapted to verbal presentation on the telephone. In both of these experiments the correspondence between the verbal report of the scale and the response card was close (i.e., all the labeled points on the scale were mentioned by the telephone interviewer). Groves and Kahn also report another case for which the correspondence was not as tight—a 100 point scale, with nine labeled points on the response card but only three points mentioned in the introduction of the question on the telephone. There were large differences in the distribution of answers between the two modes. With the response card answers tended to cluster on the labeled points of the scale. With the verbal presentation of the 100 point scale, the answers tended to cluster about points easily divisible by 10 (i.e., 10, 20,...100) and less so about points divisible by 5. It appears that in the absence of visual markers on a numeric scale that the respondent uses simple arithmetic associations to simplify the mental image of the scale. Thus, 100 points are easily divided into deciles as a first approximation.

Miller (1984) experimented with alternative ways of administering a seven point scale without the use of a response card. One method is that above, by which the respondent is asked to give a number from one to seven. Alternatively, a "two-step" procedure was used:

Now, thinking about your *health and physical condition in general,* would you say you are *satisfied, dissatisfied,* or *somewhere in the middle?*

1. Satisfied	2. In the Middle	3. Dissatisfied
How satisfied are you with your health and physical condition—completely satisfied, mostly, or somewhat?	If you had to choose, would would you say you are closer to being *satisfied* or *dissatisfied* with your or are you *right in the middle*?	How dissatisfied are *s.p* you with your help and physical condition—completely dissatisfied, mostly, or somewhat?

A split sample experiment was conducted with the two forms for five different items. The one-step version of the scales tended to produce more answers "in the middle" and fewer on the positive side of the scales. This result might be associated with the follow-up probe in the unfolding method, asking those "in the middle" respondents to indicate whether they lean in the positive or negative direction. This step may force some of those respondents away from the middle category. The one-step scale also produced less missing data; some persons do not answer the follow-up probes, and interviewers reported respondent annoyance with the unfolding method. The one-step scales demonstrated higher intercorrelations, mostly because of reduced variance in their distributions.

The final set of investigations concerns mode effects in the measurement of income, although the measurement issue really concerns the use of unfolding scales with sequential pairwise comparisons. In face to face interview situations, income is generally measured with the assistance of response cards listing 10 to 20 lettered income categories (e.g., "A. $0-$2500," "B. $2501-$4000"). Without such an opportunity in the telephone mode, one technique has been to ask respondents a series of questions concerning whether they had incomes above or below a particular value, then, depending on their answer, to ask whether their income fell above or below another value. This continues sequentially until the respondent's income is classified into one of the desired final categories. Locander and Burton (1976) and Monsees and Massey (1979) examined different ways of doing this. In a split sample design the latter work measured income in three ways, by (1) starting at the top category and asking if the respondent made less than that amount (less than $25,000) and descending down the remaining categories until the respondent said "no"; (2) starting at the bottom category and asking if the respondent made more than that amount (more than $5000) and ascending the remaining categories until the respondent said "no"; and (3) asking whether the respondent's family made more or less than the midpoint income ($15,000) and, depending on the answer, ascending or descending in a continuous fashion to determine whether the family made more or less than the amounts. Table 11.3 presents the reported income distributions. The first three columns compare the three versions of the question on the telephone. The results show larger proportions of families being categorized in the first income categories mentioned by the interviewer. That is, in the version asking first whether the family earned more than $5000, more families are categorized as earning less than that amount than in the results using the other procedures. Similarly, in the descending scale, asking first whether the family earned less than $25,000, many more people report earning more than in the ascending scale. The procedure that starts in the middle of the income distribution resembles the descending method.

Which of the three methods is the best? To answer that question we need some criterion data; ideally, this would be some record-based actual income amount for each family. The only estimate we have for this population is from another method, using a response card in the face to face National Health Interview Survey. This survey has a much higher response rate (near 95 percent of all sample families) than did the telephone survey data of Monsees and Massey (in the high 70s). Thus, comparing income distributions confounds nonresponse and response error differences. The last column of Table 11.3 presents the results from the National Health Interview Survey. These results are somewhat different from all the telephone survey data and suggest a bias on the telephone toward overrepresenting high income reports. (Recall that this was the finding in Chapter 4 on nonresponse in telephone surveys.) The telephone measurement procedure that yields a distribution closest to that of the National Health Interview Survey is the one based on ascending questioning. Because of the confounding of nonresponse and response error, it is difficult to decide what inference should be drawn from this comparison.

11.5.3 Mode Effects and Interviewer Effects

The comments in Section 11.5.1 on social desirability effects review one type of hypothesis about differences between modes in interviewer effects. There are, however, other types of interviewer effects, some of them conceptualized as fixed effects of interviewer attributes, while others are variable errors due to interviewer differences. With regard to variable interviewer effects, the reader is referred to Chapter 8 on interviewer variance estimates. These results suggest that levels of interviewer effects on centralized telephone surveys might be somewhat lower than those in face to face surveys. There are no results apparently available to

Table 11.3 Percentage of Families in Various Income Categories by Form of Question and Mode of Interview

| | Percentage of Respondent Families | | | |
| | Telephone Survey | | | |
Income Range	Decreasing Order	Ascending Order	Unfolding Order	Face to Face Survey
Less than $5,000	9.3%	22.7%	8.1%	16.2%
$5,000–$9,999	13.6	21.4	18.0	20.3
$10,000–$14,999	23.4	19.1	17.6	19.2
$15,000–$24,999	28.5	25.0	28.8	25.3
$25,000 or more	25.2	11.9	27.5	19.2
Total	100.0%	100.0%	100.0%	100.0%
Missing data as percentage of all interviews	13.5	11.4	18.1	8.9
Total interviews	428	388	295	18,680

Source: M. Monsees and J. Massey, "Adapting Procedures for Collecting Demographic Data in a Personal Interview to a Telephone Interview," *Proceedings of the Social Statistics Section of the American Statistical Association*, 1979, pp. 130–135, Table 4.

demonstrate what portion of the reduced interviewer variability is due to the medium of communication and what portion to the centralization of the interviewing.[5]

The interviewer effects that were viewed as fixed effects in Chapter 8 included the gender and race of interviewers. In that chapter it was shown that female telephone interviewers were obtaining more pessimistic consumer attitudes than male telephone interviewers. There were no similar data for face to face interviews. For race of interviewer we found that black respondents were providing less militant attitudes regarding civil rights when questioned by a white interviewer than when questioned by a black interviewer. In a telephone interview study, Cotter et al. (1982) find that white respondents express more pro-black attitudes to black interviewers than to white interviewers. On questions not related to race, there are no effects. Finally, black respondents appear to be less affected by the race of the interviewer than white respondents.

---

[5] A comparison of dispersed telephone interviewing (with interviewers calling from their homes) and centralized telephone interviewing would provide some evidence on this score.

11.6 SURVEY COSTS AND THE MODE OF DATA COLLECTION

While the first impression of relative costs of telephone and face to face interview surveys is that the telephone would be much cheaper, it is important to note that there are some cases for which that does not appear to be true. Centralized telephone facilities do offer an ease of measuring cost components that exceeds that possible in dispersed systems. With computerized assistance for cost accounting it seems possible that the same system that monitors cost of data collection during production can supply parameter estimates for cost models fitted to error models for survey design. This possibility is pursued below in a cost model structure for a single telephone survey from a centralized facility.

11.6.1 A Cost Model for a Centralized Telephone Facility

One solution to the multiple purposes of survey cost models is the use of simultaneous equations that are partially nested to provide information for different needs:

1. The top level contains separate components for some fixed costs (costs not easily distributable over individual surveys), such as supervisory activities, monitoring and feedback activities, and interviewing activities, and hardware/software and telephone costs.

2. The fixed cost equation includes building rental, carrels and telephone instruments, electricity, maintenance, and portion of director's time.

3. The supervisory activities equation includes training costs, sample administration tasks, and interviewer question handling, and perhaps a machine maintenance component.

4. Monitoring and feedback model includes training of monitor, locating case to be monitored, monitoring, and feedback.

5. Interviewer cost model includes costs of training, reviewing cases, dialing and seeking contact, household contact, interviewing, editing, and break time.

6. A cost model for CATI includes costs relating to hardware and software, as well as telephone connect charges.

Below appears a five-equation system for a telephone survey. It has been chosen to provide both simple summary tools for managers and the model parameters of more interest to survey designers. The system should also permit estimation of other parameters, not yet anticipated, by combining various components in the models.

$$\text{TOTAL} = \text{FIXED} + \text{SUPERVISION} + \text{MONITORING}$$
$$+ \text{INTERVIEWING} + \text{ADMINISTRATIVE} + \text{DEVICES}$$

$$C = F + S(\,\cdot\,) + M(\,\cdot\,) + I(\,\cdot\,) + A(\,\cdot\,) + D(\,\cdot\,)\,,$$

where C = total cost of the survey;

$\quad F$ = summary of fixed costs;

$\quad S(\,\cdot\,)$ = total costs for supervisory activities;

$\quad M(\,\cdot\,)$ = total costs for monitoring activities;

$\quad I(\,\cdot\,)$ = total costs for interviewer salaries and training;

$\quad A(\,\cdot\,)$ = total costs for clerical and administrative salaries;

$\quad D(\,\cdot\,)$ = total costs for device support, hardware and
$\qquad$ communications costs.

The components of this equation are dollar amounts that would permit managers to know not only total costs but also cost per completed interview of various sets of survey functions. The concerns in centralized facilities appear to surround the level of administrative and supervisory costs relative to interviewer costs. This equation would allow continual assessment of those costs.

It would not, however, allow one to examine numbers of hours devoted to the survey by each category. Note that it also does not have a separate parameter for training costs, because those will appear as attributes of individual staff members. Finally, these cost parameters are for the functions described, not necessarily job titles. For example, supervisors may have responsibilities for backup of the system or terminal error diagnosis. Salary costs for those kinds of activities would be placed under device costs, $D(\,\cdot\,)$.

The supervisor cost model is simpler than it would necessarily be if we wanted to concentrate on optimizing the work flow for supervisors. It does not, for example, attempt to separate components for contact with interviewers, handling sample administration tasks, and other duties.

$$S(\cdot) = \sum_{j=1}^{J} (\text{TRAINING}(\text{SUPERVISOR}, j))$$

$$+ \sum_{j=1}^{J} \text{OVERSIGHT}(j) \cdot \text{Rate}(S, j)$$

$$= \sum_{j=1}^{J} T(S, j) + \sum_{j=1}^{J} O(j) \cdot \text{Rate}(S, j) \, ,$$

where $T(S, j)$ = training costs incurred for the jth supervisor;

$O(j)$ = total number of hours devoted to supervision in the facility;

$\text{Rate}(S, j)$ = per hour salary of the jth supervisor.

There are a total of J supervisors used on the survey.

Data for the TRAINING component would be collected at time of hiring and training for supervisors. The training costs should include the costs of trainers as well as salaries paid to the supervisory trainee. This would then be viewed as a characteristic of the supervisor and would be stored in a supervisory–level data set.

The monitor cost model has a similar form to the supervisory cost model, but more disaggregation of costs to permit efficiency rates to be calculated on monitoring activities. The key ingredients for the efficiency rates will be the proportion of time the monitor is spending in actually listening to interviews or in giving feedback to interviewers. We could also later look at the ratio of time spent in monitoring to the time spent in feedback.

$$M(\cdot) = \sum_{k=1}^{K} \text{TRAINING}(\text{Monitor},k)$$

$$+ \sum_{k=1}^{K} \left(\begin{array}{c} \text{HUNTING}(k) + \text{LISTENING}(k) \\ + \text{FEEDBACK}(\text{Monitor},k) \end{array} \right) \cdot \text{Rate}(\text{Monitor},k)$$

$$= \sum_{k=1}^{K} T(M,k) + \sum_{k=1}^{K} \left(H(k) + L(k) + F(M,k) \right) \cdot \text{Rate}(M,k) ,$$

where $T(M,k) =$ cost of training the kth monitor for this survey;

$\quad\quad H(k) =$ number of minutes searching for cases to be monitored among active stations, performed by the kth monitor;

$\quad\quad L(k) =$ number of minutes listening to interviewers interacting with persons in sample units performed by the kth monitor;

$\quad\quad F(M,k) =$ number of minutes giving feedback to interviewers performed by the kth monitor;

$\text{Rate}(M,k) =$ per minute salary rate of the kth monitor.

There are a total of K monitors used on the survey. Data for this cost model would be collected either by the CATI system software examining behavior of the monitoring terminal or by the monitors themselves. The HUNTING parameter could be obtained by subtraction from the LISTENING parameter and the FEEDBACK parameter could be obtained on the monitoring forms.

The interviewing cost model is the most disaggregated because its components are so important to the overall cost and error structure of the survey.

$$I(\,\cdot\,) = \sum_{m=1}^{M} \text{TRAINING}(\text{Interviewer},m)$$

$$+ \sum_{m=1}^{M} \left(\begin{array}{c} \text{ATTEMPTS}(\text{Interviewer},m) \\ + \text{CONTACTS}(\text{Interviewer},m) \\ + \text{QUESTIONING}(\text{Interviewer},m) \\ + \text{EDIT}(\text{Interviewer},m) \\ + \text{REFATTEMPTS}(\text{Interviewer},m) \\ + \text{REFCONTACTS}(\text{Interviewer},m) \\ + \text{REFQUESTIONING}(\text{Interviewer},m) \\ + \text{REFEDIT}(\text{Interviewer},m) \\ + \text{BREAK}(m) \\ + \text{FEEDBACK}(\text{Interviewer},m) \end{array} \right) \cdot \text{Rate}(\text{Interviewer},m)$$

$$= \sum_{m=1}^{M} T(I,m)$$

$$+ \sum_{m=1}^{M} \left(\begin{array}{c} A(I,m) + C(I,m) + Q(I,m) \\ + E(I,m) + RA(I,m) + RC(I,m) + RQ(I,m) \\ + RE(I,m) + B(m) + F(I,m) \end{array} \right) \cdot \text{RATE}(I,m) \,,$$

where $T(I,m)$ = cost of training the mth interviewer for this survey;

$A(I,m)$ = total number of minutes accessing cases prior to contact (i.e., reviewing call notes, dialing number waiting for answer, recording call disposition) performed by the mth interviewer;

$C(I,m)$ = total number of minutes in contact prior to administering substantive questions to sample units performed by the mth interviewer;

$Q(I,m)$ = total number of minutes questioning for sample unit performed by the mth interviewer;

$E(I,m)$ = total number of minutes editing cases by the mth interviewer;

$RA(I,m)$ = total number of minutes accessing initial refusal cases prior to contact (i.e., reviewing call call notes, dialing number, waiting for answer, recording call disposition) performed by the mth interviewer;

$RC(I,m)$ = total number of minutes in contact prior to administering substantive questions to initial refusal sample units performed by the mth interviewer;

$RQ(I,m)$ = total number of minutes questioning for converted refusal sample units performed by the mth interviewer;

$RE(I,m)$ = total number of minutes editing converted refusal cases by the mth interviewer;

$B(m)$ = total number of minutes on break for the mth interviewer;

$F(I,m)$ = total number of minutes in feedback sessions for the mth interviewer;

$\text{Rate}(I,m)$ = per minute salary cost for the mth interviewer.

There are a total of M interviewers used on the survey.

The data for these components would be collected on a case basis but then added to an interviewer-level record as cases are released by interviewers. The ATTEMPT parameter would be measured by the amount of time elapsed between CATI screens used in contact attempts. The CONTACT parameter would be defined as prequestioning contact, that is, those interactions that are necessary to obtain information about the household prior to asking substantive questions of a respondent. The EDIT parameter would be defined as time in a case after the interview is completed. The BREAK parameter may be problematic, depending on how this is being tracked, but it would include time away from the carrel. Finally, the FEEDBACK parameter could be obtained from the monitor's data collection.

The administrative cost function should include hardware/software costs, machine maintenance, and staff time devoted to system backup, disk file manipulations, and so on. In addition, we have placed total telephone time in this model. Thus, it represents the cost of the medium of data collection which can be assigned to a single survey.

$$D(\cdot) = \sum_{j=1}^{J} D(S, j) \cdot \text{Rate}(S, j) + \sum_{p=1}^{P} D(O, p) \cdot \text{Rate}(O, p)$$

$$+ \sum_{n=1}^{N} D(T, n) \cdot \text{Rate}(T, n) \, ,$$

where $D(\cdot)$ = total cost of the hardware/software and telephone charges;

$\quad D(S, j)$ = total minutes spent by the jth supervisor in machine and terminal maintenance and error diagnosis;

$\quad D(O, p)$ = total minutes spent by the pth computer operators/programmers in machine and terminal maintenance and error diagnosis;

$\quad D(T, n)$ = total telephone connect minutes with the nth sample case;

$\quad \text{Rate}(S, j)$ = salary cost per minute of the jth supervisor;

$\quad \text{Rate}(O, p)$ = salary cost per minute of the pth computer operator/programmer;

$\quad \text{Rate}(T, n)$ = cost per minute of telephone connect time with the nth sample case.

If properly designed, this cost and effort modeling system will permit both survey managers and survey designers access to the information they need. In review of these ideas we need to pose this challenge to the set of equations and data retrieval systems. We sketch a few ideas below.

To assess efficiency of monitors, we might examine the ratio

$$\frac{L(k)}{H(k) + L(k)}$$

for each monitor working on the study.

To assess efficiency of interviewers, we might examine

$$\frac{Q(I,m)}{A(I,m) + C(I,m) + Q(I,m) + E(I,m) + B(I,m)}$$

for each interviewer.

For building a deterministic cost model for use in survey design decisions about optimal workload size for interviewers, we might use

$$\frac{\sum \big(T(I,m) + F(I,m) + B(I,m) \big) \cdot \text{Rate}(I,m)}{M}$$

$$+ \sum_{n=1}^{N} \left(\begin{array}{c} A(I,n) + C(I,n) + Q(I,n) \\ + E(I,n) + IA(I,n) + IC(I,n) \\ + IQ(I,n) + IE(I,n) \end{array} \right) \cdot \frac{\sum\limits_{m=1}^{M} \text{Rate}(I,m)}{M},$$

where the first term is the average cost per interviewer, and the second term is the average cost per case.

11.6.2 Empirical Data on Cost of Data Collection in Different Modes

All the discussion above specifies the form of cost functions in a centralized facility but does not give the reader any sense of the differences in various types of effort required in each mode. Groves and Kahn (1979), in an experimental comparison of telephone and face to face interview surveys on the national level, presented a detailed table listing the number of staff hours and magnitudes of other materials required for the two modes of data collection. This table is reproduced as Table 11.4.

The telephone survey interviews were conducted from a centralized facility; the face to face interviews by interviewers located in about 75 primary areas throughout the country. The average length of the telephone interview was about one–half of the face to face interview. The sampling technique for the telephone survey was about one–third as efficient, in terms of proportion of numbers that were working household numbers, as that of telephone surveys now used. Finally, the table contains costs of sampling and data collection only. These features

Table 11.4 Direct Costs for Components of Sampling and Data Collection Activities on the Telephone and Personal Interview Surveys

	Telephone Survey		Personal Interview Survey	
	Hours or Other Units	Costs ($)	Hours or Other Units	Costs ($)
I. Sampling				
Administrative salaries	86.0[a]	505.27[a]	362.0	3,305.03
Clerical/typing salaries	0	0	186.0	676.12
Chunking and listing	0	0	0	4,566.00
Data processing	0	450.00	0	0
Category total	86.0 hours	$955.27	548.0 hours	$8,547.15
Percentage of total	1.6%	2.5%	4.1%	10.1%
II. Pretest				
Ann Arbor field office salaries	34.0	224.59	32.0[a]	200.8[a]
Clerical/typing salaries	4.0	14.73	17.0	88.65
Supervisors salary	50.5	188.67	25.8	153.30
Interviewers				
Salary	76.4	226.06	125.7	444.13
Travel	0	0	666(mi)	93.32
Duplicating	1688(pp.)	69.40	7370(pp.)	119.62
Postage	0	0	0	14.00
Category total	164.9 hours	$723.45	200.5 hours	$1,113.10
Percentage of total	3.0%	1.9%	1.5%	1.3%
III. Training and Prestudy Work				
Interviewing supervisors salaries	37.1	155.57	667.0[a]	3,929.14[a]
Interviewers salaries	314.6	936.58	1,621.3[a]	5,285.51[a]
New interviewer training		660.00		0
Duplicating	524(pp.)	26.95	6725(pp.)	85.10
Supplies		14.78		223.86
Coding staff salaries	11.0	94.46	0	0
Coding evaluation of questionnaires	40.0	178.00	0	0
Category total	402.7 hours	$2,066.34	2,288.3 hours	$9,523.61
Percentage of total	7.4%	5.4%	16.9%	11.2%

534

Table 11.4 (Continued)

	Telephone Survey		Personal Interview Survey	
	Hours or Other Units	Costs ($)	Hours or Other Units	Costs ($)
IV. Materials				
Questionnaire	100,800(pp.)	802.25	224,000(pp.)	1,466.51
Other data collection instruments		278.70	24,840(pp.)	704.65
Data collection related materials and reporting forms		274.68	30,900(pp.)	1,211.24
General supplies	—	19.33	—	277.75
Category total		$1,374.96		$3,660.15
Percentage of total		3.6%		4.3%
V. Ann Arbor Field Office				
Administrative salaries	156.0	1,222.48	324.0	2,508.29
Clerical/typical salaries	55.0	172.26	392.4	1,651.13
Category total	211.0	$1,394.74	716.4 hours	$4,159.42
Percentage of total	3.9%	3.7%	5.3%	4.9%
VI. Field Salaries				
Supervisor salaries	648.0	2,303.64	988.5[a]	4,956.88[a]
Interviewer salaries	3,442.0	10,181.05	8,389.8[a]	27,321.04[a]
Foreign interviewers salaries	—	60.0	0	0
Category total	4,090.0 hours	$12,544.69	9,378.3 hours	$32,277.92
Percentage of total	75.5%	33.1%	69.4%	38.0%
VII. Field Staff Travel				
Supervisor travel		0		5,620.35
Interviewer travel				
Personal auto mileage	0	0	74,405(mi)	10,416.72
Other		0	—	778.04
Category total	—	0		$16,815.11
Percentage of total		0		19.8%

535

Table 11.4 (Continued)

	Telephone Survey		Personal Interview Survey	
	Hours or Other Units	Costs ($)	Hours or Other Units	Costs ($)
VIII. *Communications*				
Postage				
Telephone				3,491.03
For data collection		0		0
For other communications		15,793.60		1,756.45
Supplies for mailing		0		732.83
Category total	—	15,793.60	—	$5,980.31
Percentage of total		41.6%		7.0%
IX. *Control Function*				
Administrative salaries	247.5	830.55	8.0	91.30
Clerical/typing salaries	0	0	188.5	766.93
Printing and duplicating	0	0	1500(pp.)	24.99
Data processing		372.00[a]		0
Category total	247.5 hours	$1,202.55	196.5 hours	$883.22
Percentage of total	4.6%	3.2%	1.5%	1.0%
X. *Postinterviewing Activities*				
A. Interviewer evaluation/debriefing				
Supervisor salaries	12.0	44.52	0	0
Interviewer salaries	68.0	199.70	30.2[a]	98.62[a]
Ann Arbor administrative salaries	0	0	16.0	94.87
Ann Arbor clerical/typing salaries	0	0	12.0	37.80
Duplicating	50(pp.)	2.50	1200(pp.)	15.55
Postage		0		34.45
Category total	80.0 hours	$246.72	58.2 hours	$281.29
Percentage of total	1.5%	0.6%	0.4%	0.3%

Table 11.4 (Continued)

	Telephone Survey		Personal Interview Survey	
	Hours or Other Units	Costs ($)	Hours or Other Units	Costs ($)
B. Verification				
Ann Arbor administrative salaries	100.1[a]	384.78[a]	90.5	596.44
Ann Arbor clerical/typing salaries	0	0	9.0	35.13
Duplicating	0	9.30[a]	1150(pp.)	33.90
Supplies		24.79[a]		23.88
Postage		88.66		104.78
Telephone		0		82.25[a]
Category total	100.1 hours	$507.53	99.5 hours	$876.38
Percentage of total	1.8%	1.3%	0.7%	1.0%
C. Report to respondents				
Ann Arbor administrative salaries	3.0	22.63	4.0	28.70
Keypunching	34.2	256.23	32.8	246.00
Data processing		162.32[a]		112.94
Printing	0	555.00[a]	22,400(pp.)	251.00
Postage		133.76[a]		107.62
Category total	37.2 hours	$1,129.94	36.8 hours	$746.26
Percentage of total	0.7%	3.0%	0.3%	0.9%
OVERALL TOTAL	5,419.4 hours	$37,929.79	13,522.5 hours	$84,863.92
PRE-INTERVIEW TOTAL	3.3 hours	23.45	8.7	$54.82

[a]Costs based on estimates of those personnel involved in the work, usually necessitated by different categories of work being performed by the same personnel.
Source: R. Groves and R. Kahn, *Surveys by Telephone: A National Comparison with Personal Interview Surveys,* Orlando, Academic Press, 1979, Table 7.1.

537

should be kept in mind in drawing inferences about costs of the different modes.

The largest differences between modes concern the costs of communication and travel. Travel costs are high for the face to face interview survey (20 percent of the total budget) but nonexistent for the telephone survey. In contrast, the telephone communication costs are quite high for the telephone survey (42 percent of the budget) but much smaller for the face to face interview survey (7 percent of the budget). These are inherent differences in the mode. It is surprising how similar in size are the costs of travel for the face to face and the cost of telephone communication for the telephone survey.

One real source of cost savings in telephone surveys is the higher workload size of interviewers relative to that in face to face interview surveys. Hence, the telephone interviewers could collect approximately 1618 thirty minute interviews in 3442 hours, or 1 per 3.3 hours. The face to face interviewers collected 1548 interviews in 8390 hours or 1 per 8.7 hours. Even doubling the amount of time for the telephone interviewers (a gross exaggeration of the impact of doubling the length of the interview) would produce a cost savings for the telephone. This arises because interviewers share the entire sample in a centralized facility. Face to face interviewers are given only a small number of cases to process. They often work hours during which no one in the sample households is at home. Furthermore, face to face interviewers are paid for the hours they spend traveling to the sample areas.

The final notable difference concerns the number of staff hours required in administrative coordination of the data collection activities, for communication with the interviewers, receipt and logging of materials, and general oversight of the staff (labeled Ann Arbor office time in Table 11.4). The face to face interview survey required 716 hours but only 211 were required for the centralized telephone survey. This difference reflects the larger number of persons to be coordinated in a face to face interview survey.

Overall the telephone survey costs are approximately one-half the costs of the face to face interview survey, after adjustments are made for different interview lengths. This experience is not uniform, however, across different experiments. Tuchfarber and Klecka (1976) report telephone survey costs that are one-quarter equivalent face to face interview costs in a citywide survey of victimization experiences. Sykes and Hoinville (1985), in an experimental study in Britain, found total costs of 9 pounds on the telephone and 10.5 pounds on the face to face interview. However, their calculations assumed that the interviewer costs would be the same in the two modes (a result not found in centralized telephone interviewing). It is clear that cost differences between modes are heavily dependent on the administrative structure of the interviewing staffs.

11.7 COST AND ERROR MODELING WITH TELEPHONE AND FACE TO FACE SURVEYS

Dual frame designs were mentioned in Chapter 3 as a way to increase coverage of rare populations. The methodology also has a place in the choice of mode of data collection. A dual frame, mixed mode design would retain the cost savings of telephone interviewing and also enjoy the increased coverage of area probability sampling. The dual frame character of the design selects both an area sample and a telephone sample, and the mixed mode character conducts face to face interviews with area sample households and telephone interviews with telephone sample households. Subsequent survey estimation combines data from both frames.

The optimal allocation of survey resources to the two frames depends on the costs and errors associated with each. Unfortunately, there is often uncertainty about the magnitudes of these costs and errors. In addition, several administrative structures may be proposed for the dual frame, mixed mode design, with each structure potentially affecting the cost and error character stics of the survey. To the extent that the optimal allocation is sensitive to changes in the values of parameters in the cost or error models, greater emphasis should be placed on obtaining accurate estimates for those parameters before a complete dual frame design can be specified.

11.7.1 Dual Frame Survey Design

In mixes of telephone and area frame samples clustered samples are often used in both frames. In the area frame, the sample may be clustered to reduce the costs of interviewer travel from one sample unit to another. In telephone samples, clustering is also employed to reduce the costs of data collection. Typically, telephone numbers are randomly generated within geographically assigned number codes in order to select a sample of telephone households. Clustered selection by sets of 100 consecutive numbers is often introduced to increase the proportion of sample numbers generated that are connected to households (e.g., see Waksberg, 1978).

11.7.2 Administrative Structures for Dual Frame, Mixed Mode Designs Mode Designs[6]

Most organizations that conduct national face to face interview surveys in the United States maintain regional offices for the supervision of field interviewers. When telephone interviewing is first introduced into such an organization, several decisions regarding the administration of the new mode of data collection are required. First, a decision must be made concerning which personnel will conduct telephone interviewing. Second, the organization must decide whether interviewers will be centralized in telephone interviewing facilities or will use telephones in their own homes for interviewing.

As with most survey design issues, these decisions involve an interplay of costs and errors. The cost of telephone interviewing from interviewers' homes can be lower than that from a centralized facility if their assignments are clustered into areas near the interviewers" homes. For random digit dialing samples distributed independently of the primary areas of an area probability sample, the cost advantage of home-based telephone interviewing diminishes with low-cost long distance communication services available in the United States. The centralized facilities also offer advantages of size that permit large-scale training activities and continuous concentration on interviewing while the staff is in the facility, a feature that can reduce the number of hours spent on each completed case. The errors related to survey administration— nonresponse, response, and recording errors—might also be more effectively controlled in a centralized environment where constant supervision and monitoring are possible.

Representations of four dual frame, mixed mode survey administrative structures are shown in Figures 11.2, 11.3, 11.4, and 11.5. These structures vary by which interviewers do the work and from what locations the interviewing is conducted. In each, a central office (denoted by the large rectangle) is attached to several regional offices (denoted by triangles), which in turn supervise individual interviewers (denoted by small rectangles appended to each triangle) working in different primary areas. In addition, centralized telephone interviewing facilities (denoted by circles of various sizes) are attached to either the central or regional offices.

The administrative structure depicted in Figure 11.2 is used by many survey organizations to administer face to face surveys and thus is a potential structure for the dual frame, mixed mode survey designs of

---

[6] This section draws heavily on R. Groves and J. Lepkowski (1985).

interest. Interviewers conduct face to face interviews at selected area frame households within the primary areas where they reside; they also interview a sample of telephone households using their own home telephones. The regional office personnel supervise both modes of interviewing. The location of telephone sample households may be clustered within area frame primary sampling units, at the cost of increased variance for survey estimates. On the other hand, if such clustering is used, transfer of survey materials is facilitated when telephone sample respondents request a face to face interview.

The structure in Figure 11.3 requires each interviewer to conduct interviews in only one mode. The face to face interviewers conduct face to face interviews in the sample area segments near their homes, but all telephone interviews are conducted from one centralized facility attached to the central office. Administrative control over the centralized telephone facility lies within the central office. In contrast to the first administrative structure, transfer of survey materials from one mode to another by supervisory staff may be required. Some of these requests can be served by using the face to face interviewing staff located within reasonable distance from the sample person's residence.

Administrative structure III (Figure 11.4) is similar to II since it also has centralized interviewing facilities, but telephone interviewing is conducted at several facilities located across the country. The locations can be chosen to minimize long distance telephone charges and to increase access to labor markets with large numbers of potential interviewers. Each telephone facility hires, trains, schedules, and supervises its own work force.

The final administrative structure (Figure 11.5) also maintains centralized telephone interviewing facilities, but in order to minimize redundancy within the survey administration, facilities are attached to regional offices which supervise face to face interviewing. Such regional offices are responsible for both the face to face interviewers and the local staff of telephone interviewers.

These four structures represent a wide range of possible organizational approaches to the administration of such surveys. As the discussion has indicated, each model has somewhat different cost requirements, some needing a consideration of transfer costs, others needing to maintain centralized telephone interviewing facilities. Yet each model might have appeal to different survey organizations.

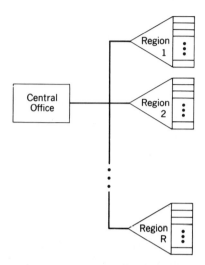

Figure 11.2 Administrative structure I: dispersed telephone interviewing.

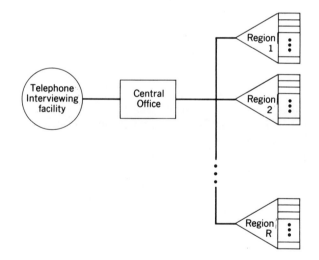

Figure 11.3 Administrative structure II: centralized telephone interviewing facility.

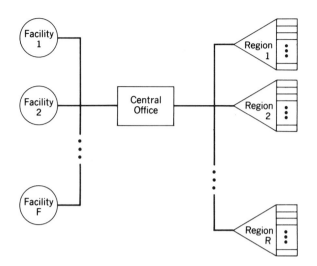

Figure 11.4 Administrative structure III: multiple centralized facilities.

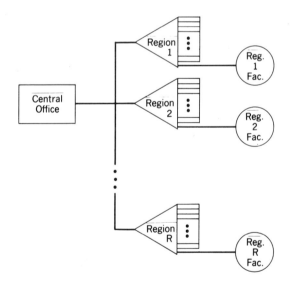

Figure 11.5 Administrative structure IV: telephone facilities attached to regional office.

11.7.3 An Example of a Dual Frame, Mixed Mode Design Simulation

The U.S. National Crime Survey (NCS) is a large continuing household sample survey designed and conducted by the U.S. Bureau of the Census for the Bureau of Justice Statistics. The primary purpose of the NCS is the measurement of self-reported crime victimization of the U.S. civilian noninstitutionalized population ages 12 and older. The NCS collects information from a stratified multistage area probability sample of approximately 14,000 housing units, yielding about 25,800 face to face interviews per month. Each person 14 years old and older is asked to complete an interview; proxy reports are taken for those 12 and 13 years old. A rotating panel design is employed in which each sample unit is visited every 6 months for a 3-1/2 year period before being replaced.

The NCS collects data on many types of victimization and on the characteristics of reported events. The types of crime covered include personal victimizations such as rape, robbery, assault, and personal larceny, as well as household victimizations such as burglary, motor vehicle theft, and household larceny. For the purposes of this investigation, five types of crime were selected to provide an opportunity to examine the problems of dual frame allocation in a multipurpose survey: total personal crimes, robbery, assault, total household crimes, and motor vehicle theft.

Separate cost models were developed for each administrative structure reviewed above. As an illustration of the nature of these cost models, consider the cost model for administrative structure III:

$$C = C_0 + C_{ro} \, \text{INT}[(m_A/32)) + 1]$$

$$+ C_{tp} \, \text{INT}[(m_A \overline{N}_A/W_A) + 1]$$

$$+ C_{tf} \, \text{INT}[(m_B \overline{N}_B/W_B)/50 + 1]$$

$$+ C_{tt} \, \text{INT}[(m_B \overline{N}_B/W_B) + 1]$$

$$+ [10 + c_A(m_B \overline{N}_B/40)]$$

$$+ c_A m_A \overline{N}_A + c_B m_B \overline{N}_B \,,$$

where C_0 = overhead or fixed costs;

C_{ro} = cost to operate a regional office;

INT[$\cdot$] = integer portion of the argument [$\cdot$];

C_{tp} = annualized cost to hire and train a face to face interviewer;

W_A = annual number of completed interviews for a face to face interviewer;

C_{tf} = cost to operate a telephone facility of 50 interviewers;

W_B = annual number of completed interviews for a telephone interviewer;

C_{tt} = annualized cost to hire and train a telephone interviewer;

c_A = remaining cost per completed face to face interview;

c_B = remaining cost per completed telephone interview;

m_A = number of clusters in the area frame;

m_B = number of clusters from the telephone frame;

$\overline{N}_A$ = number of sample households per cluster from the area frame;

$\overline{N}_B$ = number of sample households per cluster from the telephone frame.

The model is deliberately complex in order to capture as many types of cost that can be manipulated in a survey design. The individual cost components are often multiplied by a step function (i.e., INT[$\cdot$]) which reflects the fact that adding a new unit to a system is done in discrete steps, not in continuous increments (see Chapter 2, Section 2.5).

The cost models for the administrative structures vary in their components (see Table 11.5). Each cost model has a component for administrative overhead for conducting the combined survey operation and components for hiring and training costs. There is also a component for per unit costs of completing face to face and telephone interviews in each model. Structure I does not have a cost for centralized telephone interviewing facilities since interviewers conduct telephone interviews from their homes.

Groves and Lepkowski (1985) combine the cost models with error models for a dual frame estimate of victimization rates. These were used to examine the optimal allocation between the two frames. The results (shown in Table 11.6) are that the optimal allocation to the telephone frame varies from 39 percent of the interviews to 74 percent of the interviews, depending on which statistic is examined and what administrative model is used. There are greater differences among statistics than among administrative structures. For example, Figure 11.6 and 11.7 show the estimated standard error for two statistics: robbery rates in Figure 11.6 and total household crime rates in Figure 11.7. The four different lines on the curves depict the standard errors for the different administrative models. Since the administrative models are associated with different costs, different sample sizes are purchased by each. The differences are small, however. In contrast, comparing the two figures, robbery rates achieve their smallest standard errors for fixed costs when about 50 percent of the interviews come from the telephone frame; total household crime rates achieve their lowest standard errors when about 70 to 75 percent of the cases come from the telephone frame.

Furthermore, the results suggest that large gains in sampling precision are possible for some types of crime statistics in a dual frame survey design. On the other hand, the magnitude of other sources of error (e.g., response variance, response and nonresponse bias) and their effects on the allocation have not been addressed. Ideally, these nonsampling errors would be added to the error models used in determining the optimal allocations. Unfortunately, few of these errors are measurable and even fewer are routinely measured and presented with survey estimates themselves (see Lepkowski and Groves, 1986 for one attempt). For example, with respect to bias, there is little empirical evidence of systematic differences in point estimates obtained by face to face or telephone interviews, suggesting that few large differences between the frames exist with respect to bias. There are arguments, but little evidence at present, that interviewer variance is lower in centralized telephone surveys (Groves and Magilavy, 1986). There is no evidence of differences in simple response variance.

In short, it is likely that biases and nonsampling variances in the two modes might not work exclusively to the disadvantage of the telephone mode. If that is the case, the conclusions obtained by focusing on sampling variance alone, that large gains in efficiency are possible using a

Table 11.5 Components for Cost Models for Estimated Household Level Costs to Produce Annual Estimates for the National Crime Survey

Cost Component	Administrative Structure			
	I	II	III	IV
Overhead	$(0.18)C$	$(0.18)C$	$(0.18)C$	$(0.18)C$
Regional office	$(\$122,083) \cdot$ $\text{INT}[(m_A/32)+1]$	$(\$122,083) \cdot$ $\text{INT}[(m_A/32)+1]$	$(\$122,083) \cdot$ $\text{INT}[(m_A/32)+1]$	$(\$151,483) \cdot$ $\text{INT}[\text{MAX}[m_A/32,(m_B\bar{N}_B/W_B)/50,$ $(m_A\bar{N}_A+m_B\bar{N}_B)/(25W_B+16\bar{N}_A)]+1]$
Hiring and training, personal interviewers	$(\$320.94) \cdot$ $\text{INT}[\text{MAX}(m_A\bar{N}_A/W_A,$ $m_B\bar{N}_B/W_B)]$	$(\$320.94) \cdot$ $\text{INT}[(m_A\bar{N}_A/W_A)+1]$	$(\$320.94) \cdot$ $\text{INT}[(m_A\bar{N}_A/W_A)+1]$	$(\$320.94) \cdot$ $\text{INT}[(m_A\bar{N}_A/W_A)+1]$
Supplementary training, personal interviewers in centralized telephone interviewer techniques	$-^a$	$-$	$-$	$(\$41.25) \cdot$ $\text{INT}[((0.1)m_A\bar{N}_A/W_A]+1)$
Centralized telephone interviewing facility	$-$	$\$58,800+(\$13,710) \cdot$ $([\text{INT}((m_B\bar{N}_B/W_B)$ $+1)-50]/10+1)$	$(\$58,800) \cdot$ $\text{INT}[(m_B\bar{N}_B/W_B)/50+1]$	$\cdots$
Hiring and training, telephone interviewers	$-$	$(\$92.40) \cdot$ $\text{INT}[(m_B\bar{N}_B/W_B)+1]$	$(\$82.50) \cdot$ $\text{INT}[(m_B\bar{N}_B/W_B)+1]$	$(\$82.50) \cdot$ $\text{INT}[(m_B\bar{N}_B-$ $(0.6)W_B[(0.1)m_A\bar{N}_A/W_A+1]/W_B+1]$
Cases transferred from personal visit to telephone interviewing mode	$(0.5)(\$24.53)(0.05)m_B\bar{N}_B$	$[\$10+(0.5)(\$24.53)] \cdot$ $(0.05)m_B\bar{N}_B$	$[\$10+(0.5)(\$24.53)] \cdot$ $(0.05)m_B\bar{N}_B$	$(0.5)(\$24.53)(0.05)m_B\bar{N}_B$
Remaining per unit costs for personal interviewing	$(\$24.53)m_A\bar{N}_A$	$(\$24.53)m_A\bar{N}_A$	$(\$24.53)m_A\bar{N}_A$	$(\$24.53)m_A\bar{N}_A$
Remaining per unit costs for telephone interviewing	$(\$18.15)m_B\bar{N}_B$	$(\$15.12)m_B\bar{N}_B$	$(\$14.52)m_B\bar{N}_B$	$(\$13.68)m_B\bar{N}_B$

Note: The function INT $[\cdot]$ denotes the integer portion of the argument $[\cdot]$.   [a]Not applicable.
Source: R.M. Groves and J.M. Lepkowski, "Dual Frame, Mixed Mode Survey Designs," *Journal of Official Statistics*, Vol. 1, No. 3, 1985, Table A1.

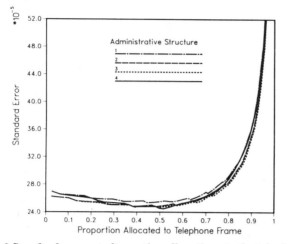

Figure 11.6 Standard error at alternative allocations to the telephone frame for robbery. From R.M. Groves and J.M. Lepkowski, "Dual Frame, Mixed Mode Survey Designs," *Journal of Official Statistics*, Vol. 1, No. 3, 1985, pp. 263–286, Figure 6.

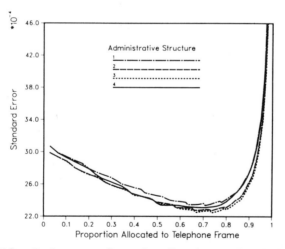

Figure 11.7 Standard error at alternative allocations to the telephone frame for total household crimes. From R.M. Groves and J.M. Lepkowski, "Dual Frame, Mixed Mode Survey Designs," *Journal of Official Statistics*, Vol. 1, No. 3, 1985, pp. 263–286, Figure 8.

dual frame design, may be repeated when a full mean square error model is used to study the allocation problem.

From the concern of sampling precision alone, it is likely that the joint use of telephone and face to face interview surveys may offer

Table 11.6 Optimal Proportion of Sample Cases Allocated to the Telephone Frame and Standard Error of the Mean Achieved at the Optimum by Administrative Structure and Type of Crime

	Administrative Structure			
	I	II	III	IV
Total Personal Crimes				
Proportion	0.63	0.65	0.66	0.65
Standard error	0.00123	0.00121	0.00120	0.00122
Robbery				
Proportion	0.45	0.48	0.39	0.51
Standard error	0.000254	0.000248	0.000247	0.000245
Assault				
Proportion	0.63	0.65	0.66	0.51
Standard error	0.000546	0.000535	0.000530	0.000538
Total Household Crimes				
Proportion	0.71	0.74	0.74	0.69
Standard error	0.00234	0.00228	0.00225	0.00231
Motor Vehicle Theft				
Proportion	0.63	0.66	0.67	0.51
Standard error	0.000503	0.000490	0.000484	0.000494

Source: R.M. Groves and J.M. Lepkowski, "Dual Frame, Mixed Mode Survey Designs," *Journal of Official Statistics*, Vol. 1, No. 3, 1985, Table 3.

advantages to U.S. household surveys in a period of rising costs of conducting interview surveys. The analysis has shown optimal allocations of 50 to 70 percent telephone in a dual frame design for the U.S. National Crime Survey, allocations which reflect the cost and sampling error advantages of telephone surveys. These results appear to favor the dual frame design over the single frame approach currently used, a design which, at present, uses telephone interviewing extensively for households being interviewed in waves subsequent to first contact. Table 11.7 summarizes the potential gains of the dual frame approach by presenting the percentage reduction in sampling variance that a dual frame design could offer when compared to a single frame design with the same total

Table 11.7 Percent Gain in Precision (or Percent Cost Saving) for the Four Dual Frame, Mixed Mode Administrative Structures Relative to a Single Frame Design by Five Types of Crimes

Victimization Rate	Single Area Frame Standard Error	Percent Gain in Variance for Fixed Cost/Percent Reduction in Cost for Fixed Variance			
		I	II	III	IV
Total personal crimes	0.00149	31.9%	34.1%	35.1%	33.0%
Robbery	0.000238	−13.9	−8.6	−7.7	−6.0
Assault	0.000648	29.0	31.8	33.1	31.1
Total household crimes	0.00279	29.7	33.2	35.0	31.4
Motor vehicle theft	0.000562	19.9	24.0	25.8	22.7

Source: R.M. Groves and J.M. Lepkowski, "Dual Frame, Mixed Mode Survey Designs," *Journal of Official Statistics*, Vol. 1, No. 3, 1985, Table 6.

cost.[7] The gains are those that occur when sampling variances of an area frame design are compared to those of a dual frame design at optimal allocation.

All types of victimization rate estimates except robbery, could achieve gains in sampling variance of 20 to 30 percent under an optimally allocated dual frame design. Robbery, however, could actually achieve a higher sampling variance under the dual frame. Alternatively, these gains in efficiency may be expressed as sample size reductions of 20 to 30 percent which could be achieved with no loss of precision for four of the five variables.

The anomalous results for robbery reflect a high element variance in the nontelephone population for this type of crime. Although the standard error of the robbery victimization rate is relatively invariant in the range of 0 to 40 percent allocation to the telephone frame, there may be other types of crime in the NCS whose standard errors will increase under dual frame allocations that are optimal for some other variables. As in all multipurpose allocation problems, some assignment of importance to various estimates must be given by the survey designer before a choice of an overall allocation can be made. No single design will be optimal for all statistics of interest, but the dual frame design appears to hold promise for many.

---

[7] The standard error for a single frame design was computed using design effects currently achieved by the single frame NCS design.

11.8 SUMMARY

The most consistent finding in studies comparing responses in face to face and telephone interviews is the *lack* of difference in results obtained through the two modes. Of course, this may be due largely to the ad hoc nature of many of the studies, whose designs did not provide for large sample sizes, and whose findings, as a result, rarely meet the criterion of statistical significance. Where differences *are* found, we are often unable to attribute them to characteristics of interaction in the two modes, or to any other single aspect of the treatments. Interviewing procedures have generally been uncontrolled in this research, and so we are unable to tell whether the performance of respondents in the telephone interview was due to the inherent nature of the medium or to the particular interviewing techniques employed within the medium.

There are, however, some theoretical principles that have gained support through replication over several studies. The faster pace of the interview leads to shorter verbal replies to questions (most evident in open questions). This may be related to the reduced cognitive processing by the respondent. This appears to be a fertile area of future research. By altering question wording, pace of the questioning, or interviewer instructions regarding probing, it may be possible to overcome this tendency. This illustrates a "reducer's" strategy.

The literature suggests that missing data on income questions are more prevalent in the telephone than in face to face interviewing. There is some evidence that these missing data levels are functions of question wording. On the whole the hypothesis that social desirability effects are larger on the telephone gets only mixed support. A review of that literature suggests that two opposing forces are confounded in the literature, one affecting the establishment of legitimacy of the survey request and another affecting the candor with which respondents answer questions.

Earlier we reviewed results that suggested reduced interviewer variance in centralized telephone interviewing. However, some attributes of interviewers that are most effectively revealed visually (e.g., race) appear to have effects in both face to face and telephone interviews. Clearly, these race effects are being communicated through voice characteristics of the interviewers. Further research might focus on other attributes of interviewer voice which affect respondent answers. This research might be a logical extension of the research on nonresponse error and voice quality in telephone surveys (see Chapter 5).

A theory of mode effects on measurement error has yet to be fully specified (see deLeeuw and van der Zouwen, 1988, for a useful start). It clearly must contain the influence of the social definition of the telephone, which will help to predict reactions to various survey uses of the medium. This definition will affect most clearly the response rates of telephone surveys for different groups. The pace of interaction on cognitive processing and attention of the respondent to the task also play a role in measurement error. The role of the physical presence of the interviewer as motivation for the respondent and as an aid for question comprehension needs to be studied. Much of the experimental literature on media effects uses highly motivated subjects, and their findings may have little inferential worth for the survey researcher.

Centralized telephone interviewing offers new possibilities for increasing the measurability and reduction of errors in survey data. Continuous supervision of interviewers can quickly alert the researcher to weaknesses in the survey procedures. Simultaneous monitoring of interviews can construct a data set which measures consistency with training guidelines. Randomized assignment of interviewers to sample cases allows estimation of interviewer variance component of measurement error.

In addition, the centralization permits closer monitoring of cost and production functions. Especially with computer–assisted telephone interviewing systems, current totals of interviewer and supervisor hours expended for different activities can be known. The current status of active sample cases can be displayed when requested by supervisors. These facilities make possible sequential designs, ones that alter response rate goals (see Chapter 5) or sample size targets while interviewing is progressing. This alteration can be guided by formal cost models. Hence, practical attempts to maximize various aspects of survey quality subject to cost constraints can be mounted.

REFERENCES

Abelson, R.P., "Beliefs are Like Possessions," *Journal for the Theory of Social Behavior,* Vol. 16, 1986, pp. 223-250.

Alba, J.W., and Hasher, L., "Is Memory Schematic?" *Psychological Bulletin,* Vol. 93, No. 2, 1983, pp. 203-231.

Alwin, D., "Approaches to the Interpretation of Relationships in the Multitrait-Multimethod Matrix," in *Sociological Methodology 1973-74,* San Francisco, Jossey-Bass, 1974, pp. 79-105.

Alwin, D., and Jackson, D., "Measurement Models for Response Errors in Surveys: Issues and Applications," in *Sociological Methodology 1980,* San Francisco, Jossey-Bass, 1980, pp. 68-113.

Amato, P.R., and Saunders, J., "The Perceived Dimensions of Help-Seeking Episodes," *Social Psychology Quarterly,* Vol. 48, No. 2, 1985, pp. 130-138.

Andersen, R., and Anderson, O., *A Decade of Health Services,* Chicago, University of Chicago Press, 1967.

Andersen, R., Kasper, J., Frankel, M.R., and Associates, *Total Survey Error,* San Francisco, Jossey-Bass, 1979.

Anderson, D., and Aitkin, M., "Variance Component Models With Binary Response: Interviewer Variability," *Journal of the Royal Statistical Society, Series A,* Vol. 47, No. 2, 1985, pp. 203-210.

Anderson, D.W., and Mantel, N., "On Epidemiologic Surveys," *American Journal of Epidemiology,* Vol. 118, No. 5, November, 1983, pp. 613-619.

Anderson, R.C., and Pichert, J.W., "Recall of Previously Unrecallable Information Following a Shift in Perspective," *Journal of Verbal Learning and Verbal Behavior,* Vol. 17, No. 1, 1978, pp. 1-12.

Andrews, F., "The Construct Validity and Error Components of Survey Measures: Estimates From a Structural Modeling Approach," Working Paper 8031, Ann Arbor, Institute for Social Research, The University of Michigan, May, 1982.

Andrews, F., "Construct Validity and Error Components of Survey Measures: A Structural Modelling Approach," *Public Opinion Quarterly,* Summer, 1984, Vol. 48, No. 2, pp. 409-422.

Argyle, M., and Dean, J., "Eye Contact, Distance and Affiliation," *Sociometry,* Vol. 28, 1965, pp. 289-304.

Aronson, S.H., "Bell's Electrical Toy: What's the Use? The Sociology of Early Telephone Use," in I. Pool (Ed.), *The Social Impact of the Telephone,* Cambridge, Mass., MIT Press, 1977.

Asher, H.B., "Some Consequences of Measurement Error in Survey Data," *American Journal of Political Science*, Vol. 18, 1974, pp. 469–485.

Bailar, B.A., "Some Sources of Error and Their Effects on Census Statistics," *Demography*, Vol. 13, No. 2, May, 1976, pp. 273–286.

Bailar, B.A., "Counting or Estimation in a Census — A Difficult Decision," *Proceedings of the Social Statistics Section, American Statistical Association*, 1983, pp. 42–47.

Bailar, B.A., and Dalenius, T. "Estimating the Response Variance Components of the U.S. Bureau of the Census' Survey Model," *Sankhya: Series B*, 1969, pp. 341–360.

Bailar, B.A., and Lanphier. C., *Development of Survey Methods to Assess Survey Practices*, Washington, DC, American Statistical Association, 1978.

Bailey, L., Moore, T.F., and Bailar, B., "An Interviewer Variance Study for the Eight Impact Cities of the National Crime Survey Cities Sample," *Journal of the American Statistical Association*, Vol. 73, March, 1978, pp. 16–23.

Ballou, J., and Del Boca, F.K., "Gender Interaction Effects on Survey Measures in Telephone Interviews," paper presented at the American Association for Public Opinion Research annual conference, 1980.

Bartlett, F.C., *Remembering: A Study in Experimental and Social Psychology*, Cambridge, Cambridge University Press, 1932.

Barton, A., "Asking the Embarrassing Question," *Public Opinion Quarterly*, Vol. 22, No. 1, 1958, pp. 67–68.

Batson, C.D., Harris, A.C., McCaul, K.D., Davis, M., and Schmidt, T., "Compassion or Compliance: Alternative Dispositional Attributions for One's Helping Behavior," *Social Psychology Quarterly*, Vol. 42, No. 4, 1979, pp. 405–409.

Belson, W., *The Design and Understanding of Survey Questions*, London, Gower, 1981.

Bem, D.J., "Self-Perception: An Alternative Interpretation of Cognitive Dissonance Phenomena," *Psychological Review*, Vol. 74, 1967, pp. 183–200.

Bendig, A.W., "Reliability and the Number of Rating Scale Categories," *The Journal of Applied Psychology*, Vol. 38, No. 1, 1954, pp. 38–40.

Benney, M., Reisman, D., and Star, S., "Age and Sex in the Interview," *The American Journal of Sociology*, Vol. 62, 1956, pp. 143–152.

Benson, P.L., and Catt, V.L., "Soliciting Charity Contributions: The Parlance of Asking for Money," *Journal of Applied Social Psychology*, Vol. 8, No. 1, 1978, pp. 84–95.

Benson, P.L., Karabenick, S.A., and Lerner, R.M., "Pretty Pleases: The Effect of Physical Attractiveness, Race, and Sex on Receiving Help," *Journal of Experimental Social Psychology*, Vol. 12, No. 5, 1976, pp. 409–415.

Benson, S., et al., "A Study of Interview Refusals," *Journal of Applied Psychology*, Vol. 35, 1951, pp. 116–119.

Benus, J., "The Problem of Nonresponse in Sample Surveys," in J.B. Lansing, et al., *Working Papers on Survey Research in Poverty Areas*, Ann Arbor, Institute for Social Research, The University of Michigan, 1971, pp. 26-59.

Bercini, D.H., and Massey, J.T., "Obtaining the Household Roster in a Telephone Survey: The Impact of Names and Placement on Response Rates," *Proceedings of the Social Statistics Section, American Statistical Association*, 1979, pp. 136-140.

Bergman, L.R., Hanve, R., and Rapp, J., "Why do Some People Refuse to Participate in Interview Surveys?" *Särtryck ur Statistisk tidskrift*, Vol. 5, 1978, pp. 341-356.

Berk, R.A., and Ray, S.C., "Selection Biases in Sociological Data," *Social Science Research*, Vol. 11, 1982, pp. 352-398.

Bethlehem, J.G., and Kersten, H.M.P., "The Nonresponse Problem," *Survey Methodology*, Vol. 7, No. 2, 1981, pp. 130-156.

Billiet, J., and Loosveldt, G., "Improvement of the Quality of Responses to Factual Survey Questions by Interviewer Training," *Public Opinion Quarterly*, Vol. 52, No. 2, 1988, pp. 190-211.

Birnbaum, Z.W., and Sirken, M.G., "On the Total Error Due to Non-Interview and to Random Sampling," *International Journal of Opinion and Attitude Research*, Vol. 4, No. 2, 1950, pp. 171-191.

Blalock, H.M., Jr., "The Cumulation of Knowledge and Generalizations to Populations," *American Sociologist*, Vol. 17, 1982, p. 112.

Bohrnstedt, G.W., "Measurement," in P.H. Rossi et al. (Eds.), *Handbook of Survey Research*, New York, Academic Press, 1983, pp. 70-122.

Boice, K., and Goldman, M., "Helping Behavior as Affected by Type of Request and Identity of Caller," *The Journal of Social Psychology*, Vol. 115, 1981, pp. 95-101.

Booker, H.S., and David, S.T., "Differences in Results Obtained by Experienced and Inexperienced Interviewers," *Journal of the Royal Statistical Society, Series A*, 1952, Part II, pp. 232-257.

Borus, M.E., "Response Error and Questioning Technique in Surveys of Earnings Information," *Journal of the American Statistical Association*, Vol. 65, No. 330, June, 1970, pp. 566-575.

Bower, G.H., Black, J.B., and Turner, T.J., "Scripts in Memory for Text," *Cognitive Psychology*, Vol. 11, 1979, pp. 177-220.

Bradburn, N.M., and Mason, W.M., "The Effect of Question Order on Responses," *Journal of Marketing Research*, November, 1964, pp. 57-61.

Bradburn, N.M., and Miles, C., "Vague Quantifiers," *Public Opinion Quarterly*, Vol. 43, No. 1, Spring, 1979, pp. 92-101.

Bradburn, N.M., Sudman, S., and Associates, *Improving Interview Method and Questionnaire Design*, San Francisco, Jossey-Bass, 1979.

Bradburn, N.M, Rips, L.J., and Shevell, S.K., "Answering Autobiographical Questions: The Impact of Memory and Inference on Surveys," *Science*, Vol. 239, 10 April, 1987, pp. 157-161.

Brehm, J.W., and Cole, H., "Effect of a Favor Which Reduces Freedom," *Journal of Personality and Social Psychology*, Vol. 3, No. 4, 1966, pp. 420-426.

Brenner, M., "Response Effects of 'Role-Restricted' Characteristics of the Interviewer," in W. Dijkstra and J. van der Zouwen (Eds.), *Response Behaviour in the Survey Interview*, New York, Academic Press, 1982.

Brewer, K.R.W., and Mellor, R.W., "The Effect of Sample Structure on Analytical Surveys," *Australian Journal of Statistics*, Vol. 15, No. 3, 1973, pp. 145-152.

Briggs, C.L., *Learning How to Ask: A Sociolinguistic Appraisal of the Role of the Interview in Social Science Research*, Cambridge, Cambridge University Press, 1986.

Brooks, C.A., and Bailar, B., *An Error Profile: Employment as Measured by the Current Population Survey*, Statistical Policy Working Paper 3, Washington, DC, Office of Federal Statistical Policy and Standards, U.S. Department of Commerce, 1978.

Brown, N.R., Rips, L.J., and Shevell, S.K., "The Subjective Dates of Natural Events in Very-Long-Term Memory," *Cognitive Psychology*, Vol. 17, 1985, pp. 139-177.

Brown, P.R., and Bishop, G.F., "Who Refuses and Resists in Telephone Surveys? Some New Evidence," University of Cincinnati, 1982.

Brunner, G.A., and Carroll, S.J., Jr., "The Effect of Prior Telephone Appointments on Completion Rates and Response Content," *Public Opinion Quarterly*, Vol. 31, 1967, pp. 652-654.

Brunner, G.A., and Carroll, S.J., Jr., "The Effect of Prior Notification on the Refusal Rate in Fixed Address Surveys," *Journal of Advertising*, Vol. 9, 1969, pp. 42-44.

Calder, B.J., Phillips, L.W., and Tybout, A.M., "The Concept of External Validity," *Journal of Consumer Research*, Vol. 9, December, 1982, pp. 240-244.

Calder, B.J., Phillips, L.W., and Tybout, A.M., "Beyond External Validity," *Journal of Consumer Research*, Vol. 10, June, 1983, pp. 112-114.

Campbell, D.T., and Fiske, D.W., "Convergent and Discriminant Validation by the Multitrait-Multimethod Matrix," *Psychological Bulletin*, Vol. 56, No. 2, 1959, pp. 81-105.

Cannell, C.F., *Reporting of Hospitalization in the Health Interview Survey*, Series 2, No. 6, National Center for Health Statistics, 1965.

Cannell, C.F., and Henson, R., "Incentives, Motives, and Response Bias," *Annals of Economic and Social Measurement*, Vol. 3, No. 2, 1974, pp. 307-317.

Cannell, C.F., Fowler, F.J., and Marquis, K.H. "A Report on Respondents' Reading of the Brochure and Letter and an Analysis of Respondents' Level of Information," Ann Arbor, Institute for Social Research, The University of Michigan, 1965.

Cannell, C.F., Lawson, S.A., and Hausser, D.L., *A Technique for Evaluating Interviewer Performance*, Ann Arbor, Institute for Social Research, The University of Michigan, 1975.

Cannell, C.F., Oksenberg, L., and Converse, J.M., "Striving for Response Accuracy: Experiments in New Interviewing Techniques," *Journal of Marketing Research*, Vol. XIV, August, 1977, pp. 306–315.

Cannell, C.F., Oksenberg, L., and Converse, J.M., *Experiments in Interviewing Techniques*, Ann Arbor, Institute for Social Research, The University of Michigan, 1979.

Cannell, C.F., Miller, P.V., and Oksenberg, L., "Research on Interviewing Techniques," in S. Leinhardt (Ed.), *Sociological Methodology, 1981*, San Francisco, Jossey-Bass, 1981.

Cannell, C.F., Groves, R.M., Magilavy, L., Mathiowetz, N., Miller, P.V., *An Experimental Comparison of Telephone and Personal Health Surveys*, National Center for Health Statistics, Technical Series 2, No. 106, 1987.

Cantril, H., and Fried, E., "The Meaning of Questions," in H. Cantril, *Gauging Public Opinion*, Princeton, NJ, Princeton University Press, 1944.

Carmines, E.C., and Zeller, R.A., *Reliability and Validity Assessment*, Sage University Paper Series on Quantitative Applications in the Social Sciences, Beverly Hills, CA, Sage Publications, Inc., 1979.

Cartwright, A., and Tucker, W., "An Attempt to Reduce the Number of Calls on an Interview Inquiry," *Public Opinion Quarterly*, Vol. 31, No. 2, 1967, pp. 299–302.

Casady, R.J., Snowden, C.B., and Sirken, M.G., "A Multiplicity Estimator for Multiple Frame Sampling," *Proceedings of the Section on Survey Research Methods, American Statistical Association*, 1981, pp. 444–447.

Cash, W.S., and Moss, A.J., *Optimum Recall Period for Reporting Persons Injured in Motor Vehicle Accidents*, Vital and Health Statistics, National Center for Health Statistics, Series 2, No. 50, 1972.

CASRO (Council of American Survey Research Organizations), Report of the CASRO Completion Rates Task Force, New York, Audits and Surveys, Inc., unpublished report, 1982.

Chai, J.J., "Correlated Measurement Errors and the Least Squares Estimators of the Regression Coefficient," *Journal of the American Statistical Association*, Vol. 66, No. 335, pp. 478–483.

Chaiken, S., "The Heuristic Model of Persuasion," in M.P. Zanna et al. (Eds.) *Social Influence: The Ontario Symposium*, Vol. 5, Hillsdale, N.J.: Erlbaum, 1987.

Champness, B.G., *The Perceived Adequacy of Four Communications Systems for a Variety of Tasks*, London, Communications Studies Group, paper No. E/72245/CH, 1972.

Champness, B.G., and Reid, A.A.L., *The Efficiency of Information Transmission: A Preliminary Comparison Between Face-to-Face Meetings and the Telephone*, Paper No. P/70240/CH, Communications Study Group, unpublished, 1970, cited in J. Short et al., *The Social Psychology of Telecommunications*, London, John Wiley & Sons, 1976.

Chromy, J., and Horvitz, D., "The Use of Monetary Incentives in National Assessment Household Surveys," *Journal of the American Statistical Association*, No. 363, September, 1978, pp. 473–478.

Cialdini, R.B., *Influence: The New Psychology of Modern Persuasion*, New York, Quill, 1984.

Clancy, K.J., and Wachsler, R.A., "Positional Effects in Shared-Cost Surveys," *Public Opinion Quarterly*, Vol. 35, No. 2, 1971, pp. 258–265.

Clark, H.H., Schreuder, R., and Buttrick, S., "Common Ground and the Understanding of Demonstrative Reference," *Journal of Verbal Learning and Verbal Behavior*, Vol. 22, 1983, pp. 245–258.

Clark, R.D., "On the Piliavin and Piliavin Model of Helping Behavior: Costs are in the Eye of the Beholder," *Journal of Applied Social Psychology*, Vol. 6, No. 4, 1976, pp. 322–328.

Cobb, S., King, S., and Chen, E., "Differences Between Respondents and Nonrespondents in a Morbidity Survey Involving Clinical Examination," *Journal of Chronic Diseases*, Vol. 6, No. 2, August, 1957, pp. 95–108.

Cochran, W.G., *Sampling Techniques*, New York, John Wiley & Sons, Inc., 1977.

Coleman, J.S., *The Adolescent Society*, New York, Free Press of Glencoe, 1961.

Collins, M., "Some Statistical Problems in Interviews with the Elderly," *SCPR Newsletter*, Winter issue 1982/83, London, Social and Community Planning Research, 1982.

Collins, M., and Butcher, B., "Interviewer and Clustering Effects in an Attitude Survey," *Journal of the Market Research Society*, Vol. 25, No. 1, 1982, pp. 39–58.

Collins, M., and Courtenay, G., "A Comparative Study of Field and Office Coding," *Journal of Official Statistics*, Vol. 1, No. 2, 1985, pp. 221–228.

Colombotos, J., "The Effects of Personal vs. Telephone Interviews on Socially Acceptable Responses," *Public Opinion Quarterly*, Vol. 29, No. 3, Fall, 1965, pp. 457–458.

Colombotos, J., et al., "Effect of Interviewer's Sex on Interview Responses," *Public Health Reports*, Vol. 83, No. 8, 1968, pp. 685–690.

Converse, J., and Schuman, H., "The Manner of Inquiry: An Analysis of Survey Question Form Across Organizations and Time," in C.F. Turner and E. Martin (Eds.), *Surveying Subjective Phenomena*, Vol. 2, New York, Russell Sage Foundation, 1984.

Converse, P., "The Nature of Belief Systems in Mass Publics," in D. Apter (Ed.), *Ideology and Discontent*, New York, Free Press, 1964.

Converse, P., "Attitudes and Nonattitudes: Continuation of a Dialogue," in E.R. Tufte (Ed.), *The Quantitative Analysis of Social Problems*, Reading, MA., Addison Wesley, 1970, pp. 168–189.

Cook, T.D., and Campbell, D.T., *Quasi-Experimentation: Design and Analysis Issues for Field Settings*, Boston, Houghton-Mifflin, 1979.

Cotter, P.R., Cohen, J., and Coulter, P.B., "Race-of-Interviewer Effects in Telephone Interviews," *Public Opinion Quarterly*, Vol. 46, No. 2, Summer, 1982, pp. 278–284.

Couch, A., and Keniston, K., "Yeasayers and Naysayers: Agreeing Response Set as a Personality Variable," *Journal of Abnormal and Social Psychology*, Vol. 60, No. 2, 1960, pp. 151–174.

Cowan, C., Murphy, L., and Weiner, J., "Effects of Supplemental Questions on Victimization Rates From the National Crime Surveys," *Proceedings of the Survey Research Methods Section, American Statistical Association*, 1978, pp. 277–282.

Craik, F.I.M., "Age Differences in Human Memory," in J. Birren and K. W. Shaie (Eds.) *Handbook of the Psychology of Aging*, New York, Van Nostrand Reinhold, 1977.

Cunningham, M.R., Steinberg, J., and Grev, R., "Wanting and Having to Help: Separate Motivations for Positive Mood and Guilt-Induced Helping," *Journal of Personality and Social Psychology*, Vol. 38, 1980, pp. 181–192.

Davies, M.F., "Cooperative Problem Solving: An Exploratory Study," Paper No. E/71159/ DV, London, Communications Studies Group, 1971.

Dawes, R.M., *Rational Choice in an Uncertain World*, San Diego, Harcourt, Brace, Jovanovich, 1988.

Dawes, R.M., and Pearson, R.W., "The Effects of Theory Based Schemas on Retrospective Data," unpublished, Carnegie–Mellon University, 1988.

DeJong, W., "Consensus Information and the Foot-in-the-Door Effect," *Personality and Social Psychology Bulletin*, Vol. 7, No. 3, 1981, pp. 423–430.

DeJong, W., and Funder, D., "Effect of Payment for Initial Compliance: Unanswered Questions About the Foot-in-the-Door Phenomenon," *Personality and Social Psychology Bulletin*, Vol. 3, 1977, pp. 662–665.

deLeeuw, E.D., and van der Zouwen, J., "Data Quality in Telephone and Face to Face Surveys: A Comparative Meta-Analysis," in R.M. Groves et al. (Eds.), *Telephone Survey Methodology*, New York, John Wiley & Sons, 1988.

DeMaio, T.J., "Refusals: Who, Where, and Why?" *Public Opinion Quarterly*, 1980, pp. 223–233.

DeMaio, T.J., "Social Desirability and Survey Measurement: A Review," in C.F. Turner and E. Martin (Eds.), *Surveying Subjective Phenomena*, Russell Sage, Vol. 2, pp. 257–282, 1984.

DeMaio, R., Marquis, K., McDonald, S., Moore, J., Sedlacek, D., and Bush, C., "Cognitive and Motivational Bases of Census and Survey Response," *Proceedings of the Second Annual Research Conference*, U.S. Bureau of the Census, 1986, pp. 271-295.

Deming, W.E., "On Errors in Surveys," *American Sociological Review*, Vol. 9, No. 4, August, 1944, pp. 359-369.

Deming, W.E., "On the Distinction Between Enumerative and Analytic Surveys," *Journal of the American Statistical Association*, Vol. 48, 1953, pp. 224-255.

Deming, W.E., "On a Probability Mechanism to Attain an Economic Balance Between the Resultant Error of Response and the Bias of Nonresponse," *Journal of the American Statistical Association*, Vol. 48, No. 264, 1953, pp. 743-772.

Dijkstra, W., "How Interviewer Variance Can Bias the Results of Research on Interviewer Effects," *Quality and Quantity*, Vol. 17, pp. 179-187.

Dillman, D., *Mail and Telephone Surveys*, New York, John Wiley & Sons, 1978.

Dillman, D., personal communication, 1988.

Dillman, D., Gallegos, J.G., and Frey, J.H., "Reducing Refusal Rates for Telephone Interviews," *Public Opinion Quarterly*, Vol. 40, No. 1, Spring, 1976, pp. 66-78.

Dodge, R.W., "Victim Recall Pretest," unpublished memorandum, Washington, DC, U.S. Bureau of the Census, 1970, also excerpted in R.G. Lehnen and W.G. Skogan, *National Crime Survey: Working Papers*, Vol. I, Bureau of Justice Statistics, 1981.

Dohrenwend, B.S., "An Experimental Study of Payments to Respondents," *Public Opinion Quarterly*, Vol. 34, 1970, pp. 621-624.

Dohrenwend, B.S., and Dohrenwend, B.P., "Sources of Refusals in Surveys," *Public Opinion Quarterly*, Vol. 32, No. 1, 1968, pp. 74-83.

Dumouchel, W.H., and Duncan, G.J., "Using Sample Survey Weights in Multiple Regression Analyses of Stratified Samples," *Journal of the American Statistical Association*, Vol. 78, No. 383, 1983, pp. 535-543.

Duncan, D.O., Featherman, D.L., and Duncan, B., *Socioeconomic Background and Achievement*, New York, Seminar Press, 1972.

Duncan, G.J., and Hill, D.H., "An Investigation of the Extent and Consequences of Measurement Error in Labor Economic Survey Data," *Journal of Labor Economics*, Vol. III, No. 4, 1985, pp. 508-532.

Dunkelberg, W., and Day, G., "Nonresponse Bias and Callbacks in Sample Surveys," *Journal of Marketing Research*, Vol. X, May, 1973, pp. 160-168.

Ebbinghaus, H., *Memory: A Contribution to Experimental Psychology*, New York, Teacher's College, 1913, original work published 1885.

Ekman, P., "Differential Communication of Affect by Head and Body Cues," *Journal of Personality and Social Psychology*, Vol. 2, 1965, pp. 726-35.

Erdos, P.L., "How to Get Higher Returns from Your Mail Surveys," *Printer's Ink*, Vol. 258, Issue 8, 1957, pp. 30–31.

Erikson, R.S., "The SRC Panel Data and Mass Political Attitudes," *British Journal of Political Science*, Vol. 9, 1979, pp. 89–114.

Fay, R.E., Passel, J.S., and Robinson, J.G., *The Coverage of Population in the 1980 Census*, Evaluation and Research Report PHC80-E4, Washington, DC, U.S. Department of Commerce, 1988.

Feather, J., *A Study of Interviewer Variance*, Saskatoon, Canada, Department of Social and Preventive Medicine, University of Saskatchewan, 1973.

Fellegi, I.P., "Response Variance and Its Estimation," *Journal of the American Statistical Association*, Vol. 59, December, 1964, pp. 1016–1041.

Fellegi, I.P., and Sunter, A.B., "Balance Between Different Sources of Survey Errors — Some Canadian Experiences," *Sankhya, Series C*, Vol. 36, Pt. 3, 1974, pp. 119–142.

Ferber, R., "The Effect of Respondent Ignorance on Survey Results," *Journal of the American Statistical Association*, Vol. 51, No. 276, 1956, pp. 576–586.

Ferber, R., and Sudman, S., "Effects of Compensation in Consumer Expenditure Studies," *Annals of Economic and Social Measurement*, Vol. 3, No. 2, 1974, pp. 319–331.

Fischhoff, B., "Perceived Informativeness of Facts," *Journal of Experimental Psychology: Human Perception and Performance*, Vol. 3, No. 2, 1977, pp. 349–358.

Forbes, G.B., and Gromoll, H.F., "The Lost Letter Technique as a Measure of Social Variables: Some Exploratory Findings," *Social Forces*, Vol. 50, September, 1971, pp. 113–115.

Fowler, F.J., Jr., and Mangione, T.W., *The Value of Interviewer Training and Supervision*, Final Report to the National Center for Health Services Research, Grant #3-R18-HS04189, 1985.

Fox, J.A., and Tracy, P.E., *Randomized Response: A Method for Sensitive Surveys*, Sage Series in Quantitative Applications in the Social Sciences, No. 58, Beverly Hills, CA, Sage Publications, 1986.

Frankel, M.R., *Inference from Survey Samples: An Empirical Investigation*, Ann Arbor, Institute for Social Research, The University of Michigan, 1971.

Freedman, J.L., and Fraser, S.C., "Compliance Without Pressure: The Foot-in-the-Door Technique," *Journal of Personality and Social Psychology*, Vol. 4, 1966, pp. 195–202.

Furst, L.G., and Blitchington, W.P., "The Use of a Descriptive Cover Letter and Secretary Pre-letter to Increase Response Rate in a Mailed Survey," *Personnel Psychology*, Vol. 32, 1979, pp. 155–159.

Gallup, G.H., *A Guide to Public Opinion Polls*, Princeton, Princeton University Press, 1948.

Gauld, A., and Stephenson, G.M., "Some Experiments Relating to Bartlett's Theory of Remembering," *British Journal of Psychology*, Vol. 58, No. 1 and 2, 1967, pp. 39-49.

Gergen, K.J., and Back, K.W., "Communication in the Interview and the Disengaged Respondent," *Public Opinion Quarterly*, Vol. 30, 1966, pp. 385-398.

Glenn, N.G., "Aging, Disengagement, and Opinionation," *Public Opinion Quarterly*, Vol. 33, No. 1, 1969, pp. 17-33.

Goldberger, A.S., "Linear Regression After Selection," *Journal of Econometrics*, Vol. 15, No. 2, 1981, pp. 357-366.

Goodstadt, M.S., "Helping and Refusal to Help: A Test of Balance and Reactance Theories," *Journal of Experimental Social Psychology*, Vol. 7, 1971, pp. 610-622.

Gove, W., and Geerken, M., "Response Bias in Surveys of Mental Health: An Empirical Investigation," *American Journal of Sociology*, Vol. 82, 1977, pp. 1289-1317.

Goyder, J., *The Silent Minority: Nonrespondents in Sample Surveys*, Boulder, Colorado, Westview Press, 1987.

Groves, R.M., "An Empirical Comparison of Two Telephone Sample Designs," *Journal of Marketing Research*, Vol. 15, 1978, pp. 622-631.

Groves, R.M., "On the Mode of Administering a Questionnaire and Responses to Open-Ended Items," *Social Science Research*, Vol. 7, 1978, pp. 257-271.

Groves, R.M., and Fultz, N.H., "Gender Effects Among Telephone Interviewers in a Survey of Economic Attitudes," *Sociological Methods and Research*, Vol. 14, No. 1, August, 1985, pp. 31-52.

Groves, R.M., and Kahn, R.L., *Surveys by Telephone*, New York, Academic Press, 1979.

Groves, R.M., and Lepkowski, J.M., "Cost and Error Modeling for Large-Scale Telephone Surveys," *First Annual Research Conference, Proceedings, U.S. Bureau of the Census*, Washington, DC, U.S. Bureau of the Census, 1985, pp. 330-357.

Groves, R.M., and Lepkowski, J.M., "Dual Frame, Mixed Mode Survey Designs," *Journal of Official Statistics*, Vol. 1, No. 3, 1985, pp. 263-286.

Groves, R.M., and Magilavy, L.J., "Increasing Response Rates to Telephone Surveys: A Door in the Face for Foot-in-the-Door?" *Public Opinion Quarterly*, 1981, Vol. 45, pp. 346-358.

Groves, R.M., and Magilavy, L.J., "Measuring and Explaining Interviewer Effects in Centralized Telephone Surveys," *Public Opinion Quarterly*, Vol. 50, No. 2, 1986, pp. 251-266.

Groves, R.M., and Robinson, D., *Final Report on Callback Algorithms on CATI Systems*, Washington, DC, report to the U.S. Bureau of the Census, 1982.

Groves, R.M., Miller, P.V., and Handlin, V., *Telephone Survey Methodology: A Review*, report submitted to the Bureau of Justice Statistics, Washington, DC, 1980.

Guest, L., "A Comparison of Two-Choice and Four-Choice Questions," *Journal of Marketing Research*, Vol. II, 1962, pp. 32–34.

Haas, D.F., "Survey Sampling and the Logic of Inference in Sociology," *The American Sociologist*, Vol. 17, May, 1982, pp. 103–111.

Hansen, M.H., and Hurwitz, W.N., "The Problem of Nonresponse in Sample Surveys," *The Journal of the American Statistical Association*, December, 1958, pp. 517–529.

Hansen, M.H., Hurwitz, W.N., and Madow, W.G., *Sample Survey Methods and Theory*, Vols. I and II, New York, John Wiley & Sons, 1953.

Hansen, M.H., Hurwitz, W.N., and Bershad, M.A., "Measurement Errors in Censuses and Surveys," *Bulletin of the International Statistical Institute*, Vol. 38, No. 2, 1961, pp. 359–374.

Hansen, M.H., Hurwitz, W.N., and Pritzker, L., "Response Variance and Its Estimation," *Journal of the American Statistical Association*, Vol. 59, 1964, pp. 1016–1041.

Hansen, M.H., Madow, W.G., and Tepping, B.J., "An Evaluation of Model-Dependent and Probability-Sampling Inferences in Sample Surveys," *Journal of the American Statistical Association*, Vol. 78, No. 384, December, 1983, pp. 776–793.

Hanson, R.H., and Marks, E.S., "Influence of the Interviewer on the Accuracy of Survey Results," *Journal of the American Statistical Association*, Vol. 53, No. 283, September, 1958, pp. 635–655.

Harahush, T.W., and Fernandez, I., "The Coverage of Housing Units — Results from Two Census Pretests," *Proceedings of the Social Statistics Section, American Statistical Association*, 1978, pp. 426–431.

Harkins, S.G., and Petty, R.E., "The Effects of Source Magnification of Cognitive Effort on Attitudes: An Information Processing View," *Journal of Personality and Social Psychology*, Vol. 40, 1981, pp. 401–413.

Harlan, W., et al., *Dietary Intake and Cardiovascular Risk Factors, Part II. Serum Urate, Serum Cholesterol, and Correlates*, Series 11, No. 227, Washington, DC, National Center for Health Statistics, 1983.

Harris, M.B., and Meyer, F.W., "Dependency, Threat, and Helping," *The Journal of Social Psychology*, Vol. 90, 1973, pp. 239–242.

Hartley, H.O., "Contribution to Discussion of Paper by F. Yates," *Journal of the Royal Statistical Association: Series A*, Vol. 109, 1946, pp. 12–43.

Hartley, H.O., "Multiple Frame Surveys," *Proceedings of the Social Statistics Section, American Statistical Association*, 1962, pp. 203–206.

Hartley, H.O., "Multiple Frame Methodology and Selected Applications," *Sankhya, Series C*, Vol. 36, Pt. 3, 1974, pp. 99–118.

Hasher, L., and Griffin, M., "Reconstructive and Reproductive Processes in Memory," *Journal of Experimental Psychology*, Vol. 4, No. 4, 1978, pp. 318–330.

Hasher, L., Attig, M.S., and Alba, J.W., "I Knew it All Along: Or, Did I?", *Journal of Verbal Learning and Verbal Behavior*, Vol. 20, 1981, pp. 86-96.

Hastie, R., and Carlston, D., "Theoretical Issues in Person Memory," in R. Hastie et al. (Eds.), *Person Memory: The Cognitive Basis of Social Perception*, Hillsdale, NJ, Lawrence Erlbaum, 1980, pp. 1-53.

Hatchett, S., and Schuman, H., "Race of Interviewer Effects Upon White Respondents," *Public Opinion Quarterly*, Vol. 39, No. 4, 1975-76, pp. 523-528.

Hawkins, D.F., "Estimation of Nonresponse Bias," *Sociological Methods and Research*, Vol. 3, 1975, pp. 461-488.

Hawkins, L., and Coble, J., "The Problem of Response Error in Interviews," in J.B. Lansing et al. (Eds.), *Working Papers on Survey Research in Poverty Areas*, Ann Arbor, Institute for Social Research, The University of Michigan, 1971.

Heckman, J.J., "Sample Selection Bias as a Specification Error," *Econometrica*, Vol. 47, No. 1, January, 1979, pp. 153-161.

Heckman, J.J., "Sample Selection Bias as a Specification Error," in J.P. Smith (ED.), *Female Labor Supply: Theory and Estimation*, Princeton, NJ, Princeton University Press, 1980.

Herriot, R.A., "Collecting Income Data on Sample Surveys: Evidence From Split-Panel Studies," *Journal of Marketing Research*, Vol. 14, 1977, pp. 322-329.

Herzog, A.R., and Rodgers, W.L., "Survey Nonresponse and Age," unpublished manuscript, Ann Arbor, Institute for Social Research, The University of Michigan, 1981.

Herzog, A.R., and Rodgers, W.L., "Age and Response Rates to Interview Sample Surveys," *Journal of Gerontology: Social Sciences*, Vol. 43, No. 6, 1988, pp. S200-S205.

Herzog, A.R., and Rodgers, W.L., "Age Differences in Memory Performance and Memory Ratings in a Sample Survey," *Psychology of Aging*, 1989.

Hess, I., and Pillai, R.K., "Nonresponse Among Family Heads and Wives in SRC Surveys," Ann Arbor, Survey Research Center, The University of Michigan, mimeograph, 1962.

Hill, C.T., Rubin, Z., Peplau, L.A., and Willard, S.G., "The Volunteer Couple: Sex Differences, Couple Commitment, and Participation in Research on Interpersonal Relationships," *Social Psychology Quarterly*, Vol. 42, No. 4, 1979, pp. 415-420.

Hill, D.H., "Home Production and the Residential Electric Load Curve," *Resources and Energy*, Vol. 1, 1978, pp. 339-358.

Hochstim, J.R., "A Critical Comparison of Three Strategies of Collecting Data From Households," *Journal of the American Statistical Association*, Vol. 62, 1967, pp. 976-989.

Hofstadter, D.R., *Gödel, Escher, Bach: An Eternal Golden Braid*, New York, Vintage Books, 1980.

Holahan, C.J., "Effects of Urban Size and Heterogeneity on Judged Appropriateness of Altruistic Responses: Situational vs. Subject Variables," *Sociometry*, Vol. 40, No. 4, 1977, pp. 378–382.

Horowitz, I.A., "Effects of Choice and Locus of Dependence on Helping Behavior," *Journal of Personality and Social Psychology*, Vol. 8, 1968, pp. 373–376.

Horvath, F.W., "Forgotten Unemployment: Recall Bias in Retrospective Data," *Monthly Labor Review*, March, 1982, pp. 40–44.

Horvitz, D.G., and Koch, G.G., "The Effect of Response Errors on Measures of Association," in N. Johnson and H. Smith, Jr. (Eds.), *New Developments in Survey Sampling*, New York, John Wiley & Sons, 1969.

Horvitz, D.G., Shah, B.U., and Simmons, W.R., "The Unrelated Questions Randomized Response Model," *Proceedings of the Social Statistics Section, American Statistical Association*, 1967, pp. 65–72.

House, J., and Wolf, S., "Effects of Urban Residence and Interpersonal Trust and Helping Behavior," *Journal of Personality and Social Psychology*, Vol. 36, No. 9, 1978, pp. 1029–1043.

Hulicka, I.M., "Age Differences in Retention as a Function of Interference," *Journal of Gerontology*, Vol. 22, 1967, pp. 180–184.

Hyman, H., *Interviewing in Social Research*, Chicago, University of Chicago Press, 1954.

Hyman, H.H., and Sheatsley, P.B., "The Current Status of American Public Opinion," in J.C. Payne (Ed.), *The Teaching of Contemporary Affairs*, National Council of Social Studies, 1950.

Jackson, J.M., and Latane, B., "Strength and Number of Solicitors and the Urge Toward Altruism," *Personality and Social Psychology Bulletin*, Vol. 7, No. 3, September, 1981, pp. 415–422.

Jackson, J.S., *The Impact of a Favor and Dependency of the Reinforcing Agent on Social Reinforcer Effectiveness*, unpublished doctoral dissertation, Detroit, Wayne State University, 1972.

Johnson, W.R., Sieveking, N.A., and Clanton, E.S., III, "Effects of Alternative Positioning of Open-Ended Questions in Multiple-Choice Questionnaires," *Journal of Applied Psychology*, Vol. 59, 1974, pp. 776–778.

Jones, E.E., and Nisbett, R.E., "The Actor and the Observer: Divergent Perceptions of the Causes of Behavior," in E.E. Jones et al. (Eds.), *Attribution: Perceiving the Causes of Behavior*, Morristown, N.J., General Learning Press, 1972.

Jones, R.A., "Volunteering to Help: The Effects of Choices, Dependence, and Anticipated Dependence," *Journal of Personality and Social Psychology*, Vol. 14, No. 2, 1970, pp. 121–129.

Jordan, L.A., Marcus, A.C., and Reeder, L.G., "Response Styles in Telephone and Household Interviewing: A Field Experiment," *Public Opinion Quarterly*, Vol. 44, No. 2, Summer, 1980, pp. 210-222.

Joreskog, K.G., and Sorbom, D., *LISREL—Analysis of Linear Structural Relationships by the Method of Maximum Likelihood, User's Guide, Version VI*, Chicago, National Educational Resources, 1984.

Kahn, R.L., and Cannell, C.F., *The Dynamics of Interviewing*, New York, John Wiley & Sons, 1957.

Kahneman, D., and Tversky, A., "Subjective Probability: A Judgment of Representativeness," *Cognitive Psychology*, Vol. 3, 1971, pp. 430-454.

Kalton, G., *Compensating for Missing Survey Data*, Ann Arbor, Institute for Social Research, The University of Michigan, 1983a.

Kalton, G., "Models in the Practice of Survey Sampling," *International Statistical Review*, Vol. 51, 1983b, pp. 175-188.

Kalton, G., Collins, M., and Brook, L., "Experiments in Wording Opinion Questions," *Journal of the Royal Statistical Society, Series C*, Vol. 27, No. 2, 1978, pp. 149-161.

Kalton, G., Roberts, J., and Holt, D., "The Effects of Offering a Middle Response Option with Opinion Questions," *The Statistician*, Vol. 29, No. 1, 1980, pp. 65-78.

Katosh, J.P., and Traugott, M.W., "The Consequences of Validated and Self-Reported Voting Measures," *Public Opinion Quarterly*, Vol. 45, No. 4, 1981, pp. 519-535.

Kemsley, W.F.F., "Family Expenditure Survey. A Study of Differential Response Based on a Comparison of the 1971 Sample with the Census," *Statistical News*, No. 31, November, 1975, pp. 31.16-31.21.

Kemsley, W.F.F., "National Food Survey — A Study of Differential Response Based on a Comparison of the 1971 Sample with the Census," *Statistical News*, Nov. 1976, No. 35, pp. 35.18-35.22.

Kenkel, W.F., "Sex of Observer and Spousal Roles in Decision Making," *Marriage and Family Living*, May, 1961, pp. 185-186.

Kintsch, W., and Van Dijk, T.A., "Toward a Model of Text Comprehension and Production," *Psychological Bulletin*, Vol. 85, No. 5, 1978, pp. 363-394.

Kish, L., "Studies of Interviewer Variance for Attitudinal Variables," *Journal of the American Statistical Association*, Vol. 57, March, 1962, pp. 92-115.

Kish, L., *Survey Sampling*, New York, John Wiley & Sons, 1965.

Kish, L., "Optima and Proxima in Linear Sample Designs," *Journal of the Royal Statistical Society, Series A (General)*, Vol. 139, Part 1, 1976, pp. 80-95.

Kish, L., "Chance, Statistics, and Statisticians," *Journal of the American Statistical Association*, Vol. 73, No. 361, 1978, pp. 1-6.

Kish, L., "Populations for Survey Sampling," *Survey Statistician*, No. 1, February, 1979, pp. 14-15.

Kish, L., and Frankel, M.R., "Balanced Repeated Replications for Standard Errors," *Journal of the American Statistical Association*, Vol. 65, 1970, pp. 1071-1094.

Kish, L., and Frankel, M.R., "Inference From Complex Samples," *Journal of the Royal Statistical Society, Series B*, Vol. 36, No. 1, 1974, pp. 1-37.

Kish, L., and Hess, I., "On Noncoverage of Sample Dwellings," *Journal of the American Statistical Association*, Vol. 53, June, 1958, pp. 509-524.

Kish, L., Groves, R., and Krotki, K., "Sampling Errors for Fertility Surveys," *Occasional Papers, World Fertility Survey*, London, World Fertility Survey, 1976.

Kleinke, C.L., "Effects of Dress on Compliance to Requests in a Field Setting," *The Journal of Social Psychology*, Vol. 101, 1977, pp. 223-224.

Klevmarken, N.A., "Econometric Inference From Survey Data," mimeo, Department of Statistics, University of Gothenburg, Sweden, August, 1983.

Kmenta, J., *Elements of Econometrics*, New York, Macmillan Publishing Co., 1971.

Koch, G.G., "An Alternative Approach to Multivariate Response Error Model for Sample Survey Data With Applications to Estimators Involving Subclass Means," *Journal of the American Statistical Association*, Vol. 68, 1973, pp. 906-913.

Koo, H.P., et al., "An Experiment on Improving Response Rates and Reducing Call Backs in Household Surveys," *Proceedings of the Social Statistics Section, American Statistical Association*, 1976, pp. 491-494.

Koons, D.A., *Quality Control and Measurement of Nonsampling Error in the Health Interview Survey*, Washington, DC, National Center for Health Statistics, Series 2, Number 54, DHEW Publication No. HSM 73-1328, March, 1973.

Kormendi, E., "The Quality of Income Information in Telephone and Face to Face Surveys," in R.M. Groves et al. (Eds.), *Telephone Survey Methodology*, New York, John Wiley & Sons, 1988.

Korns, A., "Coverage Issues Raised by Comparisons Between CPS and Establishment Employment," *Proceedings of the Social Statistics Section, American Statistical Association*, 1977, pp. 60-69..

Korte, C., and Kerr, N., "Response to Altruistic Opportunities in Urban and Nonurban Settings," *The Journal of Social Psychology*, Vol. 95, 1975, pp. 183-184.

Korte, C., Ypma, I., and Toppen, A., "Helpfulness in Dutch Society as a Function of Urbanization and Environmental Input Level," *Journal of Personality and Social Psychology*, Vol. 32, No. 6, 1975, pp. 996-1003.

Kristiansson, K.E., "A Non-Response Study in the Swedish Labour Force Surveys," in Methodological Studies From the Research Institute for Statistics on Living Conditions, National Central Bureau of Statistics, Stockholm, Sweden, No. 10E, 1980.

Krosnick, J.A., and Alwin, D.F., "An Evaluation of a Cognitive Theory of Response-Order Effects in Survey Measurement," *Public Opinion Quarterly*, Vol. 51, No. 2, 1987, pp. 201-219.

Kviz, F.J., "Toward a Standard Definition of Response Rate," *Public Opinion Quarterly*, Vol. 41, 1977, pp. 265-267.

Landis, J.R., et al., "Feminist Attitudes as Related to Sex of the Interviewer," *Pacific Sociological Review*, Vol. 16, No. 3, July, 1973, pp. 305-314.

Lansing, J.B., Withey, S.B., and Wolfe, A.C., *Working Papers on Survey Research in Poverty Areas*, Ann Arbor, Institute for Social Research, The University of Michigan, 1971.

Latane, B., and Darley, J.M., "Group Inhibition of Bystander Intervention in Emergencies," *Journal of Personality and Social Psychology*, Vol. 10, 1968, pp. 215-221.

Laurent, A., "Effects of Question Length on Reporting Behavior in the Survey Interview," *Journal of the American Statistical Association*, Vol. 67, No. 338, June, 1972, pp. 298-305.

Law Enforcement Assistance Administration, *San Jose Methods Test of Known Crime Victims*, Statistics Technical Report No. 1, Washington, DC, June, 1972.

Lazerwitz, B., "A Sample of a Scattered Group," *Journal of Marketing Research*, Vol. 1, February, 1964, pp. 68-71.

Lepkowski, J.M., and Groves, R.M., "A Mean Squared Error Model for Dual Frame, Mixed Mode Survey Design," *Journal of the American Statistical Association*, Vol. 81, No. 396, 1986, pp. 930-937.

Lessler, J.T., Bercini, D., Tourangeau, R., and Salter, W., "Results of the Cognitive/ Laboratory Studies of the 1986 NHIS Dental Care Supplement," *Proceedings of the Survey Research Methods Section, American Statistical Association*, 1985, pp. 460-463.

Lessler, J.T., Kalsbeek, W.D., and Folsom, R.E., "A Taxonomy of Survey Errors, Final Report," Research Triangle Institute project 255U-1791-03F, Research Triangle Park, NC, Research Triangle Institute, 1981.

Lindstrom, H., "Non-Response Errors in Sample Surveys," *Urval*, No. 16, 1983.

Linton, M., "Transformations of Memory in Everyday Life," in U. Neisser (Ed.), *Memory Observed*, New York, W.H. Freeman and Co., 1982, pp. 77-91.

Locander, W.B., and Burton, J.P., "The Effect of Question Form on Gathering Income Data By Telephone," *Journal of Marketing Research*, Vol. 13, 1976, pp. 189-192.

Loftus, E.F., *Eyewitness Testimony*, Cambridge, MA, Harvard University Press, 1979.

Loftus, E.F., and Loftus, G.R., "On the Permanence of Stored Information in the Human Brain," *American Psychologist*, Vol. 35, No. 5, May, 1980, pp. 409-420.

Long, J.S., *Confirmatory Factor Analysis*, Beverly Hills, Sage Publications, 1983.

Lord, C.G., "Schemas and Images as Memory Aids: Two Modes of Processing Social Information," *Journal of Personality and Social Psychology*, Vol. 38, No. 2, 1980, pp. 257-269.

Lord, F., and Novick, M.R., *Statistical Theories of Mental Test Scores*, Reading, MA, Addison-Wesley, 1968.

Lynch, J.G., Jr., "On the External Validity of Experiments in Consumer Research," *Journal of Consumer Research*, Vol. 9, December, 1982, pp. 225-239.

Madow, W.G., *Interview Data on Chronic Conditions Compared With Information Derived From Medical Records*, Vital and Health Statistics, Series 2, No. 23, Washington, DC, National Center for Health Statistics, 1967.

Madow, W.G., Olkin, I., and Rubin, D.B., *Incomplete Data in Sample Surveys*, New York, Academic Press, Vols. 1-3, 1983.

Mahalanobis, P.C., "Recent Experiments in Statistical Sampling in the Indian Statistical Institute," *Journal of the Royal Statistical Society*, Vol. 109, 1946, pp. 325-378.

Malinvaud, E., *Statistical Methods of Econometrics*, Chicago, Rand McNally, 1966.

Marckwardt, A.M., "Response Rates, Callbacks, and Coverage: The WFS Experience," *World Fertility Survey Scientific Reports*, No. 55, London, World Fertility Survey, April, 1984.

Marcus, A., and Telesky, C., "Nonparticipants in Telephone Follow-up Interviews," *Proceedings of the Health Survey Methods Conference*, Washington, DC, National Center for Health Services Research, 1984, pp. 128-134.

Markus, G.B., "Stability and Change in Political Attitudes: Observed, Recalled, and 'Explained'," *Political Behavior*, Vol. 8, 1986, pp. 21-44.

Markus, H., and Zajonc, R.B., "The Cognitive Perspective in Social Psychology," in G. Lindzey and E. Aronson (Eds.), *Handbook of Social Psychology*, 3rd Edition, New York, Random House, 1985.

Marquis, K., "Purpose and Procedure of the Tape Recording Analysis," in J. Lansing, S. Withey, and A. Wolfe (Eds.), *Working Papers on Survey Research in Poverty Areas*, Ann Arbor, Institute for Social Research, The University of Michigan, 1971.

Marquis, K., *Record Check Validity of Survey Responses: A Reassessment of Bias in Reports of Hospitalizations*, Rand Corporation, 1978.

Marquis, K., "Survey Response Rates: Some Trends, Causes, and Correlates," paper presented at the 2nd Biennial Health Survey Methods Research Conference, Washington, DC, National Center for Health Services Research, 1979.

Marquis, K.H., and Cannell, C.F., *A Study of Interviewer-Respondent Interaction in the Urban Employment Surveys*, Ann Arbor, Institute for Social Research, The University of Michigan, 1969.

Marsh, C., *The Survey Method: The Contribution of Surveys to Sociological Explanation*, London, Allen and Unwin, 1982.

Martin, E., *Report on the Development of Alternative Screening Procedures for the National Crime Survey*, Washington, DC, Bureau of Social Science Research, 1986.

Matarazzo, J.D., Weitman, M., Saslow, G., and Wiens, A.N., "Interviewer Influence on Durations of Interviewee Speech," *Journal of Verbal Learning and Verbal Behavior*, Vol. 1, 1963, pp. 451–458.

Matell, M.S., and Jacoby, J., "Is There an Optimal Number of Alternatives for Likert Scale Items? Study I: Reliability and Validity," *Educational and Psychological Measurement*, Vol. 31, 1971, pp. 657–674.

Matell, M.S., and Jacoby, J., "Is There an Optimal Number of Alternatives for Likert Scale Items? Effects of Testing Time and Scale Properties," *Journal of Applied Psychology*, Vol. 56, No. 6, 1972, pp. 506–509.

Mathiowetz, N., and Cannell, C., "Coding Interviewer Behavior as a Method of Evaluating Performance," *Proceedings of the American Statistical Association*, 1980, pp. 525–528.

Mathiowetz, N., and Groves, R.M., "The Effects of Respondent Rules on Health Survey Reports," *American Journal of Public Health*, Vol. 75, 1985, pp. 639–644.

McCarthy, P.J., and Snowden, C.B., "The Bootstrap and Finite Population Sampling," Washington, DC, National Center for Health Statistics, Series 2, No. 95, U.S. Department of Health and Human Services, 1985.

McGowan, H., "Telephone Ownership in the National Crime Survey," unpublished memorandum, Washington, DC, U.S. Bureau of the Census, 1982.

McGuire, W.J., "The Nature of Attitudes and Attitude Change," in G. Lindzey and E. Aronson (Eds.), *The Handbook of Social Psychology*, Second Edition, Reading, MA, Addison-Wesley, 1969.

Mehrabian, A., *Silent Messages*, Belmont, Wadsworth, 1968.

Mercer, J.R., and Butler, E.W., "Disengagement of the Aged Population and Response Differentials in Survey Research," *Social Forces*, Vol. 46, No. 1, 1967, pp. 89–96.

Merrens, M.R., "Nonemergency Helping Behavior in Various Sized Communities," *The Journal of Social Psychology*, Vol. 90, 1973, pp. 327–328.

Miller, J.D., Cisin, I.H., Courtless, T., and Harrell, A.V., *Valid Self-Reports of Criminal Behavior: Applying the 'Item Count Approach,'* Final Report Grant No. 82-IJ-CX-0045, Washington, DC, National Institute of Justice, 1986.

Miller, P.V., "Alternative Question Forms for Attitude Scale Questions in Telephone Interviews," *Public Opinion Quarterly*, Vol. 48, No. 4, 1984, pp. 766-668.

Miller, P.V., and Groves, R., "Measuring Reporting Accuracy of Criminal Victimization Through Record Checks," *Public Opinion Quarterly*, Vol. 49, Fall, 1985, pp. 366-380.

Monsees, M.L., and Massey, J.T., "Adapting Procedures for Collecting Demographic Data in a Personal Interview to a Telephone Interview," *Proceedings of the Section on Survey Research Methods, American Statistical Association*, 1979, pp. 130-135.

Mook, D.G., "In Defense of External Invalidity," *American Psychologist*, April, 1983, pp. 379-387.

Morgenstern, R., and Barrett, N., "The Retrospective Bias in Unemployment Reporting by Sex, Race, and Age," *Journal of the American Statistical Association*, Vol. 69, 1974, pp. 355-357.

Morley, I.E., and Stephenson, G.M., "Interpersonal and Interparty Exchange: A Laboratory Simulation of an Industrial Negotiation at the Plant Level," *British Journal of Psychology*, Vol. 60, 1969, pp. 543-545.

Morton-Williams, J., "The Use of 'Verbal Interaction Coding' for Evaluating a Questionnaire," *Quality and Quantity*, Vol. 3, 1979, pp. 59-75.

Munsat, S., *The Concept of Memory*, New York, Random House, 1966.

Nathan, G., "An Empirical Study of Response and Sampling Errors for Multiplicity Estimates with Different Counting Rules," *Journal of the American Statistical Association*, Vol. 71, No. 356, December, 1976, pp. 808-815.

National Academy of Sciences, *Panel on Privacy and Confidentiality as Factors in Survey Response*, Washington, DC, 1979.

Nealon, J., "The Effects of Male vs. Female Telephone Interviewers," Washington, DC, U.S. Department of Agriculture, Statistical Research Division, June, 1983.

Neter, J., Maynes, E.S., and Ramanathan, R., "The Effect of Mismatching on the Measurement of Response Errors," *Journal of the American Statistical Association*, Vol. 60, No. 312, 1965, pp. 1005-1027.

Neyman, J., "On the Two Different Aspects of the Representative Method: The Method of Stratified Sampling and the Method of Purposive Selection," *Journal of the Royal Statistical Society*, Vol. 97, 1934, pp. 558-606.

Neyman, J., "Contribution to the Theory of Sampling Human Populations," *Journal of the American Statistical Association*, Vol. 33, 1938, pp. 101-116.

Nisbett, R.E., and Ross, L., *Human Inference: Strategies and Shortcomings of Social Judgment*, Englewood Cliffs, NJ, Prentice-Hall, 1980.

Nisbett, R.E., Caputo, C., Legant, P., and Marecek, J., "Behavior as Seen by the Actor and as Seen by the Observer," *Journal of Personality and Social Psychology*, Vol. 27, No. 2, 1973, pp. 154-164.

O'Muircheartaigh, C.A., "Response Errors in an Attitudinal Sample Survey," *Quality and Quantity*, Vol. 10, 1976, pp. 97-115.

O'Neil, M., "Estimating the Nonresponse Bias due to Refusals in Telephone Surveys," *Public Opinion Quarterly*, 1979, pp. 218-232.

Oksenberg, L., *Analysis of Monitored Telephone Interviews*, Report to the U.S. Bureau of the Census for JSA 80-23, Ann Arbor, Survey Research Center, The University of Michigan, May, 1981.

Oksenberg, L., Coleman, L., and Cannell, C.F., "Interviewers' Voices and Refusal Rates in Telephone Surveys," *Public Opinion Quarterly*, Vol. 50, No. 1, 1986, pp. 97-111.

Passel, J.S., and Robinson, J.G., "Revised Estimates of the Coverage of the Population in the 1980 Census Based on Demographic Analysis: A Report on Work in Progress," *Proceedings of the Social Statistics Section, American Statistical Association*, 1984, pp. 160-165.

Pearce, P.L., and Amato, P.R., "A Taxonomy of Helping: A Multidimensional Scaling Analysis," *Social Psychology Quarterly*, Vol. 43, No. 4, 1980, pp. 363-371.

Pepper, S., and Prytulak, L.S., "Sometimes Frequently Means Seldom: Context Effect in the Interpretation of Quantitative Expressions," *Journal of Research in Personality*, Vol. 8, 1974, pp. 95-101.

Pettigrew, T.F., *A Profile of the Negro American*, New York, Van Nostrand, 1964.

Petty, R.E., "The Role of Cognitive Responses in Attitude Change Processes," in R. Petty et al. (Eds.) *Cognitive Responses in Persuasion*, Hillsdale, NJ, Lawrence Erlbaum, 1981.

Petty, R.E., and Cacioppo, J.T., *Attitudes and Persuasion: Classic and Contemporary Approaches*, Dubuque, IA, W.C. Brown, 1981.

Petty, R.E., and Cacioppo, J.T., *Communication and Persuasion: Central and Peripheral Routes to Attitude Change*, New York, Springer-Verlag, 1986.

Pfefferman, D., and Nathan, G., "Regression Analysis of Data from a Cluster Sample," *Journal of the American Statistical Association*, Vol. 76, 1981, pp. 681-689.

Platek, R., "Some Factors Affecting Non-response," *Survey Methodology*, Vol. 3, No. 2, 1977, pp. 191-214.

Politz, A., and Simmons, W., "An Attempt to Get the 'Not at Homes' into the Sample Without Callbacks," *Journal of the American Statistical Association*, Vol. 44, No. 245, March 1949, pp. 9-16.

Presser, S., and Schuman, H., "The Measure of a Middle Position in Attitude Surveys," *Public Opinion Quarterly*, Vol. 44, 1980, pp. 70-85.

Prufer, P., and Rexroth, M., "Zur Anwendung der Interaction-Coding Technik," *Zumanachrichten*, No. 17, Zentrum fuer Umfragen, Methoden und Analysen e.V., Mannheim, W. Germany, 1985.

Robins, L.N., "The Reluctant Respondent," *Public Opinion Quarterly*, Vol. 27, No. 2, 1963, pp. 276-286.

Rogers, T.F., "Interviews by Telephone and in Person: Quality of Responses and Field Performance," *Public Opinion Quarterly*, Vol. 40, No. 1, Spring, 1976, pp. 51-65.

Rorer, L.G., "The Great Response-Style Myth," *Psychological Bulletin*, Vol. 63, 1965, pp. 129-156.

Rosenthal, R., *Experimenter Effects In Behavior Research*, New York, Appleton-Century-Crofts, 1966.

Rosenthal, R., and Rosnow, R.L., *The Volunteer Subject*, New York, Wiley-Interscience, 1975.

Ross, L., Lepper, M., and Hubbard, M., "Perseverance in Self-Perception and Social Perception," *Journal of Personality and Social Psychology*, Vol. 32, 1975, pp. 880-892.

Royall, R.M., "The Linear Least-Squares Prediction Approach to 2-Stage Sampling," *Journal of the American Statistical Association*, Vol. 71, 1970, p. 657.

Royall, R.M., and Cumberland, W.G., "Variance Estimation in Finite Population Sampling," *Journal of the American Statistical Association*, Vol. 73, 1978, p. 351.

Royall, R.M., and Herson, J., "Robust Estimation in Finite Populations I," *Journal of the American Statistical Association*, Vol. 68, 1973a, p. 880.

Royall, R.M., and Herson, J., "Robust Estimation in Finite Populations II: Stratification on a Size Variable," *Journal of the American Statistical Association*, Vol. 68, 1973b, p. 890.

Royston, P., Bercini, D., Sirken, M., and Mingay, D., "Questionnaire Design Laboratory," *Proceedings of the Section on Survey Research Methods, American Statistical Association*, 1986, pp. 703-707.

Rubin, D.B., *Multiple Imputation for Nonresponse in Surveys*, New York, John Wiley & Sons, 1986.

Rumenik, D.K., et al., "Experimenter Sex Effects in Behavioral Research," *Psychological Bulletin*, Vol. 84, No. 5, 1977, pp. 852-877.

Rust, K.F., *Techniques for Estimating Variances for Sample Surveys*, Ph.D. dissertation, Ann Arbor, The University of Michigan, 1984.

Rustemeyer, A., "Measuring Interviewer Performance in Mock Interviews," *Proceedings of the Social Statistics Section, American Statistical Association*, 1977, pp. 341-346.

Schaeffer, N.C., "Evaluating Race-of-Interviewer Effects in a National Survey," *Sociological Methods and Research*, Vol. 8, No. 4, May, 1980, pp. 400-419.

Schaps, E., "Cost, Dependency, and Helping," *Journal of Personality and Social Psychology*, Vol. 21, No. 1, 1972, pp. 74–78.

Schopler, J., and Matthews, M.W., "The Influence of the Perceived Causal Locus of Partner's Dependence on the Use of Interpersonal Power," *Journal of Personality and Social Psychology*, Vol. 2, No. 4, 1965, pp. 609–612.

Schuman, H., and Converse, J.M., "The Effects of Black and White Interviewers on Black Responses in 1968," *Public Opinion Quarterly*, Vol. 35, No. 1, 1971, pp. 44–68.

Schuman, H., and Ludwig, J., "The Norm of Even-Handedness in Surveys as in Life," *American Sociological Review*, Vol. 48, 1983, pp. 112–120.

Schuman, H., and Presser, S., "Question Wording as an Independent Variable in Survey Analysis," *Sociological Methods and Research*, Vol. 6, 1977, pp. 151–170.

Schuman, H., and Presser, S., *Questions and Answers in Attitude Surveys*, New York, Academic Press, 1981.

Schuman, H., Presser, S., and Ludwig, J., "Context Effects on Survey Responses to Questions About Abortion," *Public Opinion Quarterly*, Vol. 45, 1981, pp. 216–223.

Schuman, H., Kalton, G., and Ludwig, J., "Context and Contiguity in Survey Questionnaires," *Public Opinion Quarterly*, Vol. 47, No. 1, 1983, pp. 112–115.

Schuman, H., Steeh, C., and Bobo, L., *Racial Attitudes in America: Trends and Interpretations*, Cambridge, Mass., Harvard University Press, 1985.

Schwartz, S.H., "Elicitation of Moral Obligation and Self-Sacrificing Behavior: An Experimental Study of Volunteering to be a Bone Marrow Donor," *Journal of Personality and Social Psychology*, Vol. 15, No. 4, 1970, pp. 283–293.

Schwartz, S.H., and Clausen, G.T., "Responsibility, Norms, and Helping in an Emergency," *Journal of Personal and Social Psychology*, Vol. 16, 1970, pp. 299–310.

Schwartz, S.H., and Howard, J.A., "Explanations of the Moderating Effect of Responsibility Denial on the Personal Norm-Behavior Relationship," *Social Psychology Quarterly*, Vol. 43, No. 4, 1980, pp. 441–446.

Shank, R., and Abelson, R., *Scripts, Plans, Goals, and Understanding*, Hillsdale, NJ, Lawrence Erlbaum Associates, 1977.

Short, J.A., *Cooperation and Competition in an Experimental Bargaining Game Conducted Over Two Media*, London, Communications Studies Group, Paper No. E/71160/SH, 1971.

Short, J.A., *The Effects of Medium of Communications on Persuasion, Bargaining and Perceptions of the Other*, London, Communication Studies Group, Paper No. E/73100/SH, 1973.

Short, J., Williams, E., and Christie, B., *The Social Psychology of Telecommunications*, London, John Wiley & Sons, Ltd., 1976.

Shotland, R.L., and Heinold, W.D., "Bystander Response to Arterial Bleeding: Helping Skills, the Decision-Making Process, and Differentiating the Helping Response," *Journal of Personality and Social Psychology*, Vol. 49, No. 2, 1985, pp. 347–356.

Shotland, R., and Stebbins, C.A., "Emergency and Cost as Determinants of Helping Behavior and the Slow Accumulation of Social Psychological Knowledge," *Social Psychology Quarterly*, Vol. 46, No. 1, 1983, pp. 36–46.

Silver, B.D., Anderson, B.A., and Abramson, P.R., "Who Overreports Voting?" *American Political Science Review*, Vol. 80, No. 2, 1986, pp. 613–624.

Singer, E., and Kohnke-Aguirre, L., "Interviewer Expectation Effects: A Replication and Extension," *Public Opinion Quarterly*, Vol. 43, No. 2, 1979, pp. 245–260.

Sirken, M., "Household Surveys With Multiplicity," *Journal of the American Statistical Association*, March, 1970, pp. 257–266.

Sirken, M.G., Graubard, B.J., and LaValley, R.W., "Evaluation of Census Population Coverage by Network Surveys," *Proceedings of the Survey Research Methods Section, American Statistical Association*, 1978, pp. 239–244.

Skinner, C.J., "Design Effects of 2-Stage Sampling," *Journal of the Royal Statistical Society*, Vol. 48, No. 1, 1986, pp. 89–99.

Slocum, W.L., Empey, L.T., and Swanson, H.W., "Response to Questionnaires and Structured Interviews," *American Sociological Review*, Vol. 21, 1956, pp. 221–225.

Smith, A.D., "Age Differences in Encoding, Storage, and Retrieval," in L.W. Poon et al., *New Directions in Memory and Aging: Proceedings of the George A. Talland Memorial Conference*, Hillsdale, NJ, Lawrence Erlbaum, 1980.

Smith, S.M., "Remembering in and out of Context," *Journal of Experimental Psychology: Human Learning and Memory*, Vol. 5, No. 5, 1979, pp. 460–471.

Smith, T.W., "Sex and the GSS," General Social Survey Technical Report No. 17, Chicago, National Opinion Research Center, 1979.

Smith, T.W., "Nonattitudes: A Review and Evaluation," in C.F. Turner and E. Martin (Eds.), *Surveying Subjective Phenomena*, New York, Russell Sage Foundation, 1984, pp. 215–255.

Smithson, M., and Amato, P., "An Unstudied Region of Helping: An Extension of the Pearce-Amato Cognitive Taxonomy," *Social Psychology Quarterly*, Vol. 45, No. 2, 1982, pp. 67–76.

Smullyan, R., *5000 B.C. and Other Philosophical Fantasies*, New York, St. Martin's, 1983.

Snow, C.P., *The Two Cultures and a Second Look*, New York, Cambridge Press, 1959.

Steeh, C., "Trends in Nonresponse Rates," *Public Opinion Quarterly*, Vol. 45, 1981, pp. 40–57.

Steinberg, J., "A Multiple Frame Survey for Rare Population Elements," *Proceedings of the Social Statistical Section, American Statistical Association*, 1965, pp. 262-266.

Stember, H., and Hyman, H., "How Interviewer Effects Operate Through Question Form," *International Journal of Opinion and Attitude Research*, Vol. 3, No. 4, 1949-1950, pp. 493-512.

Stephenson, C.B., "Probability Sampling With Quotas: An Experiment," *Public Opinion Quarterly*, Vol. 43, No. 4, 1979, pp. 477-496.

Stevens, J.A., and Bailar, B., "The Relationship Between Various Interviewer Characteristics and the Collection of Income Data," *Proceedings of the Social Statistics Section, American Statistical Association*, 1976, pp. 785-790.

Stokes, S.L., "Estimation of Interviewer Effects in Complex Surveys With Application to Random Digit Dialing," *Proceedings of the Second Annual Research Conference of the U.S. Bureau of the Census*, Washington, DC, U.S. Bureau of the Census, 1986, pp. 21-31.

Strenta, A., and DeJong, W., "The Effect of a Prosocial Label on Helping Behavior," *Social Psychology Quarterly*, Vol. 44, No. 2, 1981, pp. 142-147.

Stryker, S., and Statham, A., "Symbolic Interactionism and Role Theory," in G. Lindzey and E. Aronson (Eds.), *The Handbook of Social Psychology*, 3rd Edition, New York, L. Erlbaum Associates., 1985.

Stuart, A., *Basic Ideas of Scientific Sampling*, London, Griffin, 1962.

Sudman, S., *Reducing the Cost of Surveys*, Chicago, Aldine, 1967.

Sudman, S., and Bradburn, N., "Effects of Time and Memory Factors on Response in Surveys," *Journal of the American Statistical Association*, Vol. 68, No. 344, 1973, pp. 805-815.

Sudman, S., and Bradburn, N., *Response Effects in Surveys*, Chicago, Aldine, 1974.

Sudman, S., and Ferber, R., "A Comparison of Alternative Procedures for Collecting Consumer Expenditure Data for Frequently Purchased Products," *Journal of Marketing Research*, Vol. 11, 1974, pp. 128-135.

Sudman, S., et al., "Modest Expectations: The Effects of Interviewers' Prior Expectations on Responses," *Sociological Methods and Research*, Vol. 6, No. 2, November, 1977, pp. 171-182.

Sykes, W., and Collins, M., "Comparing Telephone and Face-to-Face Interviewing in the UK," *Survey Methodology*, Vol. 13, No. 1, 1987, pp. 15-28.

Sykes, W., and Collins, M., "Effects of Mode of Interview: Experiments in the UK," in R.M. Groves et al. (Eds.), *Telephone Survey Methodology*, New York, John Wiley & Sons, 1988.

Sykes, W., and Hoinville, G., *Telephone Interviewing on a Survey of Social Attitudes: A Comparison With Face-to-face Procedures*, London, Social and Community Planning Research, 1985.

Tenebaum, M., "Results of the CES Coverage Question Evaluation," Washington, DC, U.S. Bureau of the Census, memorandum, June 21, 1971.

Thomas, G., and Batson, C.D., "Effect of Helping Under Normative Pressure on Self-Perceived Altruism," *Social Psychology Quarterly*, Vol. 44, No. 2, 1981, pp. 127–131.

Thompson, W.C., Cowan, C.L., and Rosenbaum, D.L., "Focus of Attention Mediates the Impact of Negative Affect on Altruism," *Journal of Personality and Social Psychology*, Vol. 38, 1980, pp. 291–300.

Thornberry, O.T., and Massey, J.T., "Trends in United States Telephone Coverage Across Time and Subgroups," in R.M. Groves et al. (Eds.), *Telephone Survey Methodology*, New York, John Wiley & Sons, 1988.

Thorndyke, P.W., "Cognitive Structures in Comprehension and Memory of Narrative Discourse," *Cognitive Psychology*, Vol. 9, 1977, pp. 77–110.

Tice, D.M., and Baumeister, R.F., "Masculinity Inhibits Helping in Emergencies: Personality Does Predict the Bystander Effect," *Journal of Personality and Social Psychology*, Vol. 49, No. 2, 1985, pp. 420–428.

Tobin, J., "Estimation of Relationships for Limited Dependent Variables," *Econometrica*, Vol. 26, No. 1, pp. 24–36.

Tuchfarber, A.J., Jr., and Klecka, W.R., *Random Digit Dialing: Lowering the Cost of Victim Surveys*, Washington, DC, Police Foundation, 1976.

Tucker, C., "Interviewer Effects in Telephone Surveys," *Public Opinion Quarterly*, Vol. 47, No. 1, 1983, pp. 84–95.

Tulving, E., "Episodic and Semantic Memory," in E. Tulving and W. Donaldson (Eds.), *Organization of Memory*, New York, Academic Press, 1972.

Tulving, E., and Thomson, D.M., "Encoding Specificity and Retrieval Processes in Episodic Memory," *Psychological Review*, Vol. 80, No. 5, 1973, pp. 352–373.

Turner, C.F., and Krauss, E., "Fallible Indicators of the Subjective State of the Nation," *American Psychologist*, Vol. 33, 1978, pp. 456–470.

Tversky, A., and Kahneman, D., "Judgment Under Uncertainty: Heuristics and Biases," *Science*, Vol. 185, 1974, pp. 1124–1131.

U.S. Bureau of the Census, *Standards for Discussion and Presentation of Errors in Data*, Washington, DC, Technical Paper 32, 1974.

U.S. Bureau of the Census, *The Current Population Survey: Design and Methodology*, Washington, DC, Technical Paper 40, 1977.

U.S. Department of Commerce, *Glossary of Nonsampling Error Terms*, Washington, DC, Statistical Policy Working Paper 4, 1978.

U.S. Department of Commerce, *Money Income and Poverty Status of Families and Persons in the United States: 1984*, Current Population Reports, Series P-60, No. 149, Washington, DC, 1985.

Underwood, B.J., "Interference and Forgetting," *Psychological Review*, Vol. 64, 1957, pp. 119-122.

Uranowitz, S.W., "Helping and Self-Attributions: A Field Experiment," *Journal of Personality and Social Psychology*, Vol. 31, No. 5, 1975, pp. 852-854.

Valentine, C.A., and Valentine, B.L., *Missing Men*, report to the U.S. Bureau of the Census, Washington, DC, U.S. Bureau of the Census, mimeo, 1971.

Vigderhous, G., "Scheduling Telephone Interviews: A Study of Seasonal Patterns," *Public Opinion Quarterly*, Vol. 45, pp. 250-259.

Waksberg, J., "Optimum Size of Segment—CPS Redesign," Washington, DC, U.S. Bureau of the Census, internal memorandum, 1970.

Waksberg, J., "Sampling Methods for Random Digit Dialing," *Journal of the American Statistical Association*, Vol. 73, No. 361, 1978, pp. 40-46.

Ward, J.C., Russick, B., and Rudelius, W., "A Test of Reducing Callbacks and Not-at-Home Bias in Personal Interviews by Weighting At-Home Respondents," *Journal of Marketing Research*, Vol. XXII, February, 1985, pp. 66-73.

Warner, S.L., "Randomized Response: A Survey Technique for Eliminating Evasive Answer Bias," *Journal of the American Statistical Association*, Vol. 60, 1965, pp. 63-69.

Weaver, C.N., Holmes, S.L., and Glenn, N.D., "Some Characteristics of Inaccessible Respondents in a Telephone Survey," *Journal of Applied Psychology*, Vol. 60, No. 2, 1975, pp. 260-262.

Weber, D., and Burt, R.C., *Who's Home When*, Washington, DC, U.S. Bureau of the Census, 1972.

Weeks, M., Jones, B.L., Folsom, R.E., Jr., and Benrud, C.H., "Optimal Times to Contact Sample Households," *Public Opinion Quarterly*, Vol. 44, No. 1, Spring, 1980, pp. 101-114.

Weiss, C.H., "Validity of Welfare Mothers' Interview Responses," *Public Opinion Quarterly*, Vol. 32, No. 4, 1968, pp. 622-626.

Werner, O., and Schoepfle, E.M., *Systematic Fieldwork*, Vol. 1 and 2, Beverly Hills, Sage Publications, 1987.

Wiener, M., and Mehrabian, A., *Language Within Language: Immediacy, a Channel in Verbal Communication*, New York, Appleton-Century-Crofts, 1968.

Wilcox, J.B., "The Interaction of Refusal and Not-at-home Sources of Nonresponse Bias," *Journal of Marketing Research*, Vol. XIV, 1977, pp. 592–597.

Wilks, S.S., *Mathematical Statistics*, New York, John Wiley & Sons, 1962.

Williams, E., "A Summary of the Present State of Knowledge Regarding the Effectiveness of the Substitution of Face-to-Face Meetings by Telecommunicated Meetings," Paper No. P/74294/WL, London, Communications Studies Group, 1974.

Williams, E., "Experimental Comparison of Face-to-Face and Mediated Communication: A Review," *Psychological Bulletin*, Vol. 84, 1977, pp. 963–976.

Williams, J.A., Jr., "Interviewer-Respondent Interaction: A Study of Bias in the Information Interview," *Sociometry*, Vol. 27, 1964, pp. 338–352.

Wilson, C., and Williams, E., "Watergate Words: a Naturalistic Study of Media and Communications," unpublished paper, London, Communications Studies Group, 1975.

Wiseman, F., and McDonald, P., "Noncontact and Refusal Rates in Consumer Telephone Surveys," *Journal of Marketing Research*, Vol. XVI, November, 1979, pp. 478–484.

Wolter, K., *Introduction to Variance Estimation*, New York, Springer-Verlag, 1985.

Woltman, H.F., Turner, A.G., and Bushery, J.M., "A Comparison of Three Mixed-Mode Interviewing Procedures in the National Crime Survey," *Journal of the American Statistical Association*, Vol. 75, No. 371, September, 1980, pp. 534–543.

Wright, T., and Tsao, H.J., "A Frame on Frames: An Annotated Bibliography," in T. Wright (Ed.), *Statistical Methods and the Improvement of Data Quality*, New York, Academic Press, 1983, pp. 25–72.

Yates, F., "Systematic Sampling," *Philosophical Transactions of the Royal Society* (London), (A), Vol. 241, 1948, pp. 345–377.

Yekovich, F.B., and Thorndyke, P.W., "An Evaluation of Alternative Functional Models of Narrative Schemata," *Journal of Verbal Learning and Verbal Behavior*, Vol. 20, 1981, pp. 454–469.

Young, I., "Understanding the Other Person in Mediated Interactions," Paper No. E/74266/YN, London, Communications Studies Group, 1974.

Young, I., "A Three Party Mixed-Media Business Game: A Progress Report on Results to Date," Paper No. E/75189/YN, London, Communications Studies Group, 1975.

Zuckerman, M., Siegelbaum, H., and Williams, R., "Predicting Helping Behavior: Willingness and Ascription of Responsibility," *Journal of Applied Psychology*, Vol. 7, No. 4, 1977, pp. 295–299.

INDEX